Periodic Table of the Elements

1	2	3	4	5	6	7	8	9	10	11	12	13	14	15	16	17	18
H 1, 1.0079																	**He** 2, 4.003
Li 3, 6.941	**Be** 4, 9.012											**B** 5, 10.81	**C** 6, 12.01	**N** 7, 14.01	**O** 8, 16.00	**F** 9, 19.00	**Ne** 10, 20.18
Na 11, 22.99	**Mg** 12, 24.31											**Al** 13, 26.98	**Si** 14, 28.09	**P** 15, 30.97	**S** 16, 32.07	**Cl** 17, 35.45	**Ar** 18, 39.95
K 19, 39.098	**Ca** 20, 40.08	**Sc** 21, 44.96	**Ti** 22, 47.90	**V** 23, 50.94	**Cr** 24, 52.00	**Mn** 25, 54.94	**Fe** 26, 55.85	**Co** 27, 58.93	**Ni** 28, 58.69	**Cu** 29, 63.55	**Zn** 30, 65.39	**Ga** 31, 69.72	**Ge** 32, 72.61	**As** 33, 74.92	**Se** 34, 78.96	**Br** 35, 79.90	**Kr** 36, 83.80
Rb 37, 85.47	**Sr** 38, 87.62	**Y** 39, 88.91	**Zr** 40, 91.22	**Nb** 41, 92.91	**Mo** 42, 95.94	**Tc** 43, (98.91)	**Ru** 44, 101.1	**Rh** 45, 102.9	**Pd** 46, 106.4	**Ag** 47, 107.9	**Cd** 48, 112.4	**In** 49, 114.8	**Sn** 50, 118.7	**Sb** 51, 121.8	**Te** 52, 127.6	**I** 53, 126.9	**Xe** 54, 131.3
Cs 55, 132.9	**Ba** 56, 137.3	**Lu** 71, 175.0	**Hf** 72, 178.5	**Ta** 73, 180.9	**W** 74, 183.8	**Re** 75, 186.2	**Os** 76, 190.2	**Ir** 77, 192.2	**Pt** 78, 195.1	**Au** 79, 197.0	**Hg** 80, 200.6	**Tl** 81, 204.4	**Pb** 82, 207.2	**Bi** 83, 209.0	**Po** 84, (209)	**At** 85, (210)	**Rn** 86, (222)
Fr 87, (223)	**Ra** 88, (226)	**Lr** 103, (256)															

Lanthanides

La 57, 138.9	**Ce** 58, 140.1	**Pr** 59, 140.9	**Nd** 60, 144.2	**Pm** 61, (144.9)	**Sm** 62, 150.4	**Eu** 63, 152.0	**Gd** 64, 157.2	**Tb** 65, 158.9	**Dy** 66, 162.5	**Ho** 67, 164.9	**Er** 68, 167.3	**Tm** 69, 168.9	**Yb** 70, 173.0

Actinides

Ac 89, (227.0)	**Th** 90, 232.0	**Pa** 91, 231.0	**U** 92, 238.0	**Np** 93, 237.0	**Pu** 94, (244.1)	**Am** 95, (243)	**Cm** 96, (247)	**Bk** 97, (247)	**Cf** 98, (251)	**Es** 99, (252)	**Fm** 100, (257)	**Md** 101, (258)	**No** 102, (259)

(previously printed on inside cover)

Useful Conversion Factors

$1.0 \text{ atm} = 760 \text{ Torr} = 1.013 \text{ bar} = 1.013 \times 10^5 \text{ Pa}$

$4.184 \text{ J} = 1.0 \text{ cal}$

$2.54 \text{ cm} = 1.0 \text{ inch}$

$1.0 \text{ m}^3 = 1000 \text{ L}$

$1.0 \text{ pound} = 0.4536 \text{ kg}$

$1.0 \text{ Debye} = 3.336 \times 10^{-30} \text{ C m}$

Fundamental Physical Constants in SI units*

Quantity	Symbol	Value
Speed of light in vacuum	c	$2.998 \times 10^8 \text{ m s}^{-1}$
proton charge	e	$1.602 \times 10^{-19} \text{ C}$
Faraday's constant	$\mathscr{F}$	96480 C mol^{-1}
Plank's constant	h	$6.626 \times 10^{-34} \text{ J s}$
Boltzmann's constant	k	$1.381 \times 10^{-23} \text{ J K}^{-1}$
electron mass	m_e	$9.109 \times 10^{-31} \text{ kg}$
proton mass	m_p	$1.673 \times 10^{-27} \text{ kg}$
Avogadro's number	N_A	$6.022 \times 10^{23} \text{ mol}^{-1}$
Gas constant	R	$8.314 \text{ J K}^{-1} \text{ mol}^{-1}$
permittivity of vacuum	ε_o	$8.854 \times 10^{-12} \text{ C}^2 \text{ N}^{-1} \text{ m}^{-2}$

* to four significant figures; the constants are known to much higher accuracy

(previously printed on inside cover)

Principles of Thermodynamics

UNDERGRADUATE CHEMISTRY

A Series of Textbooks

Edited by

J. J. LAGOWSKI

Department of Chemistry
The University of Texas at Austin

1. Modern Inorganic Chemistry, *J. J. Lagowski*
2. Modern Chemical Analysis and Instrumentation, *Harold F. Walton and Jorge Reyes*
3. Problems in Chemistry, Second Edition, Revised and Expanded, *Henry O. Daley, Jr., and Robert F. O'Malley*
4. Principles of Colloid and Surface Chemistry, *Paul C. Hiemenz*
5. Principles of Solution and Solubility, *Kozo Shinoda, translated in collaboration with Paul Becher*
6. Physical Chemistry: A Step-by-Step Approach, *M. K. Kemp*
7. Numerical Methods in Chemistry, *K. Jeffrey Johnson*
8. Polymer Chemistry: An Introduction, *Raymond B. Seymour and Charles E. Carraher, Jr.*
9. Principles of Colloid and Surface Chemistry, Second Edition, Revised and Expanded, *Paul C. Hiemenz*
10. Problems in Chemistry, Second Edition, Revised and Expanded, *Henry O. Daley, Jr., and Robert F. O'Malley*
11. Polymer Chemistry: An Introduction, Second Edition, *Raymond B. Seymour and Charles E. Carraher, Jr.*
12. Polymer Chemistry: An Introduction, Third Edition, Revised and Expanded, *Raymond B. Seymour and Charles E. Carraher, Jr.*
13. Seymour/Carraher's Polymer Chemistry: An Introduction, Fourth Edition, Revised and Expanded, *Charles E. Carraher, Jr.*
14. Seymour/Carraher's Polymer Chemistry: Fifth Edition, Revised and Expanded, *Charles E. Carraher, Jr.*
15. Principles of Thermodynamics, *Myron Kaufman*

Additional Volumes in Preparation

Seymour/Carraher's Polymer Chemistry: Sixth Edition, Revised and Expanded, *Charles E. Carraher, Jr.*

Principles of Thermodynamics

Myron Kaufman

Emory University
Atlanta, Georgia

CRC Press
Taylor & Francis Group
Boca Raton London New York

CRC Press is an imprint of the
Taylor & Francis Group, an **informa** business

CRC Press
Taylor & Francis Group
6000 Broken Sound Parkway NW, Suite 300
Boca Raton, FL 33487-2742

First issued in paperback 2019

ISBN-13: 978-0-8247-0692-0 (hbk)
ISBN-13: 978-0-367-39588-9 (pbk)

**Visit the Taylor & Francis Web site at
http://www.taylorandfrancis.com**

**and the CRC Press Web site at
http://www.crcpress.com**

Dedicated to my beloved brother Stuart,

whose brilliant life ended too soon.

Preface

Once upon a time there were Giants, with names like Joule, Maxwell, Carnot, Clausius, and Thomson (Lord Kelvin). They lived during a time, called the Industrial Revolution, when labor-saving machines were being developed to greatly expand the productive capabilities of humankind. In the never-ending attempts to improve the performance of these machines, the Giants were led to profound considerations of the fundamental limits of energy conversions. Starting with some simple observations, they developed the science of thermodynamics. This science deals with energy and its capabilities and transformations. Later, Boltzmann—another Giant—connected thermodynamics to the microscopic world of atoms and molecules, while Prigogine extended thermodynamics to deal with non-equilibrium systems.

The purpose of *Principles of Thermodynamics* is to convey the powerful ideas of the Giants to advanced undergraduates, beginning graduate students, and interested scientific readers at an appropriate mathematical level, enlightening this audience to the wide variety of problems for which a thermodynamic perspective is useful. In this volume, I have chosen to express the laws of thermodynamics in terms of simple principles, self-evident from everyday experience. For example, the second law—the cornerstone of any presentation of thermodynamics—is stated as "in any real process there is net degradation of energy." I believe this approach is much more comprehensible than that based on machines, used by the Giants.

Since mathematics is the language of thermodynamics, there are many equations in this book. However, the mathematics used is no more complicated than necessary. Facility with differentiation and integration at the level of a first-year course in calculus is assumed and a few relationships from multivariable calculus are used repeatedly. All the reader has to know about this subject, however, is presented in Appendix A. Although the mathematically advanced reader can skim over this, it remains as a handy reference for any question that arises on multivariable calculus.

Principles of Thermodynamics should be accessible to scientifically literate persons who are either learning the subject on their own or reviewing the material. At Emory University, this volume forms the basis of the first semester of a one-year sequence in physical chemistry. Problems and questions are included at the end of each chapter. Essentially, the questions test whether the students understand the material, and the problems test whether they can use the derived results. More difficult problems are indicated by an asterisk. Some problems, marked with an M, involve numerical calculations that are most easily performed with the use of a computer program such as Mathcad or Mathematica. A brief survey of some of these numerical methods is included in Appendix B, for cases in which the programs are unavailable or cumbersome to use.

Thermodynamics deals with relations between properties of materials and changes of these properties during processes. Some knowledge of specific properties is thus necessary before beginning a discussion of thermodynamics. This is the purpose of Chapter 1, which deals with some of the properties of gases and other materials. In Chapter 2, after defining terms and introducing the zeroth and the first law, conservation of energy is applied to a number of processes. In Chapter 3, the quality of energy is used as the basis for introducing entropy and the second law, which determines the direction of spontaneous processes and equilibrium. In Chapter 4, entropy is placed on an absolute basis with the third law, which involves low-temperature systems. The advantages of analyzing processes using free-energy functions are then introduced.

Chapter 5 gives a microscopic-world explanation of the second law, and uses Boltzmann's definition of entropy to derive some elementary statistical mechanics relationships. These are used to develop the kinetic theory of gases and derive formulas for thermodynamic functions based on microscopic partition functions. These formulas are applied to ideal gases, simple polymer mechanics, and the classical approximation to rotations and vibrations of molecules.

In Chapters 6, 7, and 8, the thermodynamic framework is successively applied to phase transformations of single-component systems, chemical reactions, and ideal solutions. Included are discussions of the thermodynamics of open systems, the phase rule, and colligative properties. Chapter 9 gives the framework for discussing nonideal multicomponent systems and describes a variety of phase diagrams of such systems. In Chapter 10, the discussion is extended to ionized systems, including galvanic cells. Chapter 11 deals with surface effects in both single- and multicomponent systems, including adsorption. Finally, in Chapter 12 the thermodynamics of open systems is applied to systems at steady state undergoing dissipative process. Although several applications of this material are considered, the aim is to give the reader the tools needed to approach the extensive literature on this subject. The material in Chapter 12 is not covered in the physical chemistry sequence, but is assigned as outside reading for outstanding students.

Principles of Thermodynamics is both compact and rigorous; almost all results are "derived." Most of all, this book tries to convey the beauty of one of the most impressive triumphs of the human mind—the application of deductive reasoning from a few simple postulates, resulting in the development of a myriad of relationships useful in just about every branch of science.

Many thanks are given to Professor C. G. Trowbridge of Emory University and Professor Wentao Zhu of Tsinghua University, Beijing, for reading parts of the manuscript. More than anyone, I thank my wife, June, for her encouragement and understanding throughout the protracted period it took to write this book.

Myron Kaufman

Contents

Preface   *v*

1   Introduction and Background   **1**

   1.1   Introduction   2

   1.2   The Ideal Gas   4

   1.3   Thermal Expansion Coefficient and Isothermal
        Compressibility   6

   1.4   A Simple Model of the Ideal Gas   8

   1.5   Real Gases: The van der Waals Equation   12

   1.6   Real Gases: Other Equations   15

   1.7   Condensation and the Critical Point   19

   1.8   Gas Mixtures   24

   1.9   Equations of State of Condensed Phases   26

   1.10 Pressure Variations in Fluids   30

        Questions   31

        Problems   32

        Notes   34

2   Thermodynamics: The Zeroth and First Laws   **36**

   2.1   The Nature of Thermodynamics   37

2.2   Systems   37
2.3   Equilibrium   38
2.4   Properties   39
2.5   Processes   40
2.6   Heat and the Zeroth Law of Thermodynamics   40
2.7   Work   43
2.8   Internal Energy   49
2.9   The First Law   49
2.10   Heat Capacities   51
2.11   The Joule Process   58
2.12   The Joule-Thomson Process   59
2.13   Reversible Adiabatic Expansion of an Ideal Gas   62
2.14   A Simple Heat Engine   64
       Questions   67
       Problems   69
       Notes   70

3   The Second Law of Thermodynamics   **71**
3.1   The Second Law   71
3.2   Entropy Changes in Some Simple Processes   79
3.3   Heat Diagrams   84
3.4   General Analysis of Thermal Devices   85
       Questions   88
       Problems   90
       Notes   92

4   The Third Law and Free Energies   **93**
4.1   Absolute Zero and the Third Law of Thermodynamics   94
4.2   Absolute Entropies   97
4.3   Helmholtz and Gibbs Free Energies   98
4.4   Partial Derivatives of Energy-like Quantities   101
4.5   Heat Capacities   106
4.6   Generalization to Additional Displacements   106
4.7   Standard States   107
4.8   Entropy of Mixing of Ideal Gases   109
4.9   Thermodynamics of Stretching Rubbers   110
       Questions   112
       Problems   114
       Notes   116

5 Statistical Mechanics **117**
  5.1 The Microscopic World 118
  5.2 The Joule Process 119
  5.3 Distribution Among Energy States 124
  5.4 Thermodynamic Functions from the Partition Function 128
  5.5 System Partition Functions 130
  5.6 Velocity Distributions 132
  5.7 A Steady-State Example 135
  5.8 Thermodynamic Functions of the Monatomic Ideal Gas 136
  5.9 Energy of Polyatomic Ideal Gases 138
  5.10 Configurations of a Polymer Chain 141
  5.11 Theory of Ideal Rubber Elasticity 144
  Questions 146
  Problems 146
  Notes 148

6 Phase Transformations in Single-Component Systems **150**
  6.1 Thermodynamics of Open Systems 151
  6.2 Entropy Change for Open Systems 153
  6.3 Phases and Phase Transformations 154
  6.4 General Criterion for Equilibrium in a Multiphase System 155
  6.5 Phase Equilibrium Conditions 161
  6.6 Vapor Pressure When a Gas Is Not Ideal 164
  6.7 Equation of State for the Two-Phase Region 166
  6.8 Effect of Inert Gas on Phase Equilibria 168
  6.9 Condensed-Phase Equilibria 169
  6.10 Equilibrium Between Three Phases 170
  6.11 Phase Diagrams 171
  6.12 Mesomorphic Phases 173
  Questions 175
  Problems 176
  Notes 178

7 Chemical Reactions **179**
  7.1 Nomenclature 180
  7.2 Thermochemistry 181
  7.3 Calorimetry 185
  7.4 Estimating the Thermodynamics of Reactions 187
  7.5 Chemical Equilibrium 190
  7.6 Direction of Chemical Reactions 192
  7.7 Concentration Dependence of Free Energy 192

7.8   Equilibrium Constants   193
7.9   General Considerations Involving Multicomponent and
      Multiphase Equilibrium: The Gibbs Phase Rule   196
7.10 Concentrations at Equilibrium   199
7.11 Temperature Dependence of the Equilibrium Constant   204
7.12 Equilibrium Constant in Terms of Partition Function   206
      Questions   207
      Problems   209
      Notes   211

8 Ideal Solutions   **212**
8.1   Measures of Concentration   213
8.2   Partial Molar Quantities   214
8.3   Measurement of Partial Molar Quantities   217
8.4   The Ideal Solution   220
8.5   The Ideally Dilute Solution   222
8.6   Freezing-Point Depression, Boiling-Point Elevation and
      Osmotic Pressure   226
8.7   Distribution of a Solute Between Two Solvents   231
8.8   Phase Diagram of a Binary Ideal Solution   232
      Questions   240
      Problems   242
      Notes   244

9 Nonideal Solutions   **245**
9.1   Activity Coefficients   245
9.2   Excess Thermodynamic Functions   248
9.3   Determining Activity Coefficients   249
9.4   Equilibrium Constants   257
9.5   Phase Diagrams of Binary Nonideal Systems   258
9.6   Phase Diagrams of Ternary Systems   266
      Questions   269
      Problems   270
      Notes   272

10 Ionized Systems   **273**
10.1 Ionic Solutions   274
10.2 Mean Ionic Activity Coefficients   275
10.3 Mean Ionic Activity Coefficients from Experimental Data   276
10.4 Calculation of Mean Ionic Activity Coefficient   277
10.5 Ionic Equilibrium   282

10.6	Ion Pairs and Ion Solvation	284
10.7	Electrochemical Systems	286
10.8	Types of Electrodes	288
10.9	Electrochemical Cells	291
10.10	Thermodynamics of Electrochemical Cells	293
10.11	Standard Electrode Potentials	296
10.12	Applications of Electrochemical Cells	300
	Questions	303
	Problems	304
	Notes	306

11 Surfaces   **308**

11.1	The Surface Region	309
11.2	The Surface of the Single-Component Condensed Phase	309
11.3	Surface Tension of Single Components	311
11.4	Processes Involving One Interface	313
11.5	Processes Involving More Than One Interface	316
11.6	Thermodynamics of Immersion	320
11.7	Effect of Surface Curvature on Vapor Pressure	320
11.8	Thermodynamics of Solution Surfaces	322
11.9	Properties of Surface Films	325
11.10	Adsorption on Solids	328
11.11	Statistical Mechanics of Adsorption	334
11.12	Colloids	337
	Questions	340
	Problems	341
	Notes	342

12 Steady-State Systems   **343**

12.1	Steady-State Systems	344
12.2	Conservative and Nonconservative Properties	344
12.3	Entropy Generation in Some Simple Processes in Steady-State Systems	346
12.4	The Phenomenological Equations Relating Flows and Forces	353
12.5	Fluxes Produced by Nonconjugate Forces	356
12.6	Thermal Diffusion	357
12.7	Thermoelectric Effects	360
	Questions	361
	Problems	362
	Notes	363

Appendix A   Multivariable Calculus   **365**
   A.1   Differentials   365
   A.2   Integrals   368
   A.3   Second Derivatives   369
           Problems   370
           Note   371

Appendix B   Numerical Methods   **372**
   B.1   Solving Equations   372
   B.2   Fitting Data   374
   B.3   Numerical Integration   375
           Problems   376

Appendix C   Tables of Thermodynamic Data   **377**

Appendix D   Glossary of Symbols   **381**

Index   *387*

Principles of Thermodynamics

1

Introduction and Background

Thermodynamics is the only science about which I am firmly convinced that, within the framework of the applicability of its basic principles, it will never be overthrown.

Albert Einstein

Thermodynamics derives from consideration of energy. Because the study of thermodynamics is much more meaningful when it is applied to familiar substances, we briefly discuss the properties of various materials in this chapter. At low and moderate pressures, the ideal gas is an excellent approximation to real gases and a good introduction to the properties of materials at equilibrium. Although thermodynamics deals with bulk matter, a microscopic or molecular description of matter is often used to understand and predict bulk properties of materials. Using the kinetic theory of gases, a description of temperature and pressure on the molecular level is obtained. Because deviations from ideal gas behavior become more pronounced at high pressure and temperatures, the equations of van der Waals, Berthelot, and Redlich and Kwong are introduced to describe real gas behavior. These equations are used to discuss properties of real substances, such as condensation and critical phenomena, that are due to the interactions between molecules. Gas mixtures are discussed and the generation of pressure variation in an extended gas (the atmosphere) due to a gravitational field is calculated. The influence of pressure, temperature, and forces on the properties of solids and liquids are illustrated.

1.1 Introduction

1.1.1 Scope

With knowledge comes power, but knowledge soon becomes overwhelming; there is just too much to be known. We need principles to organize and extend our knowledge. In the study of matter and its transformations, a body of knowledge that is often called chemistry, there are a number of basic principles by which knowledge is organized. It is interesting that these principles are, in reality, only approximations.

One of these principles is the immutability of atomic species, which enables us to consider chemical transformations as rearrangements of atoms, the basic building blocks of matter. Most readers know that atomic species are really not immutable. In fact, nuclear physics deals largely with changes of one atomic species into another. Nevertheless, immutability is a very good and convenient approximation in the low-energy environment of our everyday experience, and we do not hesitate to make use of it in the science that applies to that realm.

Another principle that we use with impunity in chemistry is conservation of mass. Since Lavoisier, mass balance has been assumed in considerations of chemical transformations, even though Einstein's theory of relativity assures us that it is only an approximation. The point is that in chemistry and other fields that consider only relatively low-energy phenomena, conservation of mass is a very good approximation indeed!

The present volume deals with a number of exceedingly useful principles for organizing and extending our knowledge of matter and its transformations. These principles are based on considerations of energy, and are called the *laws of thermodynamics*. These laws are generalization of our experience and are taken as the basic postulates of thermodynamics. There are four of these laws, and at least three of these are approximations. One law, conservation of energy, is an approximation in the same sense as is conservation of mass, because, by relativity, we know that mass and energy are interconvertible. A second law is only true in a statistical sense (but in a very, very good statistical sense), and a third law allows exception. For the most part, we will be able to use these laws and the mathematical relationships that follow from them with no consideration of their approximate nature.

Thermodynamics was developed early in the industrial revolution to aid in improving engines. Originally, it dealt with transformations of heat and work. However, over time, the laws of thermodynamics have been used to deduce powerful mathematical relationships applicable to a broad range of phenomena. Thermodynamics has been found to be an infallible guide for indicating which processes can occur in nature. Processes not ruled out by thermodynamics can occur, but may occur so slowly that they can be completely neglected. The science dealing with the rates of processes is called *kinetics* and will not be discussed in this book.

Our approach to matter and its transformations will be macroscopic and phenomenological. Thus, we will consider nitrogen as a substance with a certain group of properties that can be measured and other properties that can be calculated from the measured ones using thermodynamics. It is also possible to calculate the properties of materials from their microscopic description. For example, we know that nitrogen is made up of diatomic molecules that are rotating, vibrating, translating, and occasionally colliding with other nitrogen molecules. However, except for translation, accurate description of these motions requires *quantum mechanics*, which will not be treated in this volume. We do, in Chapter 5, establish basic relations between the microscopic and macroscopic worlds, and these can be used once the requisite quantum background is obtained.

1.1.2 Plan of the Book

We begin by presenting some experimental results dealing with the properties of different types of matter. This is necessary—so that we have something to which we can apply our thermodynamic relations. A digression into the simplest molecular description of ideal gases allows us to develop some intuitive feeling for the very important bulk properties—temperature and pressure. The four laws of thermodynamics are then presented. The zeroth law allows us to define a unique temperature scale, and the first law is conservation of energy. The idea that energy has quality, as well as quantity, is then introduced. The second law is presented as the quality of energy being reduced as we use it. This reduction in quality of energy is called entropy increase. The third law deals with very low temperatures and permits us to discuss absolute values for entropy. The use of free energies greatly simplifies the application of the laws of thermodynamics. Each of the laws of thermodynamics is expressed in mathematical form, and useful relations between thermodynamic variables are derived. Extensive application of multivariable calculus is made in the text. A short presentation of this subject is given in Appendix A.

Entropy is interpreted as the number of microscopic arrangements included in the macroscopic definition of a system. The second law is then used to derive the distribution of molecules and systems over their states. This allows macroscopic state functions to be calculated from microscopic states by statistical methods.

The laws of thermodynamics are applied to a number of topics. These include elastic properties of polymeric materials and surface effects in liquids. Phase transformations (melting, boiling, and sublimation) of single-component systems are treated, followed by chemical reactions in gases. Chemical potentials are defined and used to treat these problems, and general conditions for equilibrium are developed. For treating solutions, two ideal limits are introduced and used to calculate solution properties. The concept of activity is used to treat deviations from ideal behavior, and many of the thermodynamic relations are rephrased in terms of this variable. In ionic solutions, deviation from ideal behavior is large, even at very low concentrations, and methods are introduced for treating this behavior. Electrochemical cells are treated both for their practical importance and for their ability to provide measurements of many thermodynamic parameters. Surface effects are discussed and their thermodynamics analyzed by use of the concept of a two-dimensional surface phase. Finally, we explore the thermodynamics of steady-state systems, in which matter and/or energy continually pass through the system boundaries. Such systems are important models for chemical processes and living systems.

1.2 The Ideal Gas

One of the first quantitative relations established from experimental measurements was Boyle's observation in 1662 that for most gases, under the conditions that they could conveniently be studied in his laboratory (pressures of a few atmospheres or less), volume is inversely proportional to pressure:

$$V \propto \frac{1}{P} \tag{1}$$

Boyle's measurements were made on a fixed sample of gas over a short time, so that his laboratory temperature could be considered constant. Soon after, Charles explicitly investigated the effect of temperature variation.[1] He found that if pressure was held constant, the volume varied linearly with temperature:

$$V \propto 1 + kt \tag{2}$$

where k was found to have the value $1/273.15°C$. This equation implies that if it holds to a temperature of $-273.15°C$ (it does not), the volume of gases would be zero at that temperature. Because negative volumes are not meaningful, $t = -273.15°C$ could be considered the lowest temperature attainable and was called *absolute zero*. A new temperature, T, was defined as

$$T \equiv t + 273.15 \tag{3}$$

and termed the *absolute temperature*, or the *ideal gas temperature*.[2] When Eqs. (1)–(3) are combined, we have

$$V \propto \frac{T}{P} \tag{4}$$

Additional information was supplied by Avogadro's hypothesis that the volume of a gas (at constant T and P) was proportional to the number of molecules, N, in the gas sample. Because in the 17th century, the number of molecules in a given gas sample was not known, it was customary to deal with an arbitrary number of molecules called the mole. Although the definition of the number of molecules in a mole has changed slightly over time, at present the number of atoms in exactly $12\,g$ of the isotope ^{12}C is chosen as the standard, resulting in 6.02214×10^{23} molecules/mol. This quantity is called *Avogadro's number* and given the symbol N_A. In order to deal with an equation, rather than a proportionality, we include a constant R, the *gas constant* in our final gas equation:

$$V = \frac{nRT}{P} \tag{5}$$

The value of R depends on the units chosen for V and P. Scientists have agreed to move toward adopting the SI (Système International d'Unites) units, with V in cubic meters (m^3) and P in *Pascal* (Pascal $=$ Pa $=$ Newton/m^2) and R is $8.314\,m^3\,Pa/mol\,K = 8.314\,J/mol\,K$, where the joule (J) is the SI unit of energy. Because Eq. (5) only holds exactly in the limit of zero pressure, where molecules are infinitely far apart and their interactions can be neglected, it is called the *ideal gas law*. For most gases, it is an excellent approximation for the conditions generally used in laboratory experiments. Under high-pressure conditions that occasionally hold in the laboratory and are usually employed in industrial processes, Eq. (5) is often a very poor approximation. The Pascal is a very small pressure; $100,000\,Pa$ is called a *bar*, which is very similar to an *atmosphere* ($1.0\,atm = 1.013 \times 10^5\,Pa = 1.013\,bar$). It is important to distinguish between the atmosphere, a fixed unit of pressure, roughly equal to the average atmospheric pressure at sea level, and "atmospheric pressure," which depends on the place and time being discussed.

Pressure, volume, temperature, and number of moles are *thermodynamic properties* or *thermodynamic variables* of a system—in this case, a gas sample. Their values are measured by experimenters using thermometers, pressure gauges, and other instruments *located outside the system*. The properties are of two types: those that increase proportionally with the size of the system, such as n and V, called *extensive properties*, and those defined for each small region in the system, such as P and T, called *intensive properties*. Terms that are added together or are on opposite sides of an equal sign must contain the same number of extensive variables. The quotient of two extensive variables is an intensive variable.

When a property, such as T or P, is assigned to a system, it is implied that it is uniform (has the same value) throughout the system. These conditions are called *thermal equilibrium* and *mechanical equilibrium* for temperature and pressure, respectively. In Chapter 10, we will deal with systems in which intensive properties are not uniform.

By dividing both sides of Eq. (5) by n, it can be expressed completely in terms of intensive variables:

$$\frac{V}{n} \equiv V_m = \frac{RT}{P} \tag{6}$$

where V_m is called the *molar volume*. Any extensive property divided by the number of moles is called the "molar property." We can write $V_m(P, T)$, indicating that the molar volume is a function of P and T. Additional variables (concentrations) are needed for multicomponent systems. The *density* of a gas, ρ, is its mass per volume:

$$\rho = \frac{nM}{V} = \frac{PM}{RT} \tag{7}$$

The ideal gas law is an example of an *equation of state* (i.e., an equation relating different properties of a form of matter). Such equations are not obtained from thermodynamics. They result either from empirical measurements of the related quantities or from calculations based on molecular models.

For some purposes, a graphical presentation of information is preferable to an equation. As shown in Eq. (6), the ideal gas law is a relation among three variables. In order to represent this equation in a two-dimensional plot, one variable must be held constant. The plots are called *isotherms*, *isobars*, or *isochores*, depending on whether temperature, pressure, or volume, respectively, is held constant. The plots for an ideal gas are shown in Fig. 1.

1.3 Thermal Expansion Coefficient and Isothermal Compressibility

Several partial derivatives of state functions are used so often that they have been given names. The *thermal expansion coefficient*, α, is defined as

$$\alpha \equiv \frac{1}{V} \left(\frac{\partial V}{\partial T} \right)_P \tag{8}$$

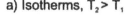

a) Isotherms, $T_2 > T_1$ b) Isobars, $P_2 > P_1$

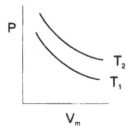
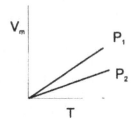

c) Isochores, $V_{m,2} > V_{m,1}$

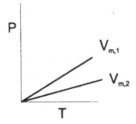

Figure 1 Isotherms, isobars, and isochores of an ideal gas.

and the *isothermal compressibility*, κ, as

$$\kappa \equiv -\frac{1}{V}\left(\frac{\partial V}{\partial P}\right)_T \qquad (9)$$

Both of these partial derivatives are divided by V to make them intensive quantities. The SI units of α are K^{-1} and those of κ are Pa^{-1}. A negative sign is used in the definition of κ, because volumes always decrease as pressure increases, and we would prefer to tabulate positive quantities.

Example 1. Find expressions for the thermal expansion coefficient and the isothermal compressibility of an ideal gas.

Solution:

$$V = \frac{nRT}{P}, \qquad \left(\frac{\partial V}{\partial T}\right)_P = \frac{nR}{P}, \qquad \left(\frac{\partial V}{\partial P}\right)_T = -\frac{nRT}{P^2}$$

$$\alpha \equiv \frac{1}{V}\left(\frac{\partial V}{\partial T}\right)_P = \frac{1}{V}\frac{nR}{P} = \frac{1}{T}, \qquad \kappa \equiv -\frac{1}{V}\left(\frac{\partial V}{\partial P}\right)_T = \frac{1}{V}\frac{nRT}{P^2} = \frac{1}{P}$$

1.4 A Simple Model of the Ideal Gas

Thermodynamics deals with relations among bulk (macroscopic) properties of matter. Bulk matter, however, is comprised of atoms and molecules and, therefore, its properties must result from the nature and behavior of these microscopic particles. An explanation of a bulk property based on molecular behavior is a *theory* for the behavior. Today, we know that the behavior of atoms and molecules is described by quantum mechanics. However, theories for gas properties predate the development of quantum mechanics. An early model of gases found to be very successful in explaining their equation of state at low pressures was the kinetic model of noninteracting particles, attributed to Bernoulli. In this model, the pressure exerted by n moles of gas confined to a container of volume V at temperature T is explained as due to the incessant collisions of the gas molecules with the walls of the container. Only the translational motion of gas particles contributes to the pressure, and for translational motion Newtonian mechanics is an excellent approximation to quantum mechanics. We will see that ideal gas behavior results when interactions between gas molecules are completely neglected.

The development first considers what happens when a single molecule hits a wall of a container. This result is then summed over all of the molecules in the container. A number of simplifications will be made in the presentation; most of these can be relaxed to give a much more complicated, but perhaps more satisfying, derivation.[3]

Consider a cubical box of side L, as shown in Fig. 2. We will first consider only the z component of motion of a single molecule of mass m. Because all interactions between particles are neglected, a molecule moving in the positive z direction will travel unimpeded until it hits the right wall of the container. It will

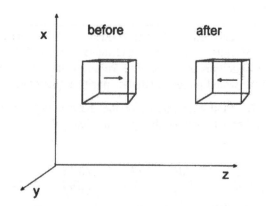

Figure 2 Elastic collision of a molecule with a wall.

be assumed that the collision with this wall is perfectly elastic, resulting in the molecule losing no translational kinetic energy. In addition, we will consider walls that are perfectly smooth and, thus, exert no forces in the x or y directions. With these assumptions, the only possible result of the collision is for the molecular velocity to reverse direction but keep the same magnitude, as shown in Fig. 2. The change of momentum of the molecule in a single collision is $-mv_z - (mv_z) = -2mv_z$. The force required to change the momentum of the molecule is provided by the right-hand wall of the box during the collision. Newton's third law equates the force to the rate of change of momentum. The force only occurs during the very short time of the collision and is highly variable. The impulsive nature of the force that occurs each time the molecule hits the right-hand wall is shown in Fig. 3.

We can also equate the time-averaged force exerted by the wall on the molecule to the time-averaged rate of change of momentum of the molecule, which is the momentum change in a single collision, times the rate of colliding with this wall. The latter quantity is molecular velocity divided by the distance the molecule travels between successive collisions with a given wall, $2L$:

$$f_{z,w \to m} = -(2mv_z)\left(\frac{v_z}{2L}\right) = -\frac{mv_z^2}{L} = -f_{z,m \to w} \tag{10}$$

The final equality results from Newton's law.[4] $f_{z,m \to w}$ is the time-averaged force exerted on the wall by a single molecule, say molecule i. To find the total force, we have to add up the contribution from each of the N molecules in the box. Dropping the $m \to w$ subscript, this sum is

$$F_z = \sum_i f_{z,i} = \frac{m}{L} \sum_i v_{z,i}^2 = \frac{mN}{L} \langle v_z^2 \rangle \tag{11}$$

time

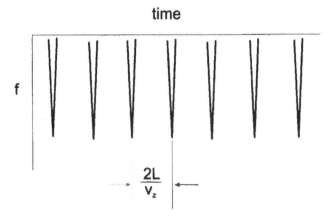

$$\frac{2L}{V_z}$$

Figure 3 Instantaneous force exerted by right-hand wall on a molecule.

using the definition of the average value of a molecular property[5]

$$\langle g \rangle \equiv \frac{1}{N} \sum_{i=1}^{N} g_i \tag{12}$$

The pressure on the right-hand wall is just the force on this wall divided by its area, L^2:

$$P = \frac{mN}{L^3} \langle v_z^2 \rangle = \frac{mN}{V} \langle v_z^2 \rangle. \tag{13}$$

The pressure due to a macroscopic number of molecules (of the order of 10^{22}) appears perfectly steady, as no pressure gauge can respond fast enough to record the individual collisions.

In considering motion in three dimensions, we can assume that the system is isotropic and the components of velocity are independent, so that

$$\langle c^2 \rangle = \langle v_x^2 \rangle + \langle v_y^2 \rangle + \langle v_z^2 \rangle = 3 \langle v_z^2 \rangle \tag{14}$$

where c is the molecular *speed*, the magnitude of the velocity. Using the average translational kinetic energy of the molecules in the box,

$$\langle \varepsilon \rangle = \frac{m}{2} \langle c^2 \rangle \tag{15}$$

$$P = \frac{mN \langle c^2 \rangle}{3V} = \frac{2N \langle \varepsilon \rangle}{3V} = \frac{2nN_A \langle \varepsilon \rangle}{3V} \tag{16}$$

Comparing Eq. (16) with Eq. (5), the ideal gas law, we see that for them to be the same,

$$\langle \varepsilon \rangle = \frac{m}{2} \langle c^2 \rangle = \frac{3}{2} \frac{RT}{N_A} = \frac{3}{2} kT \tag{17}$$

k, the ratio of the gas constant to Avogadro's number, is called the Boltzmann constant and has the value 1.38066×10^{-23} J/K. The Boltzmann constant, however, is more than just a ratio of two constants. In the microscopic world, it plays a fundamental role in the definition of entropy, a subject of major interest in this book.

Equation (17) gives us physical intuition into the microscopic meaning of temperature. Temperature is a measure of the average translational kinetic energy of molecules! However, note that it is only the absolute (or ideal gas) temperature that is directly related to the motion of molecules in this way. In Chapter 5, we will generalize this principle, showing that, under certain conditions, the absolute temperature is also a measure of the average of other types of energy (rotation and vibration) of molecules. This microscopic picture of energy will be an important adjunct to its thermodynamic definition.

From Eq. (17), we can also obtain

$$c_{rms} \equiv \sqrt{\langle c^2 \rangle} = \sqrt{\frac{3kT}{m}} = \sqrt{\frac{3RT}{M}} \tag{18}$$

where c_{rms}, the *root-mean-square velocity*,[6] is one measure of the velocity of the molecules.

Example 2. Calculate c_{rms} of N_2 at 298 K. Compare your result with the velocity of sound at ambient conditions, 340 m/s.

Solution:

$$c_{rms} = \left(\frac{3(8.314 \text{ J/mol K})(298 \text{ K}) [\text{kg m}^2/\text{s}^2]}{0.028 \text{ kg/mol} [\text{J}]} \right)^{1/2}$$

$$= 515 \frac{\text{m}}{\text{s}}$$

The speed of sound is about two-thirds of the root-mean-square molecular speed. This is reasonable because sound travels by moving molecules by reorienting, but not changing, the magnitude of their velocity. Sound can therefore travel no faster than the molecules through which it is traveling.

Note that the molecular weight of N_2 must be expressed in SI units in this calculation. Neglect of this is a very common cause of errors in such calculations. Having an intuitive feeling for the size of molecular velocities is also a good way to avoid errors.

One ramification of Eq. (18) is *Graham's law of effusion*, which deals with the rate at which gaseous molecules pass through a small hole in the wall of their enclosure (effusion). According to Graham, the rate per unit concentration is proportional to velocity and, thus, directly proportional to the square root of the absolute temperature and inversely proportional to the square root of the molecular mass.

Example 3. A small hole is punched in the wall of a container containing an equimolar mixture of H_2 and N_2. What is the initial ratio of the rate of effusion of H_2 to that of N_2?

Solution: The rate at which molecules pass through the hole is proportional to the product of their concentration in the container times their root-mean-square velocity. The concentrations are equal and molecules in the same container have the same temperature, so that only the mass dependence must be considered:

$$\frac{\text{Rate}_{H_2}}{\text{Rate}_{N_2}} = \sqrt{\frac{M_{N_2}}{M_{H_2}}} = \sqrt{\frac{28}{2}} = 3.74$$

Actually, effusion (molecules passing through a hole without colliding with each other or the walls) only holds at very low pressures. At higher pressures, many collisions of the molecules with the walls and with each other occur as the molecules pass through the hole. This process is called *diffusion*, but it has the same dependence on temperature and mass as does effusion. Gaseous diffusion has been of great importance in separating the isotopes of uranium[7] for military and electric power generating purposes. The separation is performed using gaseous UF_6 and requires many stages, as the mass differences for the different uranium-isotope variations of this compound are very small.

Before leaving this topic, we should note that Bernoulli's model was not the only theory put forth to explain gas behavior. Sir Isaac Newton, the scientific giant of the 17th and 18th centuries, proposed that the ideal gas law could be explained by postulating a repulsive force between stationary molecules. Newton's reputation was so great that there was at first little support for Bernoulli's kinetic model and, in fact, it was ridiculed when first presented before the Royal Society by Waterson. The kinetic theory, however, has proven to be so powerful in explaining many properties of gases that gradually its acceptance won out, notwithstanding Newton's reputation.

1.5 Real Gases: The van der Waals Equation

Although the ideal gas equation is both extremely simple and useful, we know that forces between molecules play an important role in the behavior of matter. What else holds molecules together in liquids and solids? As a first step in treating such forces, we will consider real gases (i.e., gases under conditions in which they do not obey the ideal gas law). It was not too long after the discovery of the ideal gas law that experimental measurements began to indicate that, for some gases (e.g., CO_2) at high pressures or low temperatures, noticeable deviations from ideal gas behavior were exhibited. A satisfying explanation of these deviations in terms of intermolecular forces was first given by van der Waals in 1873.

There are two types of intermolecular forces: *repulsive forces* and *attractive forces*. These are represented by plots of potential energy versus intermolecular separation in Fig. 4, where the forces are given by the negative of the slopes of the curves: $-dV/dr$.

The slope of the potential energy curve for repulsive forces is negative, indicating that the force is in the positive direction (i.e., tending to increase the distance between molecules). Repulsive forces are very short range. They only become important when molecules are very close to each other, but they rise quickly to very large values over a very short distance. Because of this, van der Waals treated the repulsive forces using the concept of an *excluded volume*, [i.e.,

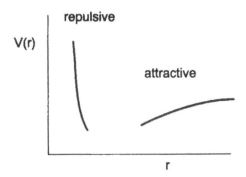

Figure 4   Attractive and repulsive forces.

a region around each molecule into which another molecule cannot penetrate (equivalent to an infinite repulsive force at the boundary of the region)]. Because the translational motion of a particle is the motion of its center of mass, it is the center of mass of one molecule that cannot enter into the excluded volume around another molecule. The van der Waals model treats gaseous molecules as attracting billiard balls.

Van der Waals reasoned that when using an ideal-gas-like treatment, the excluded volume should be subtracted from the actual volume of the container to give the free volume in which the molecules could undergo translational motion. As the excluded volume is proportional to the number of molecules (or moles) in the container, it is written as nb, where b is the excluded volume per mole, and the free volume is $V - nb$. Thus, according to van der Waals, the repulsive force modifies the ideal gas equation to

$$P = \frac{nRT}{V - nb} \tag{19}$$

and results in an increase in pressure over what would be obtained for an ideal gas.

As seen in Fig. 4, the attractive forces between molecules are weaker, but much longer range than the repulsive forces.[8] To find the form of the contribution from the attractive forces to the pressure, van der Waals noted that pressure results from collisions between molecules and the walls of the container. For a molecule colliding with a wall, attractive forces between molecules would result in a force directed toward other molecules in the container and away from the wall. The attractive forces thus result in a reduction in pressure. For any one molecule colliding with the wall, the pressure reduction should be proportional to the density of molecules pulling it back into the gas during collision (N/V). However, the rate at which molecules collide with the walls is also proportional to N/V, giving a reduction in pressure proportional to $(N/V)^2$. Writing this in terms of the number of moles gives the van der Waals equation as

$$P = \frac{nRT}{V - nb} - \frac{an^2}{V^2} \qquad (20)$$

This equation can be written more concisely in terms of molar volume, V_m:

$$P = \frac{RT}{V_m - b} - \frac{a}{V_m^2} \qquad (21)$$

Van der Waals' treatment makes no mention of three or more molecules interacting at the same time, and a billiard-ball-type of excluded volume is quite unreasonable for the "fuzzy" electron clouds required by quantum mechanics. Nevertheless, it still finds extensive use as a first correction to the ideal gas law. The equation is called *semiempirical*, in that, although it is based on physical arguments, it contains two constants, specific for each molecule, which must be evaluated by comparison with experimental data. Values for some of these constants are listed in Table 1. The numbers in such tables may vary somewhat, depending on the pressure and temperature regime in which the fitting to experimental data has been performed. Predictions of the van der Waals equation will be more accurate close to the conditions under which the constants have been determined.

If we take the billiard-ball, or hard-sphere, model literally, we can calculate the excluded volume constant, b, from the diameter of the molecular billiard balls, σ. The centers of two billiard balls, each of radius $\frac{1}{2}\sigma$, can come no closer than $r = \sigma$. Therefore, we can consider that around each molecule there is a

TABLE 1 Van der Waals' Constants of Gases

Gas	a ($N\,m^4/mol^2$)	b ($10^{-5}\ m^3/mol$)
Helium	0.0035	2.370
Neon	0.0214	1.709
Argon	0.136	3.219
Krypton	0.235	3.978
Xenon	0.425	5.105
Nitrogen	0.141	3.913
Ammonia	0.422	3.707
Ethane	0.556	6.380
Water	0.554	3.049
Carbon dioxide	0.364	4.267

Source: Data from A James, M Lord. VNR Index of Chemical and Physical Data. New York: Van Nostrand Reinhold, 1992.

sphere of radius σ and volume $4\pi\sigma^3/3$, into which no other ball can penetrate. Using this as the excluded volume for each molecule would mean that, in a collision, both partners would carry the excluded volume, which would double count the exclusion. Thus, we divide this volume by 2, getting, for b, the excluded volume per mole:

$$b = \frac{2\pi\sigma^3 N_A}{3} = 4v_{mol}N_A \tag{22}$$

where v_{mol}, the volume of a hard-sphere molecule, is

$$\frac{4\pi}{3}\left(\frac{\sigma}{2}\right)^3 = \frac{\pi}{6}\sigma^3$$

The van der Waals b constant should, therefore, be about four times the volume of 1 mol of molecules. Molecular diameters calculated from Eq. (22) are usually in rough agreement with what is expected from molecular theory.

1.6 Real Gases: Other Equations

The van der Waals equation is not the only semiempirical equation of state for gases containing two constants. Another equation is the *Berthelot equation*:

$$P = \frac{RT}{V_m - b'} - \frac{a'}{TV_m^2} \tag{23}$$

This differs from the van der Waals equation by the factor of temperature in the denominator of the second term, which makes the attractive forces relatively less important at high temperature.[9]

A third equation is the *Redlich–Kwong equation*:

$$P = \frac{RT}{V_m - b''} - \frac{a''}{T^{1/2}V_m(V_m + b)} \tag{24}$$

This equation usually gives the best fit to experimental data over a wide range of conditions, although it is difficult to give a physical explanation of the second term. We have used primed and double-primed constants for the Berthelot and Redlich–Kwong equation, respectively, in order to indicate that the constants in the different equations for the same gas are different. In the literature, the constants for all these equations are often designated as a and b. The van der Waals, Berthelot, and Redlich–Kwong equations all approach the ideal gas law at low pressure and high temperature. The van der Waals equation can be solved explicitly for P and T, but it is a cubic in V_m. Solutions for V_m can be obtained by iteration, most easily using a computer program, such as Mathcad or Mathematica. The Berthelot equation is a quadratic in T. Numerical solutions are required for both T and V_m when using the Redlich–Kwong equation.

In Fig. 5, predictions of the ideal gas law, the van der Waals equation, and the Redlich–Kwong equation are shown for ammonia at high (800 K) and low (420 K) temperatures. The high-temperature curves agree well, especially at low densities (high molar volumes). For the low-temperature curves, the ideal gas law is obviously inadequate at all but the largest molar volumes. At this temperature the van der Waals and Redlich–Kwong equations agree very well, except at the highest densities (lowest molar volumes).

A more systematic way of approaching deviations from ideal gas behavior is by means of the *compression factor*, Z, defined as the ratio of the volume of the gas to that of an ideal gas at the same temperature and pressure:

$$Z \equiv \frac{V}{V_{ig}} = \frac{PV_m}{RT} \tag{25}$$

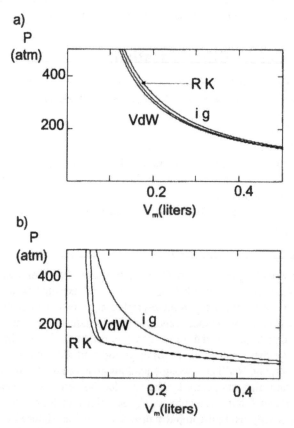

Figure 5 Ideal gas, van der Waals, and Redlich–Kwong equations of state for ammonia: (a) 800 K, (b) 420 K.

[This quantity is often called the *compressibility factor*, a name that is easy to confuse with the isothermal compressibility, defined in Eq. (9).] Z is obviously unity for an ideal gas. At the other extreme, when the pressure or density is very large, excluded volume effects become dominant, $V > V_{ig}$, and $Z > 1.0$.

Because deviations from ideal gas behavior result from intermolecular forces, which go to zero as the average distance between molecules gets very large, we expect Z to approach unity for real gases, as their molar density, $n/V = 1/V_m$, approaches zero. This suggests that it might be useful to expand Z in a series of powers of the molar density:

$$Z = 1 + B\left(\frac{1}{V_m}\right) + C\left(\frac{1}{V_m}\right)^2 + \cdots \tag{26}$$

Another useful expansion for the compression factor is in terms of pressure:

$$Z = 1 + B'P + C'P^2 + \cdots \tag{27}$$

These expansions are known as the *virial expansions*,[10] and the coefficients B, B', C, C', and so forth, which are functions of temperature, are known as the *virial coefficients*. $B(T)$ and $B'(T)$ are called the *second virial coefficients*. (The first virial coefficient is unity.)

Although the virial expansions might seem very complicated, because they contain an infinite number of terms, their power lies in the fact that usually only a few terms must be considered. As pressure is reduced and molar volume gets very large, the higher terms in the expansion become negligible and only the first two terms need be considered. Equation (26) then becomes

$$Z = \frac{V_m P}{RT} = 1 + \frac{B(T)}{V_m} \tag{28}$$

which can be rearranged to

$$V_m = \frac{RT}{P} + B(T)\frac{RT}{PV_m} \approx \frac{RT}{P} + B(T) \tag{29}$$

The final approximation results from V_m being not too different from the ideal gas molar volume (RT/P) at low and moderate pressures. Equation (29) shows that the second virial coefficient is the first correction to the ideal gas molar volume for real gases. At very low temperature, molecules are strongly influenced by attractive intermolecular forces (in fact, if the temperature is low enough, the gas will liquefy), indicating that the second virial coefficient should be negative at low temperatures. As the temperature is increased, thermal velocities become so large that the weak intermolecular attractive forces can have little effect on molecular motions. The excluded volume, resulting from the repulsive forces, should dominate under these conditions, giving a positive second virial coefficient. The typical variation of B with temperature is shown in Fig. 6. The

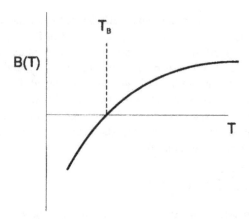

Figure 6 Variation of the second virial coefficient with temperature.

temperature at which the second virial coefficient is zero is called the *Boyle temperature*:

$$B(T_B) = 0 \tag{30}$$

At this temperature, attractive and repulsive influences on the molar volume just cancel and the gas behaves most ideally at low and moderate pressures, for which Eq. (29) is a good approximation.

Example 4. Find an expression for the Boyle temperature of a van der Waals gas.

Solution: In order to set the second virial coefficient equal to zero, we must put the van der Waals equation into the virial form (i.e., a power series in $1/V_m$). The second term in the van der Waals equation, being proportional to $(1/V_m)^2$, is already in this form. In order to put the term $RT/(V_m - b)$ into the virial form, we write it as

$$\frac{RT}{V_m - b} = \frac{RT}{V_m}\left(\frac{1}{1 - b/V_m}\right)$$

Because b, the excluded volume per mole (the actual size of the molecules) is much less than V_m, $x \equiv b/V_m$ is less than 1 and we can use the Taylor expansion[11]

$$\frac{1}{1 - x} = 1 + x + x^2 + x^3 + \cdots$$

to give

$$P = \frac{RT}{V_m}\left[1 + \left(\frac{b}{V_m}\right) + \left(\frac{b}{V_m}\right)^2 + \cdots\right] - \frac{a}{V_m^2}$$

or

$$Z = \frac{PV_m}{RT} = 1 + \left(b - \frac{a}{RT}\right)\left(\frac{1}{V_m}\right) + \cdots$$

When this is compared with Eq. (33),

$$B = b - \frac{a}{RT}$$

which is zero at T_B, giving $T_B = a/Rb$.

For most gases (except He, H_2, and Ne), T_B is above 298 K, indicating that at room temperature, $Z < 1.0$ and attractive forces dominate deviations from ideality.

1.7 Condensation and the Critical Point

Condensation refers to the formation of a liquid or solid directly from a gas. (Sometimes, forming a solid directly from a gas is called deposition.) The opposite processes are evaporation or sublimation, respectively. During all of these processes, pressure and temperature cannot be varied separately. If one is fixed, so is the other. It is common experience that water boiling at 1.0 atm pressure remains at 100°C as heat is added and its volume increases, until there is no more liquid remaining in the system. During the process of vaporization, liquid and gaseous water simultaneously exist in the system. The 1.0-atm isobar of water is shown in Fig. 7.

Liquid water has a very small molar volume and a small thermal expansion coefficient, making this part of the isobar almost horizontal. During boiling, liquid and gas coexist, and there is a vertical portion of the isobar as the relative proportions of these two phases change. The isobar acquires its gaseous form only after all of the liquid is converted to gas. As shown in Fig. 8, isotherms show corresponding behavior. If the pressure on water vapor held at 100°C is gradually increased, its molar volume at low pressure will decrease along a hyperbolic curve, similar to those shown in Fig. 1a. At 1.0 atm pressure, when liquid begins to form, the pressure will remain constant until all of the gas condenses.

The pressure at the horizontal part of the curve is known as the *vapor pressure* of the substance at the temperature of the isotherm. The curve representing liquid on the left side of this isotherm is much steeper than that representing gas on the right side of the isotherm. This is due to the much smaller

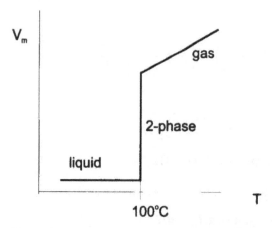

Figure 7 A 1.0-atm isobar of water, showing boiling.

isothermal compressibility, κ, for a liquid than for a gas. In the horizontal segment of this curve, both liquid and gas exist simultaneously in the system. Isobars and isotherms showing vaporization behavior have sudden changes in slope and thus cannot be represented by an analytical equation, such as those we have previously discussed.

If we look at isotherms at successively higher temperatures, we find that the horizontal portion of the isotherm gets shorter and shorter. This results mainly from the fact that at higher temperatures, gaseous molecules must be pushed closer together (to smaller molar volumes) in order for attractive forces to

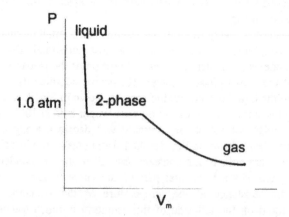

Figure 8 Isotherm, showing condensation of a gas.

overcome their increased kinetic energy. In addition, at higher temperature, thermal expansion gives the liquid a slightly higher molar volume. Eventually, a temperature is reached at which the horizontal segment of the isotherm disappears completely and only an inflection point[12] remains. This is illustrated in Fig. 9, where the width of the 2-phase region decreases at higher temperature. At the inflection point on this isotherm, liquid and gas have the same molar volume and are indistinguishable. The temperature for this limiting isotherm is known as the *critical temperature*, T_c; the inflection point on this isotherm is known as the *critical point* and it occurs at the *critical pressure*, P_c, and *critical molar volume*, $V_{m,c}$. The critical constants of a number of gases are given in Table 2.

Below the critical temperature, a phase transition occurs when compressing a gas. The formation of a liquid phase is usually first noted by the formation of droplets on the walls of the container. At temperatures above the critical temperature, a substance can be continuously compressed without a separate liquid phase forming. Under such conditions, the substance is a gas, because it continues to fill its container. However, because densities comparable to those of the liquid can be reached by such compression, it is customary to call a substance above its critical temperature a *supercritical fluid*, where the term *fluid* (from flow) refers to either liquid or gas. Supercritical fluids, with densities comparable to liquid and high thermal energy, can be exceedingly good solvents and have found use recently in processes such as decaffeination of coffee.

A substance can be characterized by its *critical constants*, T_c, $V_{m,c}$, and P_c. These constants provide information about the intermolecular potential energy of interaction in the substance, roughly related to the maximum attractive energy, the

TABLE 2 Critical Constants of Gases

Gas	P_c (10^6 Pa)	T_c (K)	$V_{m,c}$ (10^{-4} m^3/mol)
Helium	0.229	5.2	0.58
Neon	2.72	44.4	0.42
Argon	4.86	150.8	0.75
Krypton	5.50	209.4	0.92
Xenon	5.84	289.7	1.19
Nitrogen	3.40	126.3	0.90
Ammonia	11.28	405.6	0.72
Ethane	4.88	305.4	1.48
Water	22.12	647.4	0.55
Carbon dioxide	7.38	304.2	0.94

Source: Data from A James, M Lord. VNR Index of Chemical and Physical Data. New York: Van Nostrand Reinhold, 1992.

distance at which this maximum occurs, and the variation of the energy with distance, respectively.[13] The constants in an equation of state, such as that of van der Waals, can be expressed in terms of the critical constants of a gas in the following manner. Because at temperatures just below the critical temperatures, isotherms have a horizontal segment, the critical point must be a horizontal inflection point. In other words, both dP/dV_m and d^2P/dV_m^2 must be zero at the critical point. Applying these conditions to the van der Waals equation [Eq. (21)] gives

$$\frac{dP}{dV} = -\frac{RT_c}{(V_{m.c} - b)^2} + \frac{2a}{V_{m,c}^3} = 0 \quad \text{or} \quad \frac{RT_c}{(V_{m.c} - b)^2} = \frac{2a}{V_{m,c}^3} \tag{31}$$

and

$$\frac{d^2P}{dV^2} = \frac{2RT_c}{(V_{m.c} - b)^3} - \frac{6a}{V_{m,c}^4} = 0 \quad \text{or} \quad \frac{2RT_c}{(V_{m.c} - b)^3} = \frac{6a}{V_{m,c}^4} \tag{32}$$

Dividing Eq. (31) by Eq. (32) gives

$$\frac{V_{m,c} - b}{2} = \frac{V_{m,c}}{3} \quad \text{or} \quad b = \frac{V_{m,c}}{3} \tag{33}$$

Substituting this into Eq. (31) then gives

$$a = \frac{9RT_c V_{m,c}}{8} \tag{34}$$

Equations (33) and (34) are not the optimum method of determining a and b, as $V_{m,c}$ is usually measured with much less accuracy than P_c or T_c. However, substitution of these equations into the van der Waals equation [Eq. (21)], evaluated at the critical point, gives

$$V_{m,c} = \frac{3RT_c}{8P_c} \tag{35}$$

Upon substitution of this into Eq. (33) and (34), we obtain

$$a = \frac{27R^2T_c^2}{64P_c}, \qquad b = \frac{RT_c}{8P_c} \quad \text{for the van der Waals equation} \tag{36}$$

When these values of a and b are used, the van der Waals equation fits the critical point and the slope and curvature of the critical isotherm. By continuity, the equation should also be a good fit to experimental data at temperatures slightly above the critical point. Other values of the constants may provide a better fit to data at conditions far from the critical point. Because it is an analytical function, the van der Waals equation cannot reproduce the discontinuities characteristic of vaporization shown at the two-phase regions in Figs. 7–9. Equation (21) (as well as other two-parameter equations, such as those of Berthelot, and Redlich and

Kwong) oscillate in the two-phase region. A procedure for interpreting the van der Waals equation in this region will be given in Chapter 6. The Berthelot and Redlich–Kwong equations can be analyzed in a similar manner, giving the following for the characteristic constants of these equations in terms of T_c and P_c:

$$a' = \frac{27R^2 T_c^3}{64P_c}, \qquad b' = \frac{RT_c}{8P_c} \quad \text{for the Berthelot equation} \tag{37}$$

$$a'' = \frac{R^2 T_c^{5/2}}{2.34P_c}, \qquad b'' = \frac{0.260RT_c}{3P_c} \quad \begin{array}{l}\text{for the Redlich–Kwong}\\ \text{equation}\end{array} \tag{38}$$

Using Eqs. (35) (solved for R) and (36) to eliminate R, a and b from the van der Waals equation [Eq. (21)] gives

$$\left(\frac{P}{P_c}\right) = \frac{8}{3}\left(\frac{T}{T_c}\right)\left(\frac{V_m}{V_{m,c}} - \frac{1}{3}\right)^{-1} - 3\left(\frac{V_m}{V_{m,c}}\right)^{-2} \tag{39}$$

This equation is now in a very interesting form. By defining new variables $P_r \equiv P/P_c$ (the *reduced pressure*), $T_r \equiv T/T_c$ (the *reduced temperature*), and $V_{m,r} \equiv V_m/V_{m,r}$ (the *reduced molar volume*), the equation takes the form

$$P_r = \frac{8}{3}T_r\left(V_{m,r} - \frac{1}{3}\right)^{-1} - 3V_{m,r}^{-2} \tag{40}$$

This equation has a universal form that is seemingly independent of molecular species. We say "seemingly" because, in fact, we need to know the critical constants of the species in order to calculate the reduced variables P_r, T_r, and $V_{m,r}$. Equation (40) is one example of the principle of *corresponding states*, which states that different substances have closely similar properties as functions of reduced variables. The principle is usually applied by using graphs of the compression factor as a function of P_r and T_r, as shown in Fig. 10. These graphs are usually based on a large amount of experimental data, rather than the use of a single equation, such as that of van der Waals.

Example 5. Using the law of corresponding states, as expressed in Fig. 10, calculate the density (g/cm³) of ethane at 100°C and 100 atm. For ethane, $T_c = 305\,\text{K}$ and $P_c = 49\,\text{atm}$.

Solution:

$$T_r = \frac{(100 + 273)\,\text{K}}{305\,\text{K}} = 1.22, \qquad P_r = \frac{100\,\text{atm}}{49\,\text{atm}} = 2.04$$

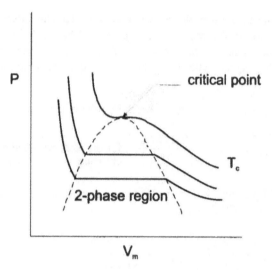

Figure 9 The critical point.

From Fig. 10, on the $T_r = 1.2$ curve at $P_r = 2.0$,

$$Z \approx 0.65 = \frac{PV}{nRT}$$

$$\rho = \frac{Mn}{V} = \frac{MP}{ZRT} = \frac{30 \text{ g}}{\text{mol}} \frac{100 \text{ atm}}{0.65} \frac{\text{mol K}}{0.082L \text{ atm}} \frac{1}{373 \text{ K}}$$

$$= 150 \frac{\text{g}}{L} = 0.15 \frac{\text{g}}{\text{cm}^3}$$

1.8 Gas Mixtures

A *solution* is defined as a mixture that is homogeneous on the macroscopic scale. Therefore, gas mixtures are solutions. In the limit of ideal gases, the total force on a wall of the container is the sum of the forces due to the collisions of the molecules of each component of the gas mixture. The total pressure is consequently the sum of the *partial pressures* of the components:

$$P = \sum_i P_i \tag{41}$$

where P_i, the partial pressure of component i in the mixture, is just the pressure that would exist in the container if it only contained the N_i molecules (n_i mol) of species i. This is *Dalton's law of partial pressures*. For an ideal gas, Eq. (41) can be written

$$P = \sum_i \frac{n_i RT}{V} = \frac{n_{tot} RT}{V} = \sum_i \frac{n_i}{n_{tot}} \frac{n_{tot} RT}{V} = \sum_i x_i P \tag{42}$$

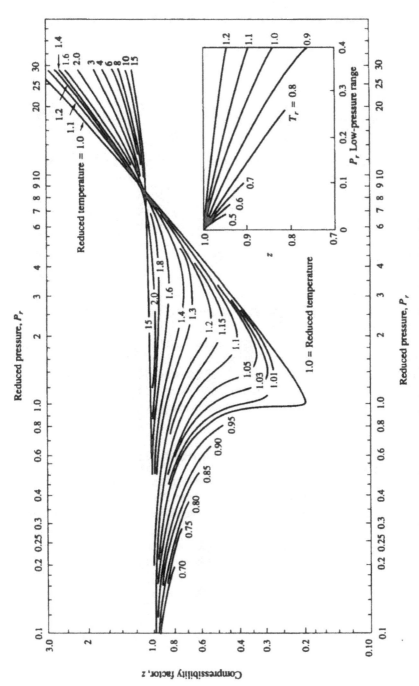

Figure 10 The law of corresponding states, $Z(T_r, P_r)$.

where $x_i = n_i/n_{tot}$ is the *mole fraction* of species i. A comparison of Eqs. (41) and (42) shows that

$$P_i = x_i P \tag{43}$$

Summing mole fractions gives

$$\sum_i x_i = \sum_i \frac{n_i}{n_{tot}} = \frac{1}{n_{tot}} \sum_i n_i = 1 \tag{44}$$

Because mole fractions must sum to unity, in a mixture of k components, only $k - 1$ mole fractions must be specified, in particular, in a *binary* (two-component) mixture, $x_B = 1 - x_A$.

Amagat's law states that the volume of a gas is the sum of the volumes that its components would have, each at the temperature and pressure of the mixture:

$$V = \sum_i V_i, \qquad V_i = \frac{n_i RT}{P} \tag{45}$$

the *partial molar volume* of component i.

For ideal gases, Amagat's law follows directly from Dalton's law (Problem 12).

Dalton's law is not very accurate under conditions at which deviations from ideal behavior are appreciable. The reason for this is that the densities at which partial pressures are calculated may be considerably different from those existing in the gas mixture. Amagat's law is usually more accurate, because the V_i's are calculated at the temperature and pressure existing in the mixture.

1.9 Equations of State of Condensed Phases

In liquids and solids, particles are very close together and repulsive forces play a much more important role than they do in gases. It is not surprising, therefore, that equations of state, such as those of van der Waals, and Redlich and Kwong, do not do a good job of predicting liquid and solid phase data. Usually, data for liquids and solid are presented in the form of thermal expansion coefficients and isothermal compressibilities. Some data for α and κ of selected liquids and solids are given in Table 3.

α and κ are usually expressed as power series in temperature and pressure, respectively:

$$\alpha \equiv \frac{1}{V}\left(\frac{\partial V}{\partial T}\right)_P = \alpha_0 + \alpha_1 T + \alpha_2 T^2 + \cdots \tag{46}$$

$$\kappa \equiv -\frac{1}{V}\left(\frac{\partial V}{\partial P}\right)_T = \kappa_0 + \kappa_1 P + \kappa_2 P^2 + \cdots \tag{47}$$

TABLE 3 Thermal Expansion Coefficients and Isothermal Compressibility of Liquids and Solids

Substance	α $(10^{-6}\ K^{-1})$	κ $(10^{-6}\ atm^{-1})$
Water	210	45
Ethanol	1400	110
Copper	50	0.77
Steel (typical)	30	0.6
Rubber (typical)	1500	5000
Glass (typical)	20	2

Source: Data from DR Lide, ed. Handbook of Chemistry and Physics. 1966; EU Condon, H Odishaw, ed. Handbook of Physics. 1958; A James, M Lord. VNR Index of Physical Data. New York: Van Nostrand Reinhold, 1992.

In using tabulated data, it should be noted that the α_n's are also functions of pressure and the κ_n's are functions of temperature.

Often for solids, *linear thermal expansion coefficients*, $\alpha_L \equiv (1/L) \times (\partial L/\partial T)_P$, are tabulated. For an *isotropic* substance (the same in all directions), we can relate α_L to the volumetric thermal expansion coefficient, defined in Eq. (8), by considering a cube with $V = L^3$:

$$\alpha_V \equiv \frac{1}{V}\left(\frac{\partial V}{\partial T}\right)_P = \frac{1}{L^3} 3L^2 \left(\frac{\partial L}{\partial T}\right)_P = 3\alpha_L \tag{48}$$

α_L for copper is $16.6 \times 10^{-6}\ K^{-1}$. Special alloys have been designed with α_L as low as 10^{-7} (e.g., Invar alloy = 64% Fe and 36% Ni). For some substances (rubbers under tension), α_L has opposite signs in different directions (see Chapter 4).

A rather accurate expression for the compressibility of most liquids is the *Tait equation*[14]:

$$\kappa = -\frac{C}{V[f(T) + P]} \tag{49}$$

where C is a constant and the function $f(T)$ is usually expressed as a power series in T.

A liquid does not have a fixed shape, so the surface area of a liquid can be easily changed. (The surface area of solids can also be changed by processes such as grinding. However, this requires a considerable amount of energy.) In condensed phases, molecules on the surface have a different environment from molecules in the bulk; therefore, a measure of the surface area is necessary to completely define the state of the system. In Chapter 11, we will discuss surface effect in liquids by use of the *surface tension*, γ, which is the extra energy per unit increase of surface area of a sample. An empirical equation of state, giving the temperature dependence of γ, is given in Chapter 11.

In liquids and gases, the pressure force at a point is transmitted equally in all directions. This is not the case in solids. The shape of a solid can be changed by applying a force, f, along a particular direction. The volume of a solid can often be considered constant during the application of a force, requiring that as the solid extends in one direction, it contracts in perpendicular directions. The fractional extension of the solid in the direction of the applied force is known as the *strain*, ε:

$$\varepsilon \equiv \frac{l - l_0}{l_0} \tag{50}$$

where l and l_0 are the length and initial length of the solid.

Because a given force is obviously more effective in producing strain when it is applied to an object of small cross-sectional area, we divide the force by the area perpendicular to its application, and obtain the *stress*, f^*:

$$f^* \equiv \frac{f}{a} \tag{51}$$

Note that stress has the same units as pressure; it is a pressure along a single direction.

Young's modulus, E, for a material is defined as

$$E \equiv \frac{df^*}{d\varepsilon} \tag{52}$$

It is the slope of the stress–strain curve. If a material is *elastic*, it recovers to its original length if the applied stress is removed. In general, E depends on the strain of the material. However, for many substances, if the stress is not too large, E is independent of strain. Over this range, the substance is said to obey *Hook's law*:

$$f^* = E\varepsilon \tag{53}$$

A typical stress–strain diagram for a metal is shown in Fig. 11. This metal follows Hook's law up to a *proportional limit* (or *yield strength*) of 2×10^9 Pa. The elastic limit, above which the metal undergoes *plastic deformation*, which is not recoverable when the stress is removed, is close to the proportional limit. The maximum stress that the metal can support is the *ultimate strength* (or *tensile strength*) of the metal, which occurs at the maximum extension of the material.

Nonmetallic substances show a wide variety of stress–strain diagrams, with each type related to the bonding of the particular material. One type, that for an elastomeric rubber, is shown in Fig. 12. Compared to a metal, the rubber can support much larger strains and only much smaller stresses. It follows Hook's law only as a limit at very small displacements. However, displacement is elastic well outside of this range.

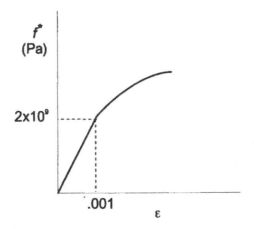

Figure 11   A typical stress–strain diagram for a metal.

Only at very low elongation is there a linear relation between f^* stress and strain. At somewhat higher strain, the curve in Fig. 11 can be represented by

$$f^* = a\varepsilon - b\varepsilon^2 \tag{54}$$

where a and b are constants.

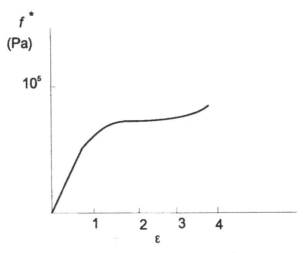

Figure 12   Stress–strain diagram for an elastomer.

1.10 Pressure Variations in Fluids

When external fields are considered, mechanical equilibrium no longer corresponds to a state of uniform pressure. This is most important for systems in a gravitational field. Equating the upward and downward forces on the volume element of fluid shown in Fig. 13 gives

$$PA = (P + dP)A + g\rho A \, dz \tag{55}$$

or

$$dP = -\rho g \, dz \tag{56}$$

A liquid can usually be considered *incompressible* (ρ constant). Integrating from $P = P_0$ at $z = 0$ to a distance h *below* the surface of the liquid ($z = -h$),

$$P \overset{\text{liq}}{=} P_0 + \rho g h \tag{57}$$

In a gas, density depends on pressure. Assuming the gas is ideal,

$$\rho = \frac{nM}{V} = \frac{PM}{RT} \tag{58}$$

giving

$$dP \overset{\text{ig}}{=} -\frac{PMg}{RT} \, dz \tag{59}$$

Separating variables and integrating,

$$\int_{P_0}^{P(h)} \frac{dP}{P} = -\int_0^h \frac{Mg}{RT} \, dz \tag{60}$$

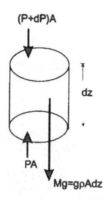

Figure 13 Fluid in a gravitational field.

Integrating with the assumptions of constant temperature and molecular weight,

$$P = P_0 \exp\left(-\frac{Mgh}{RT}\right) \tag{61}$$

This relation is known as the *barometric distribution* and is a first approximation to the pressure distribution in the atmosphere. For a more accurate representation of the pressure variation, Eq. (59) must be integrated, using the known temperature variation in the atmosphere. (The average molecular weight of atmospheric gases is, to a very good approximation, invariant throughout the troposphere and stratosphere, the two lowest regions of the atmosphere.)

Questions

1. Which of the following laws or principles: immutability of atoms, conservation of mass, and Newton's laws, applies to each of the following phenomena:

   (a) Motion of a baseball
   (b) Motion of an electron in an atom
   (c) A thermonuclear explosion

2. Give one reason why Charles' law does not hold to $-273.15°C$.

3. What is the difference between a pressure of $1.0\,atm$ and atmospheric pressure?

4. Choose a property of materials with which you have some familiarity (e.g., mass, viscosity, thermal conductivity or electrical resistance) and consider how you would measure this property. Are you inside or outside the system when you make the measurement?

5. Classify the following properties of a system as extensive or intensive: volume, pressure, energy, thermal expansion coefficient, and viscosity.

6.* If each of the following was included in the kinetic theory of gases, do you think the result [Eq. (16)] would change? Qualitatively justify your answer.

   (a) Allowing for motion of particles in three dimensions
   (b) Allowing for collisions of point particles
   (c) Allowing for inelastic collisions with the walls

7. Show that the van der Waals equation approaches the ideal gas law as $P \to 0$ and as $T \to \infty$. (Hint: As an estimate of the magnitude of V_m, use the value for the ideal gas.)

8. For a strict interpretation of the van der Waals equation, how does the potential energy of interaction of molecules have to vary as a function of intermolecular separation?

9. What are the units of a'' and b'' in the Redlich–Kwong equation?

10. On the corresponding states graph (Fig. 10), where is the critical point? Indicate the region that represents gases. Roughly sketch in the two-phase region and indicate the region that represents liquids.

11. Sketch the compression factor as a function of $1/V_m$ for

   (a) $T > T_B$
   (b) $T < T_B$

12. Give two properties of gases that are better explained using the kinetic theory of gases than Newton's idea of gas properties resulting from repulsive forces between stationary gas molecules.

13. Sketch an isochore of a substance showing condensation.

14. What is a typical ratio of the maximum strain of a rubber to that of a metal?

15.* The Tait equation predicts that the isothermal compressibility of a liquid approaches zero as the pressure becomes infinitely large. Is this reasonable in terms of the van der Waals model of real fluids?

16. Show that a perfectly elastic substance dissipates zero energy upon repeated stress–strain cycles. Draw a hysteresis diagram (a diagram that shows the states of the material during a full extension–relaxation cycle) on a figure similar to Fig. 11 for a material that relaxes along a different path than that along which it expands. Show that this leads to energy dissipation upon repeated stress–strain cycles.

Problems

1. What is the value of the gas constant in units that use pounds per square inch (psi) for pressure and cubic feet (ft^3) for volume?

2. There are about 20 drops in 1 mL of water. What is the volume of Avogadro's number of drops of water? Compare this with the volume of the Earth's oceans, 1.4×10^{18} m^3.

3. Calculate the mass of the air at standard temperature and pressure (STP) in a room of dimensions $20\,ft \times 20\,ft \times 8\,ft$. You can take $29.0\,g/mol$ as the average molecular weight of air. Compare your result with the mass of water it would require to fill the room.

4. Using the chain rule for partial derivatives [Eq. (7) of Appendix A] express $(\partial P/\partial T)_V$ in terms of the thermal expansion coefficient α and the isothermal compressibility κ.

5. What is the ratio of effusive velocities of $^{235}UF_6$ and $^{238}UF_6$? What would be the fraction of the ^{235}U isotopomer effusing from a mixture of hexafluorides that contains 0.50% ^{235}U (the percentage of the isotope in natural abundance)?

6. In a 1.0-L sample of N_2 gas (10.0 cm on a side) at 25°C and 1.0 atm pressure, what is the rms velocity of N_2 molecules? What is the average z component of velocity? Estimate how often a single molecule hits a particular wall. What is the total rate of impacts with this wall? Do you think that there is any pressure gauge that could respond to these individual impacts?

7. Using the van der Waals b constants given in Table 1, calculate the atomic diameter of the noble gases. Do atoms have a definite size?

8.* Use the van der Waals equation for N_2 to determine the following:

(a) The pressure at 25°C and a molar volume of 2.0 L/mol
(b) The molar volume at 10 atm and 300 K
(c) The molar volume at 10 atm and 100 K

Is the ratio of the results in parts (b) and (c) what you would expect? If not, why?

9.* Repeat Problem 9 using the Redlich–Kwong equation. You can calculate the Redlich–Kwong constants for N_2 from its critical constants.

10. Find the Boyle temperature for a gas obeying the Berthelot equation.

11. Show that at the Boyle temperature, $\lim_{P \to 0}(dZ/dP) = 0$ and $\lim_{P \to 0}(dZ/dV_m) = 0$.

12. Show that the requirement that Dalton's law holds for an ideal gas leads to Amagat's law also holding.

13. Use Fig. 10 to determine the molar volume of N_2 at the conditions given in Problem 8, parts (b) and (c).

14. Derive Eq. (39) from Eqs. (21), (35), and (36).

15.* For a condensed phase with thermal expansion coefficient given by Eq. (46), show that the volume can be expressed as a function of temperature by $V = V_0(1 + \beta_0 T + \beta_1 T^2 + \cdots)$, where V_0 is the volume at 0 K. Find expressions for β_0 and β_1.

16. If a round metal bar undergoes a strain of 0.001 when a force is applied at constant volume, what is the fractional change of the diameter of the bar?

17. Find Young's modulus for a rubber whose stress–strain curve obeys Eq. (54).

18. If the pressure at sea level is 760 mm Hg, what are the pressures 1.0 m and 1.0 km above sea level? Assume an altitude-independent atmospheric temperature of 25°C and an average molecular weight of the atmosphere of 29 g/mol.

19. What is the pressure 1.0 km below the surface of the ocean, assuming a depth-independent density of ocean water of 1.0 g/cm?

20.* Repeat Problem 18 at 1.0 km for a temperature that drops linearly with elevation at a rate of 6.5°C/km.

21.* Consider the following apparatus used to measure the molar volumes of coexisting liquid and gaseous phases:

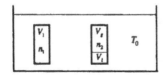

Two thick-walled glass tubes of known volumes are each filled with measured amounts of a substance. The temperature of the bath is slowly raised until the liquid phase in one of the tubes (say tube 1) just disappears. At this temperature, the volumes of the liquid and gaseous phases in the other tube are V_l and V_g. Show that the molar volumes of the coexisting liquid and gaseous phases at T_0 are

$$V_{m,g}(T_0) = \frac{V_1}{n_1}, \qquad V_{m,l}(T_0) = \frac{V_1 V_l}{n_2 V_1 - n_1 V_g}$$

22. For a Redlick–Kwong gas, find the following:

   (a) $\left(\dfrac{\partial P}{\partial T}\right)_{V_m}$

   (b) $\left(\dfrac{\partial P}{\partial V_m}\right)_T$

   (c) $\alpha = \dfrac{1}{V_m}\left(\dfrac{\partial V_m}{\partial T}\right)_P$

23. A mixture of O_2 and F_2 at 520 torr pressure is shaken with mercury. All the F_2 and none of the O_2 reacts and the pressure drops to 430 torr. What was the mole fraction of F_2 in the original mixture?

Notes

1. We will assume that Charles measured temperature with a mercury thermometer calibrated for the centigrade (Celsius) scale.
2. This is sometimes called the Kelvin temperature and measured in degrees Kelvin. More rigorously, the Kelvin temperature scale is defined by the second law of thermodynamics, but, numerically, it is the same as T. We will use the symbol K to designate absolute temperature as defined by Eq. (3).
3. The ideal gas law will also be derived from the point of view of statistical mechanics in Chapter 5.
4. If body A exerts a force on body B, then body B exerts an equal but opposite force on body A.
5. This is called the ensemble average, to distinguish it from the time average.
6. Note that the operations are performed in reverse order from which they occur in the name: square first, then average, and then square root.
7. The ^{235}U isotope undergoes fission, the ^{238}U isotope, which comprises 99.28% of naturally occurring uranium, does not.
8. The forces between neutral molecules have a much shorter range than the forces between charged particles.
9. Because molar volume usually increases with temperature, even the van der Waals equation, where the attractive forces term does not depend on temperature, predicts decreasing influence of these forces at higher temperatures.
10. From Latin, meaning "force."

11. The general formula for a Taylor expansion is

$$f(x) = f(0) + x \frac{df}{dx}\bigg|_0 + \frac{1}{2}x^2 \frac{d^2f}{dx^2}\bigg|_0 + \cdots + \frac{1}{n!}x^n \frac{d^nf}{dx^n}\bigg|_0 + \cdots$$

The terms represent successive approximations to a function at small values of its argument by the value of the function at the origin, the straight-line approximation of the function at the origin, the parabolic approximation of the function at the origin, and so forth.

12. An inflection point is where a curve changes from concave up to concave down. Mathematically, at an inflection point of a curve $f(x)$, $d^2f/dx^2 = 0$.

13. DA McQuarrie. Statistical Thermodynamics. New York: Harper & Row, 1973, Chap. 12.

14. JO Hirschfelder, CE Curtiss, RB Bird. Molecular Theory of Gases and Liquids. New York: Wiley, 1954, p 261.

2

Thermodynamics: The Zeroth and First Laws

A theory is the more impressive the greater the simplicity of its premises, the more different kinds of things it relates and the more extended its area of applicability.

Albert Einstein on thermodynamics

Facility with thermodynamics requires understanding the precise definition of many concepts, including different types of systems, properties, processes, and equilibria. We begin by considering systems isolated from their surroundings and systems into which energy, but not matter, can flow from the surroundings (closed systems). Measurements, including those of heat and work, are made on systems by observers in the surroundings. Thermodynamically, temperature is defined as the property that determines the propensity of a system to transfer heat, and the zeroth law asserts that this property is unique. In the reversible limit of real processes, a system is in equilibrium with its surroundings and processes can be represented on P versus V work diagrams. The first law states that the energy of the universe is constant in any process. The internal energy of a system, U, is a state function, and the change of U in a constant-volume process is the heat added to the system. Likewise, the change of enthalpy, H, is the heat added in a constant-pressure process. Heat capacities at constant volume and pressure are related to the rate of change of U and H, respectively, with temperature in such processes. The Joule and Joule–Thomson processes provide information on the variation of U and H with volume and pressure, respectively. The efficiency of a Carnot cycle engine (a reversible engine that transforms heat to work on a continuous basis) is analyzed using the first law of thermodynamics.

2.1 The Nature of Thermodynamics

Few endeavors reward precise thinking and penalize sloppy thinking as much as the study of thermodynamics. In this subject, we take familiar concepts, such as heat and energy, rigorously define them, and derive a multitude of relationships that are useful in many branches of science. By the above criteria of Einstein, thermodynamics is impressive indeed!

Thermodynamics is applicable to the types of substances and occurrences that we are familiar with in our daily lives or can easily create in the laboratory. As a result, it is easy to compare the predictions of thermodynamics with "real-world" behavior. Needless to say, there are impressive amounts of data that verify the conclusions of this subject. Relatively few examples of the agreement of experiment with thermodynamic theory are given in this volume. The reader is referred elsewhere for additional examples of the conformity of thermodynamic theory and experiment.[1]

Thermodynamics is not unique in dealing with energy. Energy considerations are also of paramount importance in fields such as mechanics, electricity and magnetism, and atomic and molecular structures. In these other fields, the energy of individual particles are discussed. The types of energy that they recognize are kinetic energy (the energy of motion), potential energy (the energy of position), and the energy of electric and magnetic fields. Thermodynamics is different from these other fields in that it considers the energy of entire systems, consisting of huge numbers of particles (and perhaps radiation fields). The approach of thermodynamics is totally macroscopic and its conclusions are not based on any particular model for the behavior and nature of the microscopic particles.

2.2 Systems

We live in a very complicated universe. Clearly, not even with the most powerful computers would we be able to study the details of all parts of the universe. If we are to make any progress in thinking about energy, we must focus our attention on only part of the universe. This part we call the *system*. The remainder of the universe is termed the *surroundings*.

In some cases, it will be possible to consider the system as *isolated* (i.e., not interacting with the surroundings). In order to be isolated, the boundaries of a system must be impermeable to mass and energy. Such boundaries cannot allow any interaction with external mechanical or electrical forces. For example, if there is an external pressure, the walls of the system must be rigid so that they cannot be moved by the pressure. In addition, the system must also be *adiabatic* (i.e., not allowing any energy to flow through the walls in the absence of such forces).

In most cases, we will be forced to define our system in such a way that it exchanges mass or energy with the surroundings. Systems that can exchange energy, but not mass, with the surroundings are called *closed* systems; those that can exchange both energy and mass are called *open* systems. It is not necessary to consider systems that exchange mass, but not energy, with the surroundings, because the transferring mass will bring its internal energy with it into the system. We will start by dealing exclusively with isolated and closed systems and then extend our considerations to open systems, beginning with Chapter 6.

2.3 Equilibrium

Equilibrium is a very important concept in discussions of thermodynamics. An isolated system is at equilibrium when it has no tendency to change—a condition that is called *internal equilibrium*. This implies that the system is at *mechanical equilibrium* (i.e., it has no tendency for bulk movement of material), *thermal equilibrium* [i.e., it has no tendency for transport of energy (without bulk movement of material)], and *material equilibrium* [i.e., it has no tendency for material to change form (such as by a phase transformation or a chemical reaction)].

It is somewhat more difficult to decide whether a system with no tendency to change is at equilibrium if it is not isolated (i.e., when it is closed or open). In such cases, we apply the mental test of removing the interaction between the system and its surroundings. If the system still has no tendency to change, then we say that (1) the original system was at internal equilibrium and (2) the original system was at *equilibrium with its surroundings*.

A system that relies on its surrounding to remain unchanged, such as a living organism, is at *steady state*. In the first 11 chapters of this volume, we will discuss properties of systems at equilibrium and processes in which systems change from one equilibrium state to another (each defined by appropriate constraints). In Chapter 12, we will briefly touch on systems at steady state.

We are rarely interested in systems at complete equilibrium, where all conceivable chemical (and nuclear) reactions have no tendency to occur. Usually, there are real or imagined *constraints* in our system, related to the *rates* of certain processes. For example, at most temperatures, we can completely ignore nuclear reactions. Also, in a mixture of H_2 and O_2 at low temperature, without catalysts, we may choose to ignore the formation of water, because it is very slow in reality. An imagined constraint is often convenient to use in treating chemical reactions, where the actual reaction step is separated from an additional step, in which the energy liberated in the reaction is used to heat the reaction products.

2.4 Properties

A complete description of a system, of sufficient detail to describe its macro-scopic behavior, is called the *state* of the system. A system has a particular state independent of the way in which it has arrived at that state. A system is characterized by measuring its *properties*. For systems at equilibrium, specifica-tion of a few properties will suffice to determine all other properties of the system and, thus, its state. Quantities that are determined by defining the equilibrium state are called *state functions*. After a short period of adjusting to their constraints, gases and liquids can usually be assumed to be at equilibrium for all physical changes[2] (but not necessarily for chemical reactions). For solids with strong bonds, such as high-melting-point metals, properties may depend on the previous history of a particular sample (e.g., heat treatment) and differ from the equilibrium properties for a very long time.

In thermodynamics, the observer is outside the system and properties are measured in the surroundings. For example, pressure is measured by an external observer reading a pressure gauge on the system. Volume can be determined by measuring the dimensions of the system and calculating the volume or, in the case of complex shapes, by using the system to displace a liquid from a filled container. Important thermodynamic properties have low information content (i.e., they can be expressed by relatively few numbers). The details of the shape of a system are usually not important in thermodynamics, except, sometimes, a characteristic of the shape, such as the surface-to-volume ratio, or radii of particles, may also be considered. Information only accessible to an observer within the system, such as the positions and velocities of the molecules, is not considered in thermodynamics. However, in Chapter 5 on statistical mechanics, we will learn how suitable averages of such microscopic properties determine the variables we study in thermodynamics.

As discussed in Chapter 1, properties are either *extensive* or *intensive*, depending on whether they are proportional to the size of the system or not. Thus, mass and volume are extensive properties; temperature and pressure are intensive properties. An extensive property multiplied by an intensive property remains extensive. Because the ratio of two extensive properties is intensive, dividing an extensive property by the number of moles, the mass, or the volume of a system gives an intensive property. An example of an intensive property is the density $\rho \equiv m/V$. An intensive property has a value in each region of a system. A *uniform* intensive property is constant throughout the system; if a system has all of its intensive properties uniform, it is *homogeneous*, compared to a *heterogeneous* system, in which some properties vary in the system. In heterogeneous systems, properties may vary continuously or discontinuously in space.

2.5 Processes

In thermodynamics, we are interested in *processes*, during which the state of a system changes. A process is defined by the external constraints that exist during the changes. The *path* of a process is the continuous sequence of states in which the system exists during the process. Often, a system will begin a process in some initial state in which it is in equilibrium and end the process in a final state in which it is also in equilibrium. The intermediate states of the process in general will have tendency for change and will, thus, be nonequilibrium states. Nonequilibrium states are very complicated; their properties are usually not uniform and very difficult to specify. However, the properties of the initial and final states, being state functions, are independent of the process by which the system changes between these two states. Therefore, the change of all state properties during the process are also independent of the path of the process. Sometimes, we will want to consider systems, such as engines, which undergo a *cyclic* process. In these cases, if we chose the initial and final states to be at the same place in the cycle, they are identical states with identical properties and the change in any state function for such a process is zero.

2.6 Heat and the Zeroth Law of Thermodynamics

When a system originally at equilibrium is brought into contact with its surroundings, through boundaries that do not permit any mechanical, electrical, or magnetic forces to affect the system or any transfer of matter between the system and its surroundings, two things can happen: the system can change—in this case, we say that thermal energy has transferred between the system and the surroundings. Alternatively, the system may remain unchanged, indicating that no thermal energy has been transferred. Thermal energy is the energy of random motion of particles and its transfer between the surroundings and the system is called *heat*, given the symbol *q*. In keeping with measuring thermodynamic quantities from outside the system, we define heat as follows: *Heat is the decrease in the thermal energy of the surroundings.* By this definition, our sign convention is that positive heat corresponds to thermal energy entering the system. Strictly speaking, it is redundant to talk of "heat transfer" or "heat flow," as heat already is a transfer of thermal energy. However, these phrases are so commonly used that a statement such as "there is heat between A and B" sounds awkward and "thermal energy is transferred between A and B" is ponderous. Therefore, we will not hesitate to use the terms "heat transfer" and "heat flow."

The absence of heat flow may be a result of the walls not permitting the transfer of thermal energy. Boundaries of this kind are called *adiabatic*. (Adiabatic walls are infinitely good thermal insulators.) If the walls are *nonadiabatic* (sometimes called diabatic or diathermal) and do permit heat transfer, but it does not occur, we say that the system is at *thermal equilibrium* with its surroundings.

The laws of thermodynamics are generalizations of experience. They are accepted as fundamental postulates because exceptions to them have never been observed. There are four laws of thermodynamics, labeled the zeroth, first, second, and third laws. The numbering results from the fact that the zeroth law was so intuitive that it was not realized that it was a basic postulate, which, for logical consistency, was necessary for the derivation of the other laws until after the numbering of those laws had become generally accepted. It is the zeroth law that enables us to define temperature scales. The zeroth law is a generalization of what happens when we bring two systems, each in internal thermal equilibrium, in contact through nonadiabatic walls. We will state the zeroth law as follows:

> There is a single property of systems at thermal equilibrium that determines their propensity to transfer thermal energy.

We call this property the *temperature*, and when two systems at the same temperature are brought into thermal contact through nonadiabatic walls, there is no heat transferred between them and they remain at thermal equilibrium. If they are not at the same temperature, heat is transferred spontaneously from the system at higher temperature to the system at lower temperature. All systems that are at thermal equilibrium with one another are at the same temperature.

Heat transfer between two systems is a process that occurs at the surface of each system, and it is the propensity for this surface heat transfer that is dealt with in the zeroth law. Strictly speaking, the zeroth law only deals with the direction of heat transfer and the absence of heat transfer when the systems have finally reached thermal equilibrium with each other. During the process of heat transfer, the two systems are not at thermal equilibrium and they do not each have a unique temperature. The rate of heat transfer during the process also depends on properties of the systems, such as their thermal conductivity.

When dealing with heat transferred between a system and its surroundings, we often idealize the surroundings by considering it as a *heat reservoir*. A heat reservoir is a store of thermal energy in the surroundings at a single temperature and sufficient thermal capacity so that arbitrary amounts of thermal energy can be withdrawn or added to it without noticeably changing its temperature.

Although we will not deal with the rate of heat transfer in this book, thermodynamics does require that it rise monotonically as the temperature difference between the system and its surroundings increases. In addition to considering thermal equilibrium as the condition of zero heat transfer between a system and its surroundings, it is sometimes useful to consider it as the limit of the nonequilibrium situation—as the temperature difference between the system and the surroundings approaches zero. In the limit, the rate of heat transfer approaches zero, a uniform temperature is maintained in the system, and it takes an infinite amount of time for any finite amount of heat to be transferred. Because the direction of heat transfer in this case can be reversed with the slightest change of temperature of the system or surroundings, we say that this heat transfer is *reversible*. Because in the real world we are not able to wait an infinite amount of time to observe changes in systems, we say that reversible processes are not real. However, the reversible limit of processes is of great importance in thermodynamics. It is an easily considered idealized situation against which real processes can be compared.

Of course, it is one thing to say that there is a unique property that determines the propensity to transfer heat and another thing to actually measure that property—the temperature. Usually, it is rather inconvenient to measure temperature by determining a system's tendency to transfer heat to other systems,[3] and we determine it by measuring some property that varies monotonically with temperature. The volume of a given mass of reference material (with pressure held approximately constant or kept very low), such as that of the mercury in the common household thermometer, is usually chosen for this purpose. A measurement of the length of the mercury column is equivalent to a volume measurement, because, to a very good approximation, the dimensions of the glass tube are independent of temperature. We allow the system to reach thermal equilibrium with our *thermometer*, at which point the system and thermometer have the same temperature, and then measure the volume of the reference material. It is assumed that the mass of the thermometer is so much less than that of the system that the system temperature will not be changed in the process of reaching thermal equilibrium. Two systems that give the same reading when they reach thermal equilibrium with a thermometer have the same temperature, and the zeroth law assures us that there will be no heat transferred if these two systems are brought in contact.

Temperature scales such as the centigrade scale discussed in Chapter 1 are quite arbitrary, requiring a choice of two calibration points and the material and property used to interpolate between them. We saw that we could alternatively use a single-calibration-point scale based on a broad class of materials—the ideal gases. A conceptually much simpler temperature scale, based on the second law of thermodynamics, will be discussed in Chapter 3.

2.7 Work

Heat is one way of transferring energy to or from a system; work, the transfer of energy due to a force, is another way. Work is the product of a force and the distance that the point of application of the force moves. For example, a body of mass M in the Earth's gravitational field experiences a gravitation force of magnitude Mg directed downward (g is the acceleration of gravity). If the body moves downward by a distance h, the gravitational field has done work equal to Mgh on the body. An alternative way of describing this process is to say that the body's potential energy in the gravitational field has been reduced by Mgh. These are alternative ways of describing the same process. It is important to not include both the work and the decrease of potential energy in analyzing systems.

As an example more important to thermodynamics, we consider a gas at equilibrium, confined in a cylinder topped by a movable, leak-free, and massless piston, as illustrated in Fig. 1. A mass, M, on the piston exerts a force of Mg on the piston. If the system is at *mechanical equilibrium* with its surroundings, the external and internal forces at the piston boundary are in balance. The pressure in the gas must therefore be $P = Mg/A$, where A is the area of the piston. No work is done at mechanical equilibrium, because there is no motion of the boundary, and work is the product of a force and the distance its point of application moves. If the piston does move, the work is positive if the motion is in the direction of the force. In Fig. 1, if the gas is compressed, the piston moves in the direction of the force and the force does positive work on the system.

Let us say that the mass on the piston is increased to M'. The system is no longer in mechanical equilibrium and the piston will move down, compressing

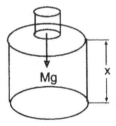

Figure 1 Gas in a cylinder.

the gas, and increasing its pressure until mechanical equilibrium is restored. With the force in the negative x direction, the work done on the gas is

$$w = F\Delta x = (-M'g)\Delta x = -\left(\frac{M'g}{A}\right)(A\Delta x) = -P_{ext}\Delta V \tag{1}$$

The symbol Δ is used generally to express the change of a state function during a process. It is equal to the value of the state function at the end of the process less its value at the beginning of the process. The work is positive, because ΔV is negative (the force and displacement are in the same direction). P_{ext} is the external pressure on the system. We can also discuss work in terms of potential energy. The potential energy of a mass in the Earth's gravitational field is Mgh, where h is measured from some arbitrary reference elevation. From this perspective, the change of mechanical potential energy in the surroundings is $M'g[(h_0 + \Delta x) - h_0] = M'g\Delta x$. We will find it most convenient to define work in the following manner:

> Work is the decrease in mechanical or electrical energy of the surroundings.

This gives $-M'g\Delta x$ for the work—the same result that was calculated in Eq. (1).

From the vantage point of the system, the above process is really very complicated. When the mass on the piston is increased, the downward force on the piston is greater than the upward force, and the piston accelerates downward. During this motion, the gas in the cylinder is unevenly compressed, turbulence is generated, and the gas pressure and temperature rise. Because of its momentum, the piston passes the point of equilibrium and compresses the gas to where its pressure is greater than the pressure caused by the external mass. There is some oscillation of the piston, causing further heating of the gas. After a while, the piston comes to rest and nonuniformity in the temperature and pressure of the gas disappear. However, from the point of view of the surroundings, the process is much simpler; we have a force moving a distance Δx or an external pressure changing the volume of a system by ΔV. If the system is not perfectly insulated, then there may also be heat transferred between the system and the surroundings. This does not affect our calculation of the work. This is the usual agenda of thermodynamics. Measurements are made in the surroundings.

> **Example 1.** A gas is confined in the bore of a cannon of length L by a shell of mass M and cross-sectional area A. Initially, the shell is at a distance L_i from the end of the cannon. Assuming 1 atm pressure, find an expression for the ejection velocity of the shell in terms of the work done by the expanding gas.
>
> *Solution:* We will define the system as the confined gas. Mechanical energy is increasing in the surroundings, due to compression of the

atmosphere and acceleration of the shell. The sum of these is $-w$, the work done by the gas on the surroundings:

$$-w = P_{\text{ext}}\Delta V + \tfrac{1}{2}Mv^2 = (1 \text{ atm})(L - L_i)A + \tfrac{1}{2}Mv^2$$

$$v = \left(\frac{2}{M}(-w - (1 \text{ atm})(L - L_i)A)\right)^{1/2}$$

Thus, not all the work done by the expanding gas goes to accelerate the shell; some goes to compress the atmosphere.

Electrical energy is considered to allow us to discuss batteries and electrochemical cells, as well as motors and resistance heating. If a charge Q is transferred to a system at an electrostatic potential (voltage) ϕ with respect to the surroundings, the work done is $Q\phi$:

$$w_{\text{elec}} = \phi Q = \phi I \Delta t \tag{2}$$

where I is the electric current which flows for time Δt. The SI unit for Q is the Coulomb and ϕ has the derived unit of the electrical potential through which unit charge must be moved to do 1.0 J of work. This unit is called a volt.

In the limit of an infinitesimal amount of motion, an infinitesimal amount of work is done. We write this as

$$\delta w = -P_{\text{ext}} \, dV \tag{3}$$

or

$$\delta w_{\text{elec}} = \phi \, dQ \tag{4}$$

The differential symbol, d, means an infinitesimal change in, and we use the symbol δ to mean an infinitesimal amount of. We make this distinction to emphasize that w is a transfer of energy through the boundaries of a system; it is not a property of the system, like V, that can change. (Note that $\int_i^f dX = X_f - X_i \equiv \Delta X$, whereas $\int \delta x = x$.) The total work done in a process is just the sum of the little bits of work done in each stage of the process. In the limit of a true infinitesimal, we must integrate Eq. (3) to get

$$w = \int \delta w = -\int P_{\text{ext}} \, dV \tag{5}$$

which gives Eq. (1), when P_{ext} is constant.

It is very important to remember that because work is measured in the surroundings, it is the external pressure that determines the work. There are, however, a number of circumstances in which P_{ext} equals P, the system pressure, and this substitution can be made in Eq. (5). These cases are both important practical processes and useful standards against which to compare such processes, and their careful discussion is necessary. One of these situations is where the system is in complete equilibrium, including mechanical equilibrium with the surroundings. In this case, because there is no unbalanced force at the boundary of the system, there is no movement and no work.

To be exact, we should say that in mechanical equilibrium, there is no movement in a finite time. We could imagine, in Fig. 1, doing work by horizontally sliding an infinite number of individual grains of sand from an infinite number of appropriately located platforms onto the piston. In the limit of infinitely small particles of sand, P_{ext} would be equal to P and we would maintain mechanical equilibrium. Do not worry if you cannot imagine just how this could be done. This is an example of a process that is impossible in practice, but does not violate any of the laws of physics. It is called by the German name, a *gedanken* (thought) process. Such a *gedanken* process is called reversible. Because the compression is infinitely slow, any process going on in the system (such as a pressure-dependent chemical reaction) will remain at equilibrium.

Assuming no friction, the above process can be reversed without the addition of energy by sliding the grains back onto the platforms. Consideration of reversible processes will be very important in our theoretical development of thermodynamics. Although reversible processes are not real, there are some real processes which occur so slowly that the system may be considered to be at equilibrium. Geological processes are usually of this type.

Replacement of P_{ext} by P, however, does not require complete equilibrium; mechanical equilibrium with the surroundings is sufficient. In many slow processes, the system pressure closely tracks the external pressure and can be substituted for it in Eq. (5). The most commonly encountered of these is the *constant-pressure process*. Because we define our constraints in the surroundings, a constant-pressure process has constant P_{ext}. If the system has a moveable boundary and *the system is initially in mechanical equilibrium with the surroundings* $(P_i = P_{ext,i})$, then P will remain equal to P_{ext} for the following two processes:

1. The process has no tendency to change the pressure of the system. For example, it may be a chemical reaction that is both thermoneutral and for which the number of moles of gaseous reactants and products are equal.
2. The process does tend to change the pressure of the system, but it occurs slowly enough so that the system boundary can move to maintain $P = P_{ext}$. Reactions occurring in the liquid or solid phase exposed to atmospheric pressure are of this sort. The boundary in this case is just the upper surface of the liquid or solid, which, having negligible momentum, responds almost instantaneously to pressure changes in the system. Similarly, gaseous reactants confined by a weightless, or very light, piston will also allow the system pressure to

remain at the constant P_{ext}, unless the reaction occurs very rapidly (such as an explosion). Because we will not usually be considering explosions, we will generally assume that in constant-pressure processes, $P = P_{ext}$. In an explosion, such a substitution would actually be meaningless, as the pressure in the system is not uniform.

In the case of a reversible process, $P_{ext} = P$ and the work is given by

$$\delta w_{rev} = -P dV, \qquad w_{rev} = -\int_i^f P \, dV \tag{6}$$

where we have explicitly shown that this integral is from some *initial* to some *final* state. By the definition of an integral, work done by the system under reversible conditions is given by the area on a P versus V diagram under the curve $P(V)$. P versus V diagrams, such as shown in Fig. 2, are known as a *work diagrams*. Only reversible processes can be shown on work diagrams.

In thermodynamic systems, pressure is usually a function of at least two other variables (e.g., $P = RT/V_m$ for an ideal gas), so the integral in Eq. (6) is not defined until more information concerning the process is given. For example, in the work diagram of Fig. 2, three different reversible *paths* going from an initial state i to a final state f are shown. This diagram emphasizes that the work done in going between two states is not specified by defining these states and, thus, work is not a state function. In the upper path, the expansion is carried out at constant pressure to state 2, followed by a reduction of pressure (by lowering the temperature) to get to the final state. The area and, thus, the reversible work done by the system is $P_i \Delta V$ ($w = -P_i \Delta V$) and is greatest in this case. The lowest

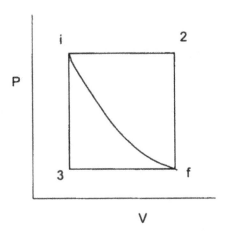

Figure 2 Work diagram for an expansion.

amount of reversible work done by the system $P_f \Delta V$ ($w = -P_f \Delta V$), is by the path through state 3, where pressure reduction precedes expansion. By taking the curved path between the initial and final state, an intermediate amount of reversible work is done.

One specific reversible expansion of an ideal gas that will be of particular interest to us is the one in which the system remains at constant temperature (by being immersed in a thermostat). Such a process is called *isothermal*. For this case, we can use Eq. (6), with $P = nRT/V$:

$$w_{rev} = -\int_i^f P \, dV \overset{ig}{=} -\int_i^f \frac{nRT}{V} \, dV \overset{isotherm}{=} -nRT \int_i^f \frac{dV}{V} = -nRT \ln\left(\frac{V_f}{V_i}\right)$$

(7)

The work is negative because the system is expanding.

In this book, when equalities are restricted to certain conditions or certain substances, we will often indicate these restrictions by notations above the equals sign. The designations "ig" (ideal gas) and "isotherm" (isothermal) have been used in Eq. (7).

The type of work that we will deal with most often in this book is work of expansion and contraction, which we will call *PV* work. Usually, the expansion or contraction is against the pressure of the atmosphere. In cases in which other types of work are involved, such as the work required to stretch an object or increase its surface area or the work of electrochemical cells or driving chemical reactions, we will usually designate these as w_{oth} or δw_{oth}. We then have

$$\delta w = -P_{ext} \, dV + \delta w_{oth}$$

(8)

or, in the reversible case,

$$\delta w_{rev} = -P \, dV + \delta w_{oth,rev}$$

(9)

The term $\delta w_{oth,rev}$ can be written in a form analogous to $-P \, dV$ (i.e., as the sum of products of changes in generalized displacement coordinates, dl_i, times the intensive forces conjugate to these displacement coordinates, L_i.

$$\delta w_{rev} = -P \, dV + \sum_i L_i \, dl_i$$

(10)

where the sum is over all forms of "other" work). One example of other work is electrical, where $L_{elec} = \phi$ (the electrical potential) and $dl_{elec} = dQ$ (the charge transferred). Another example is stretching a surface, where $L_{surf} = \gamma$ [the surface tension (discussed in Chapter 11)] and $dl_{surf} = d\sigma$ (the increase in surface area).

2.8 Internal Energy

In elementary physics, we are taught to recognize the different types of energy of simple bodies. These may include the translational and rotational kinetic energies of motion of the body as a whole, which we call K, and the potential energy due to the position of the body in a gravitational or electrical external field, V. From Newton's laws, it can be shown that energy is conserved in motion. A falling body converts its potential energy in a gravitational field into kinetic energy of motion; conversely, an upward-moving body will be brought to rest for an instant,[4] its initial kinetic energy being converted into potential energy in the gravitational field. In the presence of air, these conversions are not perfect; some energy is consumed by friction and causes a slight heating of the air and the body. We account for heating of the air by including its thermal energy in our considerations. In order to account for the energy absorbed by the body, we invoke its *internal energy, U. U* is the energy of motion or of position of one part of a body relative to other parts. Most importantly, it includes the energy of the random motions of the atoms and molecules of which the body is composed. For example, for a monatomic ideal gas, U is just the total kinetic energy of the atom discussed in Chapter 1. U is a state function; it depends only on the properties of the body, primarily its temperature. We are usually only interested in changes in the internal energy of systems and are thus free to arbitrarily fix a "minimum set" of internal energies at a particular temperature. We will see in Chapter 7 that a convenient minimum set is the internal energy of each chemical element in the state that it is found under ambient conditions. We can set these values equal to zero.

Internal energy is an extensive property of a system. If we double the size of a system, keeping intensive variables such as temperature and pressure constant, we double the system's internal energy. If we divide the internal energy of a system by the number of moles in the system, we obtain the *molar internal energy*, $U_m = U/n$, which is an intensive quantity. Other molar properties, such as the molar volume, are also indicated by the subscript m.

2.9 The First Law

The first law of thermodynamics is the recognition that

In any process, the energy of the universe is conserved.

Discussing the universe might seem grandiose, but it is just a way of reminding ourselves that we must consider both the system and the surroundings. Energy can flow between the system and surroundings and it is only the sum of these energies that is constant. The law is trivially applied to isolated systems, because, by definition, such systems cannot exchange mass or energy with the surroundings. Thus, a corollary of the first law is as follows:

The energy of an isolated system is constant in any process.

We will be more interested in closed systems, which can exchange energy with their surroundings by either heat transfer or work at the system boundary. A sign convention will be used where both heat and work are positive if they represent transfer of energy from the surroundings to the system. Because heat and work are decreases in the thermal and mechanical energy of the surroundings, they must show up as increases in the energy of the system. The mathematical form of the first law for closed systems is therefore

$$\Delta E = q + w \tag{11}$$

If there is more than one heat or work term, q and w are the net heat and work. E is the total energy of the system—the sum of its overall kinetic and potential energy plus its internal energy. Therefore, we can write

$$\Delta E = \Delta K + \Delta V + \Delta U = q + w \tag{12}$$

where ΔK is the change of the kinetic energy of motion of the system as a whole, ΔV is its change of potential energy due to its position in an external field, and ΔU is its change of internal energy. The change of the kinetic energy of the motion of the system as a whole and the change of its potential energy are usually the concern of mechanics rather than thermodynamics and we will deal only with processes in which ΔK and ΔV are both equal to zero. Our mathematical statement of the first law then becomes

$$\Delta U = q + w \tag{13}$$

This equation also applies to isolated systems, where $\Delta U = 0$ because q and w are zero.

In applying Eq. (13), the variables that determine q and w will often be continuously changing. Such change can be handled by adding up very small changes of U or, more exactly, by integrating over infinitesimal changes in U. The differential form of Eq. (13) is

$$dU = \delta q + \delta w \tag{14}$$

Although the first law might seem logically obvious to us today, it was only around 1800 that Rumford showed that a body could be heated by doing work on it, as well as by transferring heat to it. Prior to this, the *caloric* theory of heat was in vogue. Caloric was supposedly a fluid contained in bodies and transferred from hot bodies to cold bodies in the process of heating. It was not recognized that caloric was a form of energy. The amount of this fluid required to heat 1 g of water 1°C was defined as the *calorie*.[5] By the mid-19th century, Joule had measured the heat equivalent of a number of different types of mechanical energy, showing that $1\,cal = 4.184\,J$. Because many energy terms in chemistry are still given in calories, it will be worthwhile to memorize this conversion.

Example 2. What is the heat flow for an ideal gas undergoing an expansion by each of the three reversible paths shown in Fig. 2? States i and f are at the same temperature and the curved path is the isotherm between these states.

Solution: Because the initial and final states are at the same temperature, there is no change in the internal energy of the ideal gas in the overall process. Therefore, $q = -w$.

Path i-2-f: $\quad q = P_i \Delta V$

Path i-3-f: $\quad q = P_f \Delta V$

Isothermal path $\quad q = nRT \ \ln(V_f/V_i)$

Example 3. The ideal gas discussed in Example 2 could also undergo a single-stage expansion to the final state f, by suddenly reducing the pressure on the confining piston to P_f. How does this process differ from the reversible path i-3-f discussed in Example 2?

Solution: The final state in all four processes is the same. The work done is $P_f \Delta V$ ($w = -P_f \Delta V$). Therefore, $q = P_f \Delta V$, just as in path 1-3-f. There are some subtle differences, however. The single-stage expansion is not reversible and cannot be represented on a work diagram. Temperature and pressure are not uniform during the process. The system is held in a heat reservoir at a single temperature, and all the heat transferred is withdrawn from that single heat reservoir.

In the reversible process through path 1-3-f, heat transfer must be reversible. The system temperature is reduced in order to lower the pressure at constant volume. Therefore, heat must be transferred from an infinite number of heat reservoirs in the surroundings, all at a lower temperature than the final temperature of the system. We will see when we discuss entropy, in Chapter 3, that this is a very significant difference.

2.10 Heat Capacities

The heat required to heat a substance depends on just how that heating is conducted. The most common ways of heating are at constant volume or at constant pressure. In applying the first law to the process of heating a system at constant volume, no expansion work is done as $dV = 0$. Assuming that there is no other work ($\delta w_{oth} = 0$), the first law gives

$$\delta q_V = dU_V \tag{15}$$

where the subscript V indicates that the process occurs at constant volume. All of the added heat goes into the internal energy of the system. Dividing this equation by dT_V results in

$$\frac{\delta q_V}{dT_V} \equiv C_V = \frac{dU_V}{dT_V} \equiv \left(\frac{\partial U}{\partial T}\right)_V \tag{16}$$

C_V is the *heat capacity* at constant volume. We have used the partial derivative notation (see Appendix A) for dU_V/dT_V, the rate of change of internal energy with temperature when volume is held constant. We cannot use a partial derivative for $\delta q_V/dT_V$ because, as we discussed, δq_V is the *amount* of heat transferred and not the change in something, which is required for a derivative. The added heat for a finite temperature range may be found by integrating Eq. (16) in the form $\delta q_V = C_V\, dT_V$:

$$q_V = \int_{T_1}^{T_2} C_V\, dT = \int_{T_1}^{T_2} dU_V = U_2 - U_1 \equiv \Delta U \tag{17}$$

The heat capacity per unit mass, c_V, is called the *specific heat* at constant volume. The heat capacity per mole is called the *molar heat capacity* at constant volume, $C_{V,m}$. For homogeneous systems, the system heat capacity can be calculated as

$$C_V = mc_V = nC_{V,m} \tag{18}$$

For solids and liquids, heating is almost always at constant pressure. Heating such substances in constant-volume containers would build up very high stresses in the walls of the container to counter the tendency of the substances to expand as they are heated. If volume is not held constant, some expansion will occur and work can be done. Assuming $\delta w_{oth} = 0$,

$$\delta q = dU - \delta w = dU + P_{ext}\, dV = dU + P\, dV \tag{19}$$

Constant pressure means constant P_{ext}, because we define our constraints with respect to the surroundings, where we control the process. In the last step, we assume mechanical equilibrium between the system and the surroundings. This is an excellent assumption for slow processes with systems with movable boundaries, for which we will generally take $P_{ext} = P$. It is not a good assumption for processes in which there is a sudden change of constraints, such as an explosion or the removal of a "stop" which secures a piston. In such cases, the system may not even have a uniform pressure.

For a constant-pressure process, we can write Eq. (19) as

$$\delta q_P = dU + P\, dV + V\, dP \tag{20}$$

where the last term may be added, because it is zero in this case. Using the formula for the differential of a product, $d(xy) = x\, dy + y\, dx$, gives

$$\delta q_P = d(U + PV) \tag{21}$$

Defining a function, H (the *enthalpy*) by

$$H \equiv U + PV \tag{22}$$

$$\delta q_P = dH_P \tag{23}$$

If there are other work terms, the reversible forms of which are $\delta w_{oth} = L\, dl$, where l is a displacement coordinate and L is the force conjugate to the displacement coordinate, we will include additional terms of the form $-Ll$ in the definition of the enthalpy. Although not all authors adopt this extended definition, it is necessary, as shown below, to maintain the form of Eq. (23).

$$H \equiv U + PV - \sum_i L_i l_i \tag{24}$$

With δw_{oth}, the first law becomes

$$dU = \delta q - P\, dV + \sum_i L_i dl_i \tag{25}$$

This gives

$$\delta q = dH - P\, dV - V\, dP + \sum_i L_i\, dl_i + \sum_i l_i\, dL_i + P\, dV - \sum_i L_i\, dl_i$$

$$= dH - V\, dP + \sum_i l_i\, dL_i \tag{26}$$

so that at constant pressure and forces (the usual situation), the heat absorbed remains the enthalpy change.

Enthalpy plays a role in constant-pressure processes similar to that of internal energy in constant-volume processes. The heat added to a system in a constant-pressure process is the enthalpy increase of the system. Because U, P, and V (and l_i and L_i) are all state functions, H is also a state function. It is extensive. The molar enthalpy, $H_m \equiv H/n$, is intensive. Dividing Eq. (23) by dT_P gives

$$\frac{\delta q_P}{dT_P} \equiv C_P = \frac{dH_P}{dT_P} = \left(\frac{\partial H}{\partial T}\right)_P \tag{27}$$

The specific heat at constant pressure, c_p, and the molar heat capacity at constant pressure, $C_{P,m}$, are defined by

$$C_P = mc_P = nC_{P,m} \tag{28}$$

Figure 3 is a graph of molar heat capacities for a number of gaseous substances as a function of temperature. Although for most substances, heat capacities rise as temperature increases, for the noble gases, heat capacities are remarkably independent of temperature. Equation (17) of Chapter 1 gives $\frac{3}{2}kT$ for the average translational kinetic energy of ideal gas molecules. To a very good approximation, this is the only type of energy that can change in noble gas molecules at low and moderate temperatures and pressures. Multiplying by Avogadro's number, the energy per mole is $\frac{3}{2}RT$ and $C_{V,m}$ from Eq. (16), is $\frac{3}{2}R$, in agreement with Fig. 3. Molecules containing two or more atoms can also store energy as vibrational energy and rotational energy, which increases their

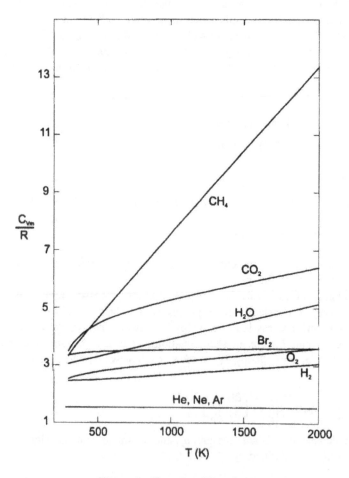

Figure 3   Heat capacities of gases.

heat capacities over that of the noble gases. Although the detailed shapes of all the curves in Fig. 3 can be calculated, such calculations involve statistical mechanics and quantum mechanics. Statistical mechanics will be treated in Chapter 5, but quantum mechanics will not be discussed in this book.

For substances with temperature-independent heat capacity, C_V can be removed from the integral in Eq. (17):

$$\Delta U = q_V = C_V(T_2 - T_1) = C_V \Delta T \qquad (29)$$

Equation (29) may be an adequate approximation for substances with temperature-dependent heat capacities if the temperature range is small enough. In addition, the substance must remain in a single phase (e.g., solid, liquid or gas) over the temperature range considered. The temperature variation of heat capacities is often represented by a multiterm empirical equation. Data for a number of gases in one often-employed representation are given in Table 1. Note that if $10^3 b = 3.26 \text{ J/K}^2 \text{ mol}; b = 3.26 \times 10^{-3} \text{ J/K}^2 \text{ mol}$.

Example 4. One reason that temperature drops as we go to high elevations is that rising air expands as it moves to regions of lower pressure. Assuming that a parcel of air behaves adiabatically and reversibly as it rises, calculate its *lapse rate*, the temperature decrease

TABLE 1 Molar Heat Capacities of Gases

Gas	$C_{P,m} = a + bT + c/T^2$		
	a (J/K mol)	$10^3 b$ (J/K^2 mol)	$10^{-5} c$ (J K/mol)
He, Ne, Ar, Kr, Xe	20.786	0	0
H_2	27.28	3.26	0.50
O_2	29.96	4.18	−1.67
N_2	28.58	3.77	−0.50
F_2	34.56	2.51	−3.51
Cl_2	37.03	0.67	−2.85
Br_2	37.32	0.50	−1.26
CO_2	44.22	8.79	−8.62
H_2O	30.54	10.29	0
NH_3	29.75	25.10	−1.55
CH_4	23.64	47.86	−1.92
SO_2	49.77	4.56	−11.05
H_2S	32.68	12.38	−1.92

Note: Valid for 298–2000 K.

Source: Data from A James, M Lord. Index of Chemical and Physical Data. New York: Van Nostrand Reinhold, 1992.

for each kilometer that the air rises. Assume dry air, so that no moisture condenses out. Consider the air to be ideal with a heat capacity $C_{P,m} = (7/2)R$.

Solution: Because we found the pressure variation in the atmosphere in Chapter 1, it is convenient to use pressure as the independent variable. Enthalpy is used when dealing with pressure. The rise of the parcel is slow enough so that it can be considered reversible:

$$dU \overset{\text{ad}}{=} -P_{\text{ext}} \, dV \overset{\text{rev}}{=} -P \, dV$$
$$dH = d(U + PV) = dU + P \, dV + V \, dP = V \, dP$$

Arbitrarily treating 1 mol of gas,

$$dH = C_{P,m} \, dT = RT \, \frac{dP}{P}$$

Using Eq. (59) of Chapter 1 for the pressure variation,

$$C_{P,m} \, dT = -Mg \, dz$$

or

$$\frac{dT}{dz} = -\frac{Mg}{C_{P,m}} = -\frac{0.029 \text{ kg/mol } (9.8 \text{ m/s}^2)1000 \text{ m/km}}{29 \text{ J/mol K}} = -9.8 \text{ K/km}$$

which is the dry, adiabatic lapse rate.

We will now derive a general relationship between C_P and C_V:

$$C_P = \left(\frac{\partial H}{\partial T}\right)_P = \left(\frac{\partial(U + PV)}{\partial T}\right)_P = \left(\frac{\partial U}{\partial T}\right)_P + P\left(\frac{\partial V}{\partial T}\right)_P \qquad (30)$$

because $(\partial P/\partial T)_P = 0$. Comparison with Eq. (16) shows that $(\partial U/\partial T)_P$ is not C_V, because P, rather than V, is held constant in the derivative. To relate $(\partial U/\partial T)_P$ to C_V, we apply Eq. (10), of Appendix A, with $F = U$, $x = T$, $y = P$, and $z = V$, to get

$$\left(\frac{\partial U}{\partial T}\right)_P = \left(\frac{\partial U}{\partial T}\right)_V + \left(\frac{\partial U}{\partial V}\right)_T\left(\frac{\partial V}{\partial T}\right)_P = C_V + \left(\frac{\partial U}{\partial V}\right)_T\left(\frac{\partial V}{\partial T}\right)_P \qquad (31)$$

giving as our final result

$$C_P = C_V + \left[P + \left(\frac{\partial U}{\partial V}\right)_T\right]\left(\frac{\partial V}{\partial T}\right)_P \qquad (32)$$

Equation (32) shows that there are two additional energies that must be supplied when heating at constant pressure compared with constant volume. The term $P(\partial V/\partial T)_P$ is the work done in expanding against the external pressure P.

Because the term $(\partial U/\partial V)_T(\partial V/\partial T)_P$ is of similar form, the quantity $(\partial U/\partial V)_T$ is called the *internal pressure*. This term represents additional energy stored *within the system*, due to the change in the average separation of the molecules when heating is performed at constant pressure. It is nonzero when forces exist between molecules.

In the Bernoulli model of the ideal gas, discussed in Chapter 1, there are no interactions between molecules. The internal energy, U, therefore does not depend on how far apart the molecules are, and $(\partial U/\partial V)_T = 0$. For an ideal gas $(\partial V/\partial T)_P = nR/P$, and

$$C_P \overset{\text{ig}}{=} = C_V + nR \quad \text{or} \quad C_{P,m} \overset{\text{ig}}{=} C_{V,m} + R \tag{33}$$

In Chapter 3, we will see how the difference in C_P and C_V can generally be obtained from the equation of state. For condensed phases, $(\partial V/\partial T)_P$ is very small, but $(\partial U/\partial V)_T$ is very large, and substantial differences between C_P and C_V can result.

Heat capacities can be measured by supplying energy to a system and observing its rise in temperature under adiabatic conditions. The energy is most easily supplied as an electrical current I passing through a known resistance for a given time:

$$w_{\text{elec}} = \phi Q = (IR)(I\Delta t) = I^2 R \Delta t \tag{34}$$

At constant volume,

$$\Delta U \overset{\text{ad}}{\underset{\text{const } V}{=}} w_{\text{elec}} \tag{35}$$

At constant pressure,

$$\Delta H = \Delta U + \Delta(PV) - \Delta(\phi Q) \overset{\text{ad}}{\underset{\text{const } P}{=}} w_{\text{elec}} - P\Delta V + P\Delta V = w_{\text{elec}} \tag{36}$$

For the last equality, we have assumed that the process begins before the current is turned on and ends after it is turned off. There is, then, no change of ϕ or Q of the system. The current is on while ϕ is not a uniform property of the system.

Example 5. When a voltage of 0.4 V is impressed on a well-insulated piece of iron of mass 44 g, exposed to the atmosphere, a current of 43 A results. After 10 s, the current is terminated and the sample is allowed to reach a uniform temperature. Its temperature is found to have risen 8.7°C. What is the heat capacity of iron? Is this value a better approximation to $C_{V,m}$ or $C_{P,m}$?

Solution: Because this system is well insulated, we will assume adiabaticity, $\delta q = 0$. In addition, we will assume that heat capacities are constant over the small temperature range of the measurements.

$$\Delta U \overset{\text{ad}}{=} w_{\text{elec}} - P_{\text{ext}}\Delta V = C_V \Delta T + \int_i^f \left(\frac{\partial U}{\partial V}\right)_T dV$$

Although $P_{\text{ext}}\Delta V$ is negligible, $(\partial U/\partial V)_T$ is very large for a solid and is not easily evaluated. Thus, the described measurement is not a good way to measure C_V:

$$\Delta H \overset{\text{const } P}{\overset{\text{ad}}{=}} w_{\text{elec}} = C_P \Delta T$$

$$C_{P,m} = \frac{(0.4 \text{ V})(43 \text{ A})(10 \text{ s})}{(8.7 \text{ K})(44 \text{ g})} \left(\frac{55.8 \text{ g}}{\text{mol}}\right) = 25.1 \frac{J}{\text{K mol}}$$

Heat capacities can be defined for processes that occur under conditions other than constant volume or constant temperature. For example, we could define a heat capacity at constant length of a sample. However, regardless of the nature of the process, the heat capacity will always be positive. This is ensured by the zeroth law of thermodynamics, which requires that as positive heat is transferred from a heat reservoir to a colder body, the temperature of the body will rise toward that of the reservoir in approaching the state of thermal equilibrium, regardless of the constraints of the heat-transfer process.

2.11 The Joule Process

The internal pressure, $(\partial U/\partial V)_T$, is so important that we should consider how it might be measured for real gases. In 1843, Joule showed that $(\partial U/\partial V)_T$ could be obtained from the simple process diagrammed in Fig. 4.

In the Joule process, a gas expands into a vacuum in a well-insulated container and the change in temperature with the change in volume of the gas is measured. The system is all of the gas, which we imagine surrounded by a flexible boundary. (The boundary expands with the gas; no gas passes through the boundary.) Because there is no opposing force as the boundary expands into vacuum, the work done is equal to zero. Assuming perfect insulation, the heat transferred to the system is also zero and $\Delta U = q + w = 0$. Thus, the Joule expansion is a constant-energy process and we measure $\Delta T_U/\Delta V_U$. If a series of measurements at different ΔV's are made, they can be extrapolated to zero ΔV_U to give $(\partial T/\partial V)_U$, which is known as the *Joule coefficient*, μ_J. From Eq. (7) of Appendix A, we can write μ_J as

$$\mu_J = \left(\frac{\partial T}{\partial V}\right)_U = -\left(\frac{\partial T}{\partial U}\right)_V \left(\frac{\partial U}{\partial V}\right)_T \tag{37}$$

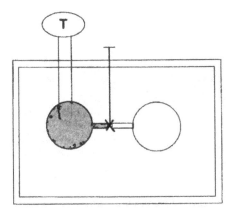

Figure 4 The Joule process.

Using Eq. (16),

$$\left(\frac{\partial U}{\partial V}\right)_T = -C_V \mu_J \tag{38}$$

The problem with the Joule experiment in practice is that although it is possible to insulate the vessel containing the gas, the vessel itself and the temperature-measuring device will be heated with the gas, greatly reducing the observed temperature change. Because the Joule coefficient is zero for ideal gases, we expect it to be small for real gases. Joule observed no temperature change when he did the experiment in the 1840s. Later, more careful measurements gave observable results, but with little precision.

2.12 The Joule–Thomson Process

An alternative, and much more accurate, method for obtaining information on the interactions between molecules is the Joule–Thomson expansion, shown in Fig. 5. This process also forms the experimental basis for much of the science of cryogenics (the study of phenomena at low temperatures), which we will discuss in Chapter 4. Industrially, cryogenic liquids, such as liquid N_2, O_2, H_2, and He, are produced by the Linde process, which uses Joule–Thomson expansions. N_2 and O_2 (and noble gases) are obtained in this process by producing and then

initial state: P_1, V_1, T_1

final state: P_2, V_2, T_2

Figure 5   The Joule–Thomson process.

fractionally distilling liquid air. O_2 and N_2 are among the top five chemicals in terms of the amount produced in the United States.

In the Joule–Thomson process, a compressed gas is allowed to rapidly expand through a porous plug. Because the apparatus is insulated and the gas expands very rapidly, no appreciable heat transfer occurs and the process is adiabatic. We will analyze the system as a quantity of gas moves through the plug. The gas is our system and it is shown in its initial and final states in Fig. 4. The process is certainly not reversible. However, the initial and final states are at internal equilibrium, with uniform temperature and pressure. The boundary of the system moves with the gas through the plug. Because of flow resistance, there is a considerable pressure drop as the gas passes through the tortuous channels of the plug. The much smaller pressure drops in the tubes upstream and downstream of the plug are neglected. Work equal to $P_1 V_1$ is done on the system by the upstream pressure, which pushes the gas through the plug. Work equal to $P_2 V_2$ is done by the system in pushing the downstream gas out of the way as it comes out of the plug. Applying the first law to this process,

$$\Delta U = U_2 - U_1 \overset{\text{ad}}{=} w = P_1 V_1 - P_2 V_2 \tag{39}$$

Rewriting, we obtain

$$U_2 + P_2 V_2 = H_2 = U_1 + P_1 V_1 = H_1 \tag{40}$$

Thus, the initial and final states of a Joule–Thomson expansion lie on a curve of constant enthalpy (*isoenthalp*) and the Joule–Thomson process occurs at constant enthalpy. The Joule–Thomson coefficient, μ_{JT}, is defined as

$$\mu_{JT} \equiv \left(\frac{\partial T}{\partial P}\right)_H \tag{41}$$

and can be determined from the limit of measurements of $\Delta T/\Delta P$ made as shown in Fig. 5. Alternatively, a single isoenthalp can be determined by expanding from an initial T_1 and the highest P_1 available to a succession of lower P_2's, measuring the resulting temperature each time. Some isoenthalps of N_2 are shown in Fig. 6. The dashed curve indicates the Joule–Thomson inversion temperature, T_i, below which $\mu_{JT} > 0$.

The Joule–Thomson coefficient is the slope of the isoenthalp and is a function of both temperature and pressure. From Eq. (23) and Eq. (7) of Appendix A, we can write

$$\mu_{JT} \equiv \left(\frac{\partial T}{\partial P}\right)_H = -\left(\frac{\partial T}{\partial H}\right)_P\left(\frac{\partial H}{\partial P}\right)_T = -\frac{1}{C_P}\left(\frac{\partial H}{\partial P}\right)_T \tag{42}$$

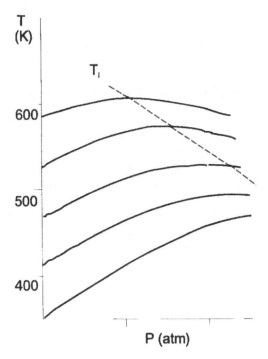

Figure 6 Isoenthalps of N_2. (Data from JR Roebuck, H Osterberg. Phys Rev 48:450, 1948.)

showing that measurements of the Joule–Thomson coefficient provide the variation of enthalpy with pressure at constant temperature. The reader should note the parallel between the Joule process for a change of volume and the Joule–Thomson process for a change in pressure.

The Joule–Thomson coefficient can be shown to be zero for ideal gases. For real gases, its magnitude is usually less than a few degrees Kelvin per atmosphere. Its sign can be either positive or negative, corresponding to the slope of the isoenthalp, as shown in Fig. 5. For gases (fluids, if we are above the critical temperature) at high temperatures, $\mu_{JT} \equiv (\partial T/\partial P)_H$ is generally negative, corresponding to heating the gas upon expansion. However, at lower temperatures and pressures, μ_{JT} is positive, corresponding to a cooling of the gas upon expansion. The temperature above which $\mu_{JT} < 0$ and below which $\mu_{JT} > 0$ is known as the *Joule–Thomson inversion temperature* at the given pressure. At this temperature, $\mu_{JT} = 0$. Most gases have $\mu_{JT} > 0$ at room temperature and moderate pressures and they cool upon expansion. Nitrogen and oxygen, for example, can be liquefied by a Joule–Thomson expansion from room temperature. For hydrogen and helium, however, the Joule–Thomson inversion temperature is lower than room temperature and these gases must be cooled (e.g., by liquid nitrogen) before they can be liquefied by further expansion. The Joule–Thomson expansion is adiabatic, but clearly nonreversible. Because it occurs in a finite time (actually quite quickly), it can be used for practical processes. A different, distinctly nonpractical, type of adiabatic process will be considered in the next section.

2.13 Reversible Adiabatic Expansion of an Ideal Gas

A reversible adiabatic expansion of an ideal gas is infinitely slow, so the system maintains internal equilibrium (mechanical, thermal, and material) and equilibrium with its surroundings. Mechanical equilibrium with the surroundings requires that the external pressure be only infinitesimally less than the internal pressure. We can therefore set $P = P_{ext}$. Thermal and material equilibria with the surroundings are not at issue, because the system is closed with adiabatic walls. A reversible adiabatic expansion is a highly idealized process! Nevertheless, it will serve as a cornerstone in our discussions of thermodynamics. Applying the first law to such a process,

$$dU = \delta q + \delta w \overset{ad}{=} - P_{ext}\, dV \overset{rev}{=} - P\, dV \tag{43}$$

For an ideal gas, U is only a function of temperature, so that we can also write

$$dU = \left(\frac{\partial U}{\partial T}\right)_V dT + \left(\frac{\partial U}{\partial V}\right)_T dV \overset{ig}{=} \left(\frac{\partial U}{\partial T}\right)_V dT = C_V\, dT \tag{44}$$

Equating these two expressions and setting $P = nRT/V$ gives

$$C_{V,m}\, dT = -RT\, \frac{dV}{V} \tag{45}$$

Separating variables,

$$C_{V,m}\, \frac{dT}{T} = -R\, \frac{dV}{V} \tag{46}$$

which can be integrated from initial conditions T_1 and V_1 to final conditions T_2 and V_2. If the ideal gas is monatomic or if the temperature range is small, $C_{V,m}$ may be taken as a constant and integration gives directly:

$$\ln\!\left(\frac{T_2}{T_1}\right) = -\frac{R}{C_{V,m}} \ln\!\left(\frac{V_2}{V_1}\right) \tag{47}$$

In the more general case, an average $C_{V,m}$, $\overline{C_{V,m}}$, must be used. [Actually, because $dT/T = d(\ln T)$, this should be an average over the logarithm of temperature.] If the final temperature is not known, $\overline{C_{V,m}}$ can be estimated by successive approximations, as in Example 6.

Example 6. Methane gas, originally at 800°C, undergoes a reversible adiabatic expansion that doubles its volume. Estimate the final temperature, considering the gas to be ideal.

Solution: For the first estimate, we will use $C_{V,m}$ at 800°C. Using data from Table 1, $C_{P,m}$ at 800°C for methane is $23.67 + 0.04786(1073) - (1.92 \times 10^5/(1073)^2 = 74.85$ J/mol K and $C_{V,m} = C_{P,m} - R = 66.53$ J/mol K. Then,

$$\ln\!\left(\frac{T_2}{T_1}\right) = -\frac{8.314}{66.53} \ln 2 = -0.0866$$

$$T_2 = T_1 \exp(-0.0866)$$
$$= (1073\text{ K})(0.917) = 984\text{ K} = 711°C$$

For a better estimate, we use $C_{V,m}$ at $\ln T$, the average of $\ln(1073)$ and $\ln(984)$. $\ln T = 6.935$ K, $T = 1028$ K, $C_{P,m} = 72.69$ J/mol K and $C_{V,m} = 64.37$ J/mol K. Then,

$$\ln\!\left(\frac{T_2}{T_1}\right) = -\frac{8.314}{64.37} \ln 2 = -0.0895$$

$$T_2 = T_1 \exp(-0.0895)$$
$$= (1073\text{ K})(0.914) = 981\text{ K} = 708°C$$

(This example is done more exactly in Problem 7.)

Assuming constant $C_{V,m}$ and setting $R = C_{P,m} - C_{V,m}$ for an ideal gas, we get

$$\ln\left(\frac{T_2}{T_1}\right) = \ln\left(\frac{V_1}{V_2}\right)^{(C_{P,m}-C_{V,m})/C_{V,m}} = \ln\left(\frac{V_1}{V_2}\right)^{\gamma-1} \tag{48}$$

where $\gamma \equiv C_{P,m}/C_{V,m}$ is called the *heat capacity ratio*. For gases, the heat capacity ratio is always greater than unity. If the logarithms of two functions are equal, the functions must be equal:

$$\frac{T_2}{T_1} = \left(\frac{V_1}{V_2}\right)^{\gamma-1} = \frac{P_2 V_2}{P_1 V_1} \tag{49}$$

or

$$P_1 V_1^{\gamma} = P_2 V_2^{\gamma} = \text{const.} \tag{50}$$

Thus, a reversible adiabatic expansion of an ideal gas occurs with constant PV^{γ}. Figure 6 shows two reversible processes of an ideal gas—a reversible isothermal process and a reversible adiabatic process—plotted on a work diagram. The pressure for the adiabatic expansion falls faster than that for the isothermal expansion (because $\gamma > 1$). The area under each curve is the amount of work that the expanding gas does on the surroundings in the process. More work is done in the isothermal process, because heat is added to the system to keep its temperature constant. Because the energy of the system remains constant with constant temperature, the added heat must end up as work done on the surroundings. In the adiabatic process, the work done on the surroundings comes exclusively from the internal energy of the system. Therefore, there is a reduction of temperature and less work is done on the surroundings in the reversible adiabatic process.

2.14 A Simple Heat Engine

A heat engine is a device that converts heat into work, the starting process for much of what is done in a modern industrial society. We have seen that this conversion can be accomplished by an isothermal expansion of a gas. If the gas is ideal, all the work done on the surroundings in such an expansion is provided by heat withdrawn from the surroundings (say, from a hot reservoir at temperature T_h). In order to continue obtaining work from a system undergoing isothermal expansion, it would have to continue expanding, and soon it would become inconveniently large. We will thus additionally require that a heat engine return to its initial state (i.e., that it undergoes a *cyclic process*). Because the system comes back to its initial state after one cycle in a cyclic process, all of the resulting change for the cycle is in the surroundings. How can we return a system (an engine) to its initial state after an isothermal expansion? One way to do this would be to perform an isothermal compression at the same temperature, T_h. In this case, however, we would just be traversing the same isotherm in the opposite direction and have to supply to the engine exactly the amount of work that we (in the surroundings) received from it in the original expansion. This would not be much of an engine at all; over one cycle, we would achieve exactly zero.[6]

One way to obtain net work over one cycle of our engine is to return it to its initial state by an isothermal expansion at a lower temperature, T_c, the temperature of a cold reservoir. The area under the lower-temperature compression on the work diagram would be lower than that under the expansion, and we would have net work. The only problem in doing this is that isotherms at different temperatures do not intersect on a work diagram. We need some way to get from the higher temperature of the expansion to the lower temperature of the compression and then back to the initial high-temperature state of the engine. This suggests that we need two additional steps to complete our cycle, as shown in Fig. 7. If these steps are adiabatic (and reversible so that they can be shown on the work diagram), we have the *Carnot cycle*, which has played a key theoretical role in the development of thermodynamics.

In Fig. 8, step I is an isothermal expansion, step II is an adiabatic expansion, step III is an isothermal compression, and step IV is an adiabatic compression. Note that zero heat is transferred in the adiabatic expansion and compression (steps II and IV) that we have added to complete the cycle, and that the work terms in these two steps exactly cancel, being equal to $\int_{T_h}^{T_c} C_V \, dT$ in the expansion and $\int_{T_c}^{T_h} C_V \, dT$ in the compression. (The work in an adiabatic

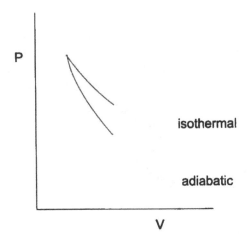

Figure 7 Reversal isothermal and reversible adiabatic expansions.

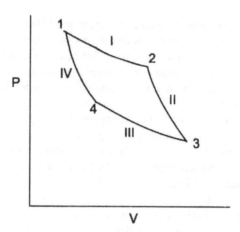

Figure 8   The Carnot cycle.

expansion comes at the expense of the internal energy of the gas.) Thus, to consider the effect on the surroundings of one cycle of the engine, we must consider only steps I and III of the Carnot cycle [i.e., the isothermal expansion (at T_h) and compression (at T_c)]. The heat and work terms in steps I and III are given in Table 2. In step III', we have made use of the fact that because V_4/V_1 and V_3/V_2 are volume ratios for adiabatic expansion and contraction between the same two temperatures, from Eq. (49) we have

$$\left(\frac{T_h}{T_c}\right) = \left(\frac{V_4}{V_1}\right)^{\gamma-1} = \left(\frac{V_3}{V_2}\right)^{\gamma-1} \qquad \text{or} \qquad \left(\frac{V_4}{V_1}\right) = \left(\frac{V_3}{V_2}\right) \tag{51}$$

The Carnot cycle engine achieves what we are looking for, a conversion of heat into work, with return of the engine to its initial state. We note, however, that in order to complete the cycle, we have paid a price. In the isothermal compression at T_c, some of the work produced in the expansion has to be used up to compress the system, finding its way into heat at the cold reservoir temperature.

TABLE 2   Heat and Work in a Carnot Cycle

Step	Heat	Work
I	$nRT_h \ln\left(\dfrac{V_2}{V_1}\right)$	$-nRT_h \ln\left(\dfrac{V_2}{V_1}\right)$
III	$-nRT_c \ln\left(\dfrac{V_3}{V_4}\right)$	$nRT_c \ln\left(\dfrac{V_3}{V_4}\right)$
III'	$-nRT_c \ln\left(\dfrac{V_2}{V_1}\right)$	$nRT_c \ln\left(\dfrac{V_2}{V_1}\right)$

An important measure of the quality of an engine is its *efficiency*, ε [the fraction of the energy that it removes from a high temperature reservoir (the heat term in step I) that it converts into work].[7] For the Carnot cycle engine to work as efficiently as possible, the heat transfers should be reversible. Thus, the heat transferred to the system in step I should be from a heat reservoir at temperature T_h, and the heat transferred from the system in step III should be to a reservoir at T_c. From Table 2, we see that the efficiency of a Carnot cycle engine is

$$\varepsilon = \frac{(-w_I) + (-w_{III})}{q_1} = \frac{[nR \ln(V_2/V_1)](T_h - T_c)}{nRT_h \ln(V_2/V_1)} = \frac{T_h - T_c}{T_h} \tag{52}$$

The Carnot cycle engine is actually the only reversible engine that we can design with two heat reservoirs. We see that because of the need to reject heat when returning the engine to its initial state, the engine cannot operate with unit efficiency. In Chapter 3, we will elevate this observation to one of the basic tenets of thermodynamics—the second law.

Questions

1. Indicate whether each of the following systems would be best described as an isolated, closed, or open system:

   (a) An empty microwave oven during a 3-min cycle.
   (b) Food is placed in the above microwave oven during the same cycle and then the cooked food is removed from the oven.
   (c) A bomb explodes in a cargo container in an airplane. (The system is the container and contents, and the analysis concludes before the walls of the system distort or rupture.)

2. Are the following isolated systems at internal equilibrium? If not, which types of internal equilibrium (thermal, mechanical, or material) do not hold?

   (a) A glass of water that has been untouched for a considerable time
   (b) A glass of room-temperature water into which some ice cubes have recently been placed
   (c) A diamond ring at ambient conditions

3. Would each of the following unchanging systems be better described as at equilibrium or at steady state?

   (a) A rock sitting at the bottom of a lake
   (b) A fish swimming in the lake
   (c) A Styrofoam cooler filled with a large amount of partially melted ice (take only the cooler, not its contents as the system)

4. Is each of the following systems homogeneous or heterogeneous?

    (a) Some crystalline sugar
    (b) An aqueous salt solution
    (c) The same solution as in part (b), after more salt is added than will dissolve
    (d) A molten solution of iron and carbon
    (e) The solution in part (d) after it has solidified into steel

5. An explosion is usually considered to be adiabatic, indicating negligible heat transfer, even though its rapidly expanding gaseous products are not at the same temperature as the surroundings, and the boundary does permit heat transfer. Can you explain this?

6. Does a real expansion produce more or less work than a reversible expansion? Does a real compression require more or less work than a reversible compression?

7. Express the definition of each of the following in mathematical form:

    (a) An adiabatic process
    (b) An isothermal process
    (c) A process in an isolated system
    (d) The Joule–Thomson expansion

8. The caloric theory of heat pictured heat as a fluid which permeates all materials. What experimental evidence clearly made this theory untenable, regardless of the properties assigned to the caloric fluid? Why?

9. Write expressions for H and dH in terms of U and dU for processes that include the following:

    (a) Electrical work
    (b) Stretching of an elastomer

10. One of the most common types of error made by students of thermodynamics is using formulas for conditions under which they do not apply. Indicate the physical conditions under which each of the following formulas holds:

    (a) $dU = \delta q - P \, dV$
    (b) $PV^\gamma = \text{const.}$
    (c) $dU = 0$ (when $dV \neq 0$)
    (d) $dH = C_P \, dT$

11. A gas expands from V_1 to V_2. For which of the following processes will the final temperature be the highest and for which will the final temperature be the lowest:

    (a) The process is reversible and adiabatic.
    (b) The process is a Joule expansion.
    (c) The process is an adiabatic irreversible expansion against the final pressure.

12. Which of the following pairs is larger:

    (a) ΔU or ΔH for an expansion against the atmosphere
    (b) $C_{P,m}$ of N_2 at $1000°C$ or at $25°C$
    (c) $C_{P,m}$ of H_2 at $500°C$ or $C_{P,m}$ of Ar at $500°C$

13. Draw a work diagram for a reversible heat engine that operates with two isothermal steps and two constant-pressure steps. How many heat reservoirs are needed to operate this engine?

14. An ideal rubber is defined as one for which $\left(\frac{\partial U}{\partial l}\right)_{T,V} = 0$. What does this definition imply for the direction of heat flow as an ideal rubber is extended? Verify your conclusion by holding a rubber band against your lips while rapidly extending it.

15. Rationalize the statement "The Carnot cycle engine is the only reversible engine that operates with just two heat reservoirs."

Problems

1. Consider the doubling of volume of 1 mol of an ideal gas at 298 K from 22.4 L (1.0 atm) to 44.8 L (0.5 atm). Figure 2 shows three paths by which this expansion can be performed. Calculate the work done on the surroundings by each path.

2. A waterfall is 200 ft high. If the water at the top has a temperature of 21.50°C, what is its temperature at the bottom?

3. A 1.0-kg iron mass falls from a height of 5.0 m into a well-insulated vessel containing 1.0 L of water. Neglecting splashing, if the original temperature of both the mass and the water was 25.00°C, what is their temperature when the final system has reached thermal equilibrium?

4. What is the temperature rise of 2.0 L of water if 2.0 A of electrical current is passed through a 200-Ω resistor in the water for 2 min? Assume the process is adiabatic and neglect the heat capacity of the resistor. The heat capacity of water is 75 J/K mol. (Note: $1.0\,\Omega A = 1.0\,V$.)

5. Find the work required to produce a strain ε in a metal following Hook's law and in a rubber following Eq. (54) of Chapter 1.

6*. (a) Derive the relation

$$\left(\frac{\partial H}{\partial P}\right)_T = \left(\frac{\partial V}{\partial P}\right)_T \left[\left(\frac{\partial U}{\partial V}\right)_T + P\right] + V$$

(b) Show that for an ideal gas

$$\left(\frac{\partial H}{\partial P}\right)_T = 0$$

7. Find a formula relating $(\partial H/\partial T)_V$ to $C_P = (\partial H/\partial T)_P$.

8. Calculate q, w, ΔU, and ΔH when 1 mol of NH_3 is heated from 300 K to 500 K at constant 1.0 atm pressure.

9. Calculate q, w, ΔU, and ΔH when 1 mol of an ideal gas expands isothermally from a pressure P_1 to a pressure P_2 against a constant opposing pressure of P_2. Put your results in terms of P_1, P_2 and T. Note that in this process, P_{ext} is constant, but P is not constant.

10. The pressure on Ar gas, originally at 25°C, is increased reversibly and adiabatically from 1.0 to 10.0 atm. What is the final temperature of the gas?

11*. (a) Using the form of $C_{P,m}$ from Table 1, integrate Eq. (46).

   (b) Using the result in part (a), find the final temperature when N_2 at 25°C expands reversibly and adiabatically from a pressure of 1.0 atm to a pressure of 0.1 atm.
   (c) Find a solution for Example 6, using the result in part (a). A program such as MathCad is useful for finding the root of the resulting equation.

12. ΔH_{vap} of water is 6.01 kJ/mol. Find ΔU_{vap} of water.

13. After finishing the measurement described in Example 5, the same iron resistor is placed in an insulated container with 75 g of an unknown substance. A current of 37 A through the resistor for 15 s raises the temperature of the system by 7.6°C. What is the specific heat of the unknown substance?

14. Give a formula for the work done in each of the following processes (use only the variables given in the problem):

   (a) A gas is heated from T_1 to T_2 at constant volume.
   (b) A battery produces a current I at a voltage ϕ for a time t.
   (c) A gas passes through a porous plug in a Joule–Thompson expansion.
   (d) An adiabatic expansion of an ideal gas in which its temperature falls from T_1 to T_2.

Notes

1. An excellent source is *Thermodynamics* by Kenneth S. Pitzer (McGraw-Hill, 1995), an update of an early classic in the field.
2. This period is much longer for liquids than for gases, and care must be taken to adequately mix liquids (and high-pressure gases) in the laboratory.
3. Although we may do something like this when we measure someone's body temperature by placing our hand on her forehead.
4. This assumes that the initial velocity is below that needed to escape from the influence of the Earth's gravitational field, the so-called escape velocity.
5. More precisely, from 14.5°C to 15.5°C at 1 atm pressure.
6. This assumes that the engine is reversible. If it were a real engine and operated at a finite speed, due to friction, we would have actually achieved a conversion of work into heat.
7. This is because we usually pay for the fuel to maintain the hot reservoir at T_h. We consider the cold reservoir as "free" (e.g., it might be a river to which we add our waste heat). Recently, it has been recognized that there are costs involved in rejecting waste heat, in the form of thermal pollution. These are not considered in the definition of thermodynamic efficiency.

3

The Second Law of Thermodynamics

All the King's horses and all the King's men
Couldn't put Humpty Dumpty together again

In addition to the quantity of energy, we must consider its quality, as indicated by its usefulness. As we use energy, its quantity remains constant, but its quality is degraded. Energy degradation is called *entropy increase*. The empirical observation that *there is always net entropy increase in real processes* is the second law of thermodynamics. Entropy, the measure of the quality of the energy of a system, is a state function. This requirement allows us to develop formulas to determine the entropy increase in both the system and its surroundings in any process. These general formulas are applied to a variety of processes involving heating, expansion, phase change, and distortion. The first and second laws of thermodynamics allow us to analyze the limits of performance of heat engines, heat pumps, and refrigerators.

3.1 The Second Law

Although the first law of thermodynamics places useful limits on the processes that can occur in nature, namely those that conserve total energy, it clearly is not the whole story regarding energy. If the only energy consideration were the first law, there would never be any energy shortages; we could just recycle energy. For example, having used a resistor to convert electrical energy into an equal amount of thermal energy, we could use a different device to convert the thermal energy back into the original amount of electrical energy. Needless to say, such a device has never been invented. As another example, if a brick falls off a table and hits the floor, energy is conserved; the initial potential energy of the brick is converted first into kinetic energy, as the brick accelerates, and then into thermal energy, as it comes to rest on the floor. The brick and floor are slightly warmer at the end of this process. However, no machine has ever been invented that could make use of this thermal energy to raise the brick back up onto the table. Of course, we could just lift up the brick, but then we would be providing the energy to return the brick to its initial state. It is obvious that many processes that occur in nature proceed in one direction, but not in the opposite direction. The generalization of this observation is the second law of thermodynamics, which tells us the direction of real processes.

It should be noted that this property of energy becoming less useful as we use it is purely a characteristic of the macroscopic realm. In the microscopic world, energy is continually transformed between kinetic and potential forms, as in a vibrating molecule, or between molecular energy and radiation, as in a molecule in a laser cavity. How microscopic systems combine to give the very different energy properties of macroscopic systems will be the subject material of Chapter 5.

What is the criterion for the direction of real processes? Because the second law cannot be proved, we must be content to state it as a crystallization of our experience. We will begin by summarizing some examples of real processes. These examples are various types of energy conversion, and we are interested in the *efficiency* of the conversions. Efficiency is defined as the ratio of the amount of the desired product energy produced to the source energy consumed, usually expressed as a percentage.

1. Mechanical potential energy (energy of position) and mechanical kinetic energy (energy of motion) can be efficiently interconverted, as in the example of the falling brick, or more apparent, in the motion of a pendulum clock, in which energy continually shifts between potential and kinetic energy. (The conservation of energy of a pendulum is even more dramatic if it operates in a vacuum, minimizing frictional losses.) Mechanical energy can also be converted into electrical energy with excellent efficiency, as is accomplished by the generators of electrical power plants (which operate with up to 99% efficiency).
2. Electrical energy can be efficiently converted into mechanical energy (with a motor).

3. Both electrical and mechanical energies can be converted with 100% efficiency into thermal energy, the former by collisions of electrons in a resistor and the latter by interactions of moving surfaces. Such *frictional processes* are what cause the disappearance of mechanical or electrical energy in systems.

4. Thermal energy cannot be completely converted into mechanical or electrical energy, as in our example of the slightly warmed brick and floor. As was shown in the discussion of the Carnot cycle engine, the conversion of thermal energy to mechanical or electrical energy is easier when the thermal energy is at high temperature than when it is at low temperature. For example, a candle flame at high temperature can boil a small amount of water, which can move a piston and raise a small weight. On the other hand, it is difficult to see how we could make use of the much larger amount of thermal energy in a lake at ambient temperature to achieve the same result.

5. The thermal energy of a high-temperature object can be transferred to a lower-temperature object by a flow of heat. Heat will not, however, spontaneously flow from a lower to a higher temperature. Of course, because we used the direction of heat flow to define temperature, this is just a reiteration of the zeroth law.

6. Chemical energy, which will be a major interest in this book, can be converted efficiently to heat (by combustion) and often also to electrical energy (as in a battery or fuel cell) or mechanical energy (as in a muscle). Considerations of chemical energy are somewhat complicated and will be deferred until Chapter 7.

The ability to convert one form of energy into other forms indicates the usefulness of the energy. The above considerations of the usefulness of different types of energy show that energy has a *quality*, as well as a *quantity*. There is a *hierarchy* of energy, indicating its quality, which is given in Fig. 1. We have listed electrical energy and mechanical energy as equivalent highest forms of energy, assuming that the slight frictional losses observed in practical interconversion of these forms are not fundamental. Thermal energy is of lower quality, and its quality is lower the lower its temperature. Chemical energy is not considered for the moment. When energy moves up on this scale (i.e., when it is converted to more useful forms), we say the energy has been *upgraded*; when it moves down on the scale, we say that the energy is *degraded*.

Having established the hierarchy of energy, we can now state the second law of thermodynamics (i.e., the principle that tells us the direction of real processes):

In any real process, there is net degradation of energy.

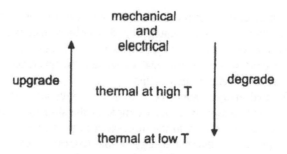

Figure 1 Hierarchy of energy.

The second law explains why we cannot continually reuse energy, as we reuse other resources, such as copper, and why energy recycling is not an effective response to energy shortages. Energy is degraded to less useful forms as we use it![1] According to the second law, processes that result in net upgrading of energy are not possible. It should be emphasized that we have not proved and cannot prove the second law, just as we did not prove the first law. These laws are a generalization of our experience, and no apparent violation of them has ever survived close scrutiny. Machines that would violate the first law are called *perpetual motion machines of the first type*, whereas those that would violate the second law are called *perpetual motion machines of the second type*. A working model of either of these types of perpetual motion machines has never been delivered to any patent office. You will be wise to keep the laws of thermodynamics in mind when someone approaches you with an opportunity to invest in the development of one of these machines!

There are two innocuous-seeming words in our statement of the second law, *real* and *net*, which are worthy of further discussion. What do we mean by a *real process*? Obviously, any process that is not "real" cannot occur, at least it cannot occur in a finite time. We have already discussed one class of nonreal processes in Chapter 2, the reversible expansions (isothermal and adiabatic) of an ideal gas. These are examples of processes in which there is complete balance between the internal and external forces on a system and, thus, there is no net driving force for the process. We can imagine the process occurring infinitely slowly, but it is not a real process. For example, in the reversible isothermal expansion, the process can be reversed by sliding the grains of sand back onto the piston, with no outside energy cost, assuming that the surfaces are frictionless. A reversible process can only be approached, not attained; it does not degrade energy. We should distinguish between truly reversible processes, which are completely at equilibrium and processes that are slow enough so that we can treat them as reversible with negligible loss of accuracy. The latter type of process is real.

The word *net* is particularly important in discussing the second law, and its neglect is the source of many errors in the application of the law. By net degradation, we mean that we must consider changes of the quality of energy that occur anywhere in the universe, as long as they are associated with the process of interest. If, in a process, a system of interest to us is exchanging energy or matter with its surroundings, there may be energy degradation in the surroundings, and this must be combined with the degradation or upgradation in the system to obtain net degradation. At this point, we will be dealing only with closed and isolated systems, so we will only be interested in energy exchange with the surroundings. In Chapter 6, we will begin considering open systems.

Before continuing, let us decide to define a new thermodynamic function that measures energy degradation. We will call this function the *entropy* and designate it by the symbol S. In terms of entropy our second law becomes

In any real process, there is net entropy increase.

Is entropy a state function? There is no reason that the quality of energy should be less a state function than the quantity of energy. When the system is completely defined, both should be known. Therefore, we assume that entropy is a state function. The state function nature of entropy is critical for the mathematical definition of this thermodynamic quantity. In addition, it allows us to consider processes in which the degradation of energy (entropy increase) is not at all apparent. For example, in a mixing process, such as results when the stopcock connecting containers of two different low-pressure gases is opened, one might think that there is no change in the quality of energy. If the pressure is low enough to neglect intermolecular interactions, the energy of the system is thermal energy (molecular kinetic energy) at the beginning and at the end of the process. Dealing with entropy as a state function, however, will allow us (in Chapter 4) to calculate the energy degradation (entropy increase) in this spontaneous process by an alternative pathway.

In order to usefully apply the second law, it will be necessary to be able to calculate both ΔS, the entropy change in the system of interest, and ΔS_{sur}, the entropy change of the surroundings. (Thermodynamic functions without the subscript "sur" can be assumed to refer to the system.) The mathematical form of our second law then becomes

$$\Delta S_{univ} = \Delta S + \Delta S_{sur} \geq 0 \qquad (1)$$

Although ΔS_{univ}, the entropy change of the universe, sounds rather formidable, in practice we will only be concerned with a few parts of the universe that interact with our system during the process. In this equation, greater than zero applies to any real process and equal to zero applies to a reversible process. In the latter, change occurs infinitely slowly, eliminating energy degradation due to friction and turbulence.

Of the two terms in Eq. (1), it is easier to calculate ΔS_{sur}, because we, the observers, are in the surroundings and can have more intimate knowledge of the surroundings than of the system. The surroundings are often simple (i.e., masses moving with known velocities or at known heights in gravitational fields or batteries or heat reservoirs) or sometimes they can be idealized as being simple, with little effect on the analysis. The system, on the other hand, is probably complicated at times during the process of interest, with nonuniform temperature, pressure, and concentrations.[2] Being outside the system, we are not privy to the instantaneous distributions of these variables. Nonetheless, we will be able to establish a method for measuring the entropy change of the system.

In order to calculate the ΔS_{sur} of a process, we note that mechanical and electrical energies are at the top of our hierarchy and we can assume that these types of energy are not degraded at all. Thus, we will not have to consider these forms of energy in calculating entropy changes. In addition, because we are presently not treating open systems, we do not have to consider chemical changes in the surroundings. This leaves only thermal energy changes in the surroundings to be considered.

The addition of thermal energy to a reservoir in the surroundings is $-q$, as our sign convention is that heat is positive when it is removed from reservoirs in the surroundings and added to the system. Thermal energy, however, is less degraded if it is at higher temperature. In order to take this into account, we can multiply the energy added to a heat reservoir by some function that decreases when the reservoir is at higher temperature. Such a function is T^{-m}, where m can have any positive value.[3] (We will find that only the value $m = 1$ is compatible with our requirement that entropy is a state function.) This gives for ΔS_{sur} the simple result.

$$\Delta S_{\text{sur}} = -\sum_i \frac{q_i}{T_i^m} \tag{2}$$

We have included the summation because there may be more than one heat reservoir in the surroundings that is involved in the process. For example, in the Carnot cycle engine, we remove heat from a hot reservoir and deposit heat in a cold reservoir.

In order to determine the entropy change of the system, $\Delta S = S_f - S_i$, we conceptualize an *entropy meter*. Although this is not a real device, we can contemplate its construction and mode of operation. Like any other measuring device, an entropy meter is located in the surroundings. It is an instrument containing various masses, batteries, heat reservoirs, and so forth. The distinctive thing about an entropy meter is its instructions, namely the following:

1. In order to measure ΔS of the system in a process, the entropy meter is not employed during the process, but, rather, in a separate measuring

step in which the system goes from the same initial state to the same final state as it does in the process of interest. Because we have agreed that ΔS must be a state function, ΔS must be identical in the actual process and the measuring step.

2. The entropy meter is used reversibly!

3. $\Delta S = -\Delta S_{met} = \sum_i (q_i / T_i^m)_{rev}$, as shown in the following analysis:

Requirement 2 guarantees that *in the measuring step*, there is zero net energy degradation (entropy increase) and we can write (with $\Delta S \equiv \Delta S_{system}$)

$$\Delta S_{univ} = \Delta S + \Delta S_{met} = 0 \tag{3}$$

or

$$\Delta S = -\Delta S_{met} = \sum_i \left(\frac{q_i}{T_i^m} \right)_{rev} \tag{4}$$

We have used Eq. (2) in Eq. (4) because the meter is in the surroundings. The subscript "rev" indicates that in the operation of the entropy meter, the work that it does as well as the heat that it transfers are all done reversibly. As in our previous discussion of the surroundings, it is only the heat reservoirs in the entropy meter that must be considered when calculating ΔS_{sur}.

It turns out that there is only one exponent, m, that will make entropy a state function. This can be seen by considering the simple engine discussed in Chapter 2—the reversible Carnot cycle engine operating on an ideal gas. Because we require that entropy be a state function, the change of entropy of the engine over one cycle of its operation must be zero. Of the four steps in the Carnot cycle, the reversible adiabatic expansion (step II) and compression (step IV) have $q_{rev} = 0$ and, thus, zero entropy change. Therefore, the entropy change of the ideal gas over one cycle is just the sum of the changes in steps I and III (or step III') of the cycle:

$$\Delta S_{cycle} = \Delta S_I + \Delta S_{III} = \frac{q_1}{T_h^m} + \frac{q_{III}}{T_c^m} = \frac{RT_h \ln(V_2/V_1)}{T_h^m} - \frac{RT_c \ln(V_2/V_1)}{T_c^m} \tag{5}$$

or

$$\Delta S_{cycle} = R \ln \left(\frac{V_2}{V_1} \right) (T_h^{1-m} - T_c^{1-m}) \tag{6}$$

which is only zero if $m = 1$.[4] Because we have developed general procedures for calculating entropy changes of the system and surroundings, these should be independent of the particular process or system that we have used to determine m, and the general formulas[5] must be

$$\Delta S_{sur} = -\sum_i \frac{q_i}{T_i}, \qquad dS_{sur} = -\sum_i \frac{\delta q_i}{T_i} \tag{7}$$

$$\Delta S = \sum_i \frac{q_{i,rev}}{T_i}, \qquad dS = \sum_i \frac{\delta q_{i,rev}}{T_i} \tag{8}$$

The differential forms of these formulas are useful for processes in which variables, especially temperature, are continuously changing.

The second law of thermodynamics denies the possibility of processes in which the only change is transfer of heat from a higher to a lower temperature. The zeroth law deals with thermal equilibrium and thus, by implication, the direction of heat transfer. However, the zeroth law applies only to heat transfer at a single interface, whereas the second law can deal with processes in which devices accomplish the heat transfer. These devices can have multiple interfaces with the heat reservoirs and can change during the process, as long as their change is cyclic.

The second law of thermodynamics has important philosophical and religious implications. Unlike many other laws of nature, which run equally well with time increasing or decreasing,[6] the second law states that entropy increases as time increases. Thus, entropy has often been called "time's arrow." Since even those who have not studied thermodynamics can usually recognize when a movie is being played backwards,[7] it appears that humans have an intuitive feeling for entropy. Thus, most moviegoers would know that something was wrong if Humpty Dumpty *spontaneously* came together again.[8]

The second law has not always been used correctly in philosophical discussions. Neglect of the surroundings has been an important error in the arguments of scientific creationists,[9] who use scientific arguments in support of religious views. The creationists claim that under evolution "theory," highly structured civilizations have evolved on Earth from the rather amorphous initial state of the planet. Increasing order is equivalent to an entropy decrease of the system, the Earth. Some creationists have argued that this is a clear violation of the second law, which requires entropy to continually increase. Therefore, they claim that the increase in order could not have occurred. The creationists theorize that Earth was created more perfect (with lower entropy) than it is today by a higher force, and its degree of organization has been decreasing with time, as required by the second law. Of course, what is neglected in this argument is that the Earth is not an isolated system. It is a steady-state system that is continually absorbing sunlight and radiating energy to space. Although the magnitudes of these two energy terms are just about equal (the temperature of the Earth has remained roughly constant over time), the qualities of the two types of energy are very different, due to their different temperatures. Over billions of years, the entropy increase of the surroundings has been more than sufficient to balance any entropy decrease that has occurred on Earth (if this, indeed, has happened) as well as to remove the large amount of entropy produced by dissipative (entropy-producing) processes that have occurred.[10]

On a cosmic scale, the second law is in agreement with the currently accepted theory of the expansion of the universe from an initial, highly located "Big Bang."

3.2 Entropy Changes in Some Simple Processes

Because it takes some practice to be able to use the recipes for calculating entropy changes in the system and surroundings, a few simple examples are presented here.

3.2.1 Reversible Adiabatic Expansion

Here, $q = 0$, giving $\Delta S_{sur} = 0$, and the process is reversible, so $q = q_{rev}$ and $q_{rev} = 0$, giving $\Delta S = 0$. Because entropy is unchanged, reversible adiabatic expansions and contractions are called *isentropic*.

3.2.2 Adiabatic Expansion

Because $q = 0$, $\Delta S_{sur} = 0$. ΔS depends on the details of the process but must be greater than zero to satisfy the second law.

3.2.3 Reversible Isothermal Expansion of an Ideal Gas

Using Eq. (5) of Chapter 2 and the fact that the energy of an ideal gas depends only on its temperature,

$$\Delta U \overset{\text{ig}}{\underset{\text{isotherm}}{=}} 0 = q - nRT \ln\left(\frac{V_f}{V_i}\right) \tag{9}$$

Because the process is reversible,

$$q_{rev} \overset{\text{rev}}{=} q = nRT \ln\left(\frac{V_f}{V_i}\right) \tag{10}$$

from which Eq. (8) gives

$$\Delta S = nR \ln\left(\frac{V_f}{V_i}\right) \tag{11}$$

and Eq. (7) gives

$$\Delta S_{sur} = -nR \ln\left(\frac{V_f}{V_i}\right) \tag{12}$$

so that $\Delta S + \Delta S_{sur} = 0$, as required for a reversible process.

3.2.4 Isothermal Expansion of an Ideal Gas

Entropy is a state function, so ΔS is the same as for the reversible isothermal expansion, calculated earlier. Because less work is done than in the reversible process and ΔU is the same in both cases, less heat is withdrawn from reservoirs in the surroundings. Therefore, the decrease of entropy of the surroundings is less than in the reversible expansion. In the limit of expansion against a vacuum (Joule process), no work is done and no heat is withdrawn from the surroundings. In this case, $\Delta S_{sur} = 0$.

3.2.5 Heating at Constant Volume

A differential change in entropy for heating at constant volume is

$$dS_V = \frac{\delta q_{rev,V}}{T} = \frac{C_V \, dT}{T} = \left(\frac{\partial S}{\partial T}\right)_V dT \tag{13}$$

which tells us that

$$\left(\frac{\partial S}{\partial T}\right)_V = \frac{C_V}{T} \tag{14}$$

Integrating Eq. (13) from the initial to the final temperature gives

$$\Delta S_V = \int_{T_i}^{T_f} \frac{C_V}{T} \, dT = C_V \ln\left(\frac{T_f}{T_i}\right) \tag{15}$$

where the last step assumes that the heat capacity can be taken as constant over the temperature range of the integration. This would hold for a monatomic gas or for other substances if the temperature interval were small enough. The entropy change of the surroundings depends on the details of the process. For example, if the heating is accomplished by placing the system in a heat bath at T_2,

$$\Delta S_{sur} = -\frac{q}{T_2} = -\frac{C_V(T_2 - T_1)}{T_2} \tag{16}$$

It can be shown that $\Delta S + \Delta S_{sur} > 0$.

3.2.6 Heating at Constant Pressure

The results are as given for constant volume heating, with C_V changed to C_P. In particular,

$$\left(\frac{\partial S}{\partial T}\right)_P = \frac{C_P}{T} \tag{17}$$

Example 1. Calculate the entropy change of 1 mol of Ar, considered an ideal gas, undergoing a change from 0°C and 1.0 atm to 50°C and 3.0 atm.

Solution: Because entropy is a state function, we can calculate ΔS by any path that takes the system from its initial to its final state. For the path, we choose isothermal compression at 0°C from 1.0 atm to 3.0 atm, followed by heating at a constant pressure of 3.0 atm from 0°C to 50°C. The C_P of a monatomic ideal gas is $(5/2)R$, independent of temperature and pressure. Because step 1 is an isothermal compression of an ideal gas, $V_2/V_1 = P_1/P_2$

$$\Delta S = \Delta S_1 + \Delta S_2 = R\ln\left(\frac{V_2}{V_1}\right) + C_P\ln\left(\frac{T_2}{T_1}\right) = R\ln\left(\frac{1}{3}\right)$$
$$+ \frac{5}{2}R\ln\left(\frac{323}{273}\right) = -1.09R + 0.42R = -5.57 \text{ J/K}$$

Satisfy yourself that the same result is obtained if the gas is first heated at constant pressure to 50°C and then isothermally compressed to 3.0 atm.

3.2.7 Phase Change

During a phase change, such as melting (often called fusion) or boiling, the system temperature remains constant at the phase equilibrium temperature, T_ϕ, while heat is transferred to effect the change. For calculating ΔS, the heat for the phase change must be transferred reversibly (i.e., from a reservoir at the phase equilibrium temperature). (Of course, because there is zero temperature differential driving the flow of heat, the transformation occurs infinitely slowly.) Assuming that the system is exposed to a constant pressure, the required heat is the enthalpy change for the phase transformation ($\Delta_\phi H$):

$$\Delta S = \frac{q_{rev}}{T_\phi} = \frac{\Delta_\phi H}{T_\phi} \tag{18}$$

The entropy change of the surroundings depends on the actual temperature of the reservoir from which the heat is transferred:

$$\Delta S_{sur} = -\frac{\Delta_\phi H}{T} \tag{19}$$

Because with an endothermic phase change, such as melting or boiling, heat will only flow into the system when $T > T_\phi$, $\Delta S + \Delta S_{sur} > 0$. For an exothermic phase change, such as condensation or freezing, heat will only flow out of the system when $T_\phi > T$, so that, once again, $\Delta S + \Delta S_{sur} > 0$. In Table 1, the phase transition temperatures and the enthalpy and entropy changes on melting and vaporization at 1 atm are given for a number of substances. Entropy changes of vaporization are much larger and show much less variation than those of melting. In fact, the approximation $\Delta_{vap} S = 85$ J/K mol (Trouton's rule) is reasonably accurate, except for very volatile or polar compounds.

Example 2. An ice tray containing 200 mL of water at 25°C is placed in a freezer at −15°C. What is ΔS and ΔS_{sur} for the resulting process? The heat capacity of ice and water are 2.0 and 4.18 J/K g, respectively. The heat of fusion of water is 6.01 kJ/mol.

Solution: To calculate ΔS for freezing, the freezing must occur at the phase equilibrium temperature (the normal freezing point of water). We break the process down into three parts:

TABLE 1 One atmosphere enthalpies and entropies of melting and vaporization

	T_m K	$\Delta_{fus}H$ kJ/mol	$\Delta_{fus}S$ J/K mol	T_b K	$\Delta_{vap}H$ kJ/mol	$\Delta_{vap}S$ J/K mol
Ar	83.8	1.12	13.4	87.3	6.43	73.6
N_2	63.1	0.71	11.2	77.4	5.57	72.0
ethane	90.4	2.86	31.6	185	14.7	79.6
benzene	279	9.95	35.7	353	30.7	87.0
CCl_4	250	3.28	13.1	350	29.8	85.1
H_2O	273	6.01	22.0	373	40.6	109
ethanol	159	5.02	31.6	351	38.6	110
AgCl	728	13.2	18.1	1820	199	109

Source: Handbook of Chem. and Phys., 76 ed., CRC Press (1995)

1. Cooling liquid water from 25°C to 0°C, with

$$\Delta H_a = 200 \text{ g} \times 4.18 \frac{J}{gK} \times (-25 \text{ K}) = -20{,}900 \text{ J}$$

$$\Delta S_a = 200 \text{ g} \times 4.18 \frac{J}{gK} \times \ln\left(\frac{273}{298}\right) = -73.2 \frac{J}{K}$$

2. Freezing water at 0°C

$$\Delta H_b = 200 \text{ g} \times \frac{mol}{18 \text{ g}} \times \left(-6010 \frac{J}{mol}\right) = -6680 \text{ J}$$

$$\Delta S_b = \frac{-6680 \text{ J}}{273 \text{ K}} = -24.5 \frac{J}{K}$$

3. Cooling ice from 0°C to −15°C

$$\Delta H_c = 200 \text{ g} \times 2.0 \frac{J}{gK}(-15 \text{ K}) = -6000 \text{ J}$$

$$\Delta S_c = 200 \text{ g} \times 2.0 \frac{g}{J \text{ K}} \ln\left(\frac{258}{273}\right) = -22.6 \frac{J}{K}$$

For the overall process,

$$\Delta H = \Delta H_a + \Delta H_b + \Delta H_c = -33.6 \text{ kJ}$$

$$\Delta S = \Delta S_a + \Delta S_b + \Delta S_c = -120.3 \frac{J}{K}$$

For the surroundings,

$$\Delta H_{sur} = +33.6 \text{ kJ}$$

$$\Delta S_{sur} = \frac{33{,}600 \text{ J}}{258 \text{ K}} = +130.2 \frac{J}{K}$$

Note that for this spontaneous process,

$$\Delta S_{univ} = \Delta S + \Delta S_{sur} = -120.3 + 130.2 = +9.9 \frac{J}{K} > 0$$

3.2.8 Isothermal Stretching of an Ideal Rubber

An ideal rubber is defined as one for which

$$\left(\frac{\partial U}{\partial l}\right)_T \overset{\text{id. rub}}{\equiv} 0 \qquad (20)$$

(a condition similar to that which applies to an ideal gas). This holds at either constant volume or constant pressure, because there is negligible change in volume as a rubber is stretched. As a result, when an ideal rubber is stretched at constant temperature,

$$\delta q \stackrel{\text{id. rub}}{=} -\delta w = -f_{\text{ext}}\, dl \tag{21}$$

If the stretching is done reversibly (very slowly),

$$\delta q_{\text{rev}} = -f\, dl \tag{22}$$

and

$$\Delta S = -\int \frac{f}{T}\, dl = -\frac{1}{T}\int f\, dl \tag{23}$$

In Chapter 5, we will see that the decrease in entropy of the rubber as it is stretched results from its component macromolecular chains adopting a more elongated configuration. The energy added as work escapes as heat to the surrounding, and in the reversible case,

$$\Delta S_{\text{sur}} = \frac{1}{T}\int f\, dl, \qquad \Delta S + \Delta S_{\text{sur}} = 0 \tag{24}$$

If the rubber is stretched rapidly, more work is done, because $f_{\text{ext}} > f$. As a result, more heat is transferred to the surrounding and the entropy increase of the surroundings is greater than in the reversible case. Because the entropy decrease of the rubber is a state function and the same in both cases, $\Delta S + \Delta S_{\text{sur}} > 0$ when the rubber is stretched rapidly.

3.2.9 Chemical Reactions

When a chemical reaction is proceeding, it is, by definition, not at equilibrium and thus not reversible. Thus, entropy changes in chemical reactions cannot be obtained from heat effects in calorimetric experiments. Entropy changes can be obtained by studying chemical equilibrium (Chapter 7) or by opposing the tendency of the reaction to proceed with an applied electric potential (Chapter 10).

3.3 Heat Diagrams

Because $\delta q_{\text{rev}} = T\, dS$,

$$q_{\text{rev}} = \int_i^f T\, dS \tag{25}$$

and areas on a graph of T versus S are equal to the heat for the state change carried out reversibly. Noting the similarity to the P versus V diagram, where areas are the work for the state change carried out reversibly, we call the T versus

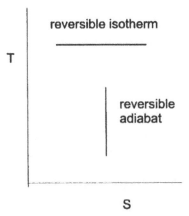

Figure 2   The heat diagram.

S diagram the *heat diagram* for the process. As shown in Fig. 2, on a heat diagram, horizontal lines are reversible isothermal processes, whereas vertical lines are reversible adiabatic (isentropic) processes.

3.4   General Analysis of Thermal Devices

Beginning with the industrial revolution, great improvements in standards of living have been achieved by the use of thermal devices, which interconvert heat and work. Three such devices, which operate with two heat reservoirs, are shown in Fig. 3. For each, the efficiency, ε, is defined as the amount of the type of energy desired divided by the amount of energy that is expended in obtaining it.

In Chapter 2, we have analyzed one particular type of heat engine, the reversible Carnot cycle engine with an ideal gas as the working substance, and found that its efficiency is $\varepsilon = 1 - T_c/T_h$. For both practical and theoretical reasons, we ask if it is possible, with the same two heat reservoirs, to design an engine that achieves a higher efficiency than the reversible Carnot cycle, ideal gas engine. What can thermodynamics tell us about this possibility?

Because we will be looking for the most efficient engine, frictional losses are set equal to zero by running the engine reversibly. The second law then becomes $\Delta S + \Delta S_{sur} = 0$. In order to keep using the engine, we will require that it operate in a cyclic manner and be unchanged over a cycle of its operation. Because the engine (the system) is unchanged, its entropy, a state function, is also unchanged over the cycle, and $\Delta S = 0$:

$$\Delta S_{sur} = -\frac{q_h}{T_h} + \frac{q_c}{T_c} = 0 \quad \text{or} \quad \frac{q_h}{q_c} = \frac{T_h}{T_c} \tag{26}$$

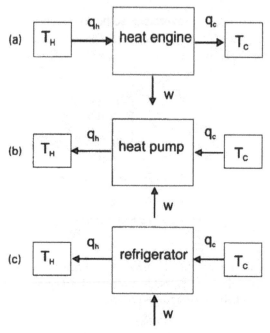

Figure 3   (a) A two-reservoir heat engine, $\varepsilon \equiv w/q_h$; (b) a heat pump, $\varepsilon \equiv q_h/w$; (c) a refrigerator, $\varepsilon \equiv q_c/w$.

Over one cycle, we also have $\Delta U = 0$, and from Fig. 3a, the first law tells us that

$$q_h - q_c - w = 0 \qquad (27)$$

The efficiency of a heat engine is defined as what we want from the engine (work) divided by what we pay for (heat from the hot reservoir):

$$\varepsilon \equiv \frac{w}{q_h} = \frac{q_h - q_c}{q_h} = 1 - \frac{T_c}{T_h} \qquad (28)$$

(i.e., the same efficiency as the Carnot cycle). Actually, a Carnot cycle engine is the only possible reversible engine that can run with two reservoirs. This follows from the requirement that heat can only be reversibly transferred to or from a single heat reservoir in an isothermal process. The steps in a two-reservoir reversible engine therefore must be either isothermal or adiabatic.

Equation (28) provides us with a means of defining a temperature scale that is as elegant as it is impractical. Our thermometer is a reversible heat engine, the efficiency of which we measure when it is operated between the temperature of interest and a reference temperature, defined as 273.15 K for an ice bath at 1 atm pressure. For temperatures above that of the ice bath, the ice bath is the cold reservoir; for temperatures colder than the ice bath, the ice bath is the hot reservoir. Such a scale requires only a single calibration point, the defined temperature of the ice bath, and is independent of the working material of the engine. Temperature is determined by Eq. (28) from the measured efficiency of the engine. Temperature measured on this thermodynamic (or Kelvin) scale is identical to that measured on the ideal gas scale discussed in Chapter 1.

As seen in Fig. 3, a heat pump is theoretically just an engine with all its energy flows reversed. Multiplying each term in Eqs. (26) and (27) by −1 does not change these equations. However, ε for a heat pump (usually called the coefficient of performance) is defined as q_h/w, because heat pumps are used to heat homes in winter by pumping in heat from the outdoors at lower temperature. This is certainly not the usual direction of heat flow, but it can be achieved by providing work, usually in the form of electrical energy. The coefficient of performance of a heat pump is, using the results of Eqs. (26) and (27),

$$\varepsilon \equiv \frac{q_h}{w} = \frac{q_h}{q_h - q_c} = \frac{T_h}{T_h - T_c} \tag{29}$$

Because, under most practical conditions, efficiencies calculated from Eq. (29) are much greater than 1, it predicts that with a heat pump, much more energy can be provided to heat a dwelling than is purchased from the local power company. This, indeed, is found to be the case. One problem that heat pumps do have, however, is providing enough capacity to keep homes warm when the outdoor temperature becomes very low (see Problem 11) and their efficiency is low. Therefore, heat pumps find most use in regions with rather mild winters. Unfortunately, their economic attractiveness in such regions is often diminished by a "capacity fee" that must be paid to the local power company for additional direct heating capacity for days when the heat pump is insufficient to comfortably heat the home. The above-calculated efficiencies assume reversible operation and are called maximum or *theoretical efficiencies*. Actual engines, heat pumps, and refrigerators will have additional losses and operate at some percentage of theoretical efficiency.

The product of the efficiency of a reversible heat engine times the coefficient of performance of a reversible heat pump, both operating between the same two heat reservoirs, is

$$\varepsilon_{eng}\varepsilon_{pump} = \frac{w}{q_h} \frac{q_h'}{w'} = 1$$

If the pump uses all of the work generated by the engine ($w = w'$), it will return to the hot reservoir just the amount of heat withdrawn by the engine. If either of these devices had an efficiency greater than we calculated, it could be used as the pump, taking the work produced by the engine and returning more heat to the hot reservoir than was used by the engine. This would represent net flow of heat from a cold to a hot reservoir, which is an upgrading of energy and violates the second law of thermodynamics.

Note that in Fig. 3c, what we call a refrigerator is the device that removes heat from the enclosure that stores food and pumps it into the surroundings. The box that stores the food is the cold reservoir. The analysis also applies to an air conditioner. In the analysis of the refrigerator, we get the same result as Eq. (26), but the first law gives

$$q_c - q_h + w = 0 \tag{30}$$

Because the useful quantity for a refrigerator is the amount of heat that it removes from the enclosure, we define its coefficient of performance as

$$\varepsilon \equiv \frac{q_c}{w} = \frac{q_c}{q_h - q_c} = \frac{T_c}{T_h - T_c} \tag{31}$$

a quantity which also is usually considerably greater than 1.

Example 3. What is the maximum rate that a heat pump that uses 1 kW of electrical power can supply heat to a house at 25°C when the outside temperature is 10°C?

Solution: Because we are interested in the maximum heat flow, we assume that the heat pump is reversible. Its coefficient of performance is

$$\varepsilon = \frac{298}{298 - 283} = 19.9 = \frac{q_h}{1.0 \text{ kW}}, \quad q_h = 19.9 \text{ kW} = 19.9 \, \frac{\text{kJ}}{\text{s}}$$

Questions

1. This chapter is based on the postulate that energy has quality, as well as quantity. List two other properties that have quality as well as quantity. List two properties that have only quantity.

2. For the freezing of water supercooled to −2.0°C, indicate whether each of the following is greater than, equal to, or less than zero: ΔS, ΔS_{sur}, and ΔS_{univ}.

3. Which of the following may be real processes? For those that cannot be real, indicate whether they violate the first or second law of thermodynamics.

   (a) A "super" heat conductor is discovered that allows heat to flow from a refrigerator to the surrounding room at a higher temperature.
   (b) A marble in a beaker of water suddenly jumps out of the water, while the water cools.
   (c) The average temperature of the Earth is increased by a huge volcanic eruption, which deposits a large amount of molten lava on the Earth's surface.
   (d) The average temperature of the Earth is increased by a large asteroid that hits the earth.

4. Are any of the following devices perpetual motion machines of the first or second kinds? Which kind are they and why?

(a) An engine operates by drawing water from a reservoir up a wick by capillary action, followed by the water falling from the top of the wick and turning a water wheel as it returns to the reservoir.

(b) An engine operates by a rotating magnet being pulled into a magnetic field. When the magnet reaches the field, a mass shifts on the magnet and gravity rotates the magnet out of the field.

(c) An engine operates by using the relatively warm surface waters of the ocean to evaporate a volatile compound and drive a piston. The resulting gas is passed to the colder lower water of the ocean, where it is compressed and reliquefied.

(d) A ship is propelled across the ocean in summer using heat from the ocean. (Take the ocean temperature as 18°C and air temperature as 25°C.)

(e) The same as part (d), but in winter, when the water temperature is 10°C and the air temperature is 2°C.

5. Draw a Carnot cycle on a heat diagram.

6. Sketch a curve representing reversible heating at constant pressure on a heat diagram.

7. Explain why a constant-pressure expansion cannot be included in a reversible engine operating with any finite number of heat reservoirs.

8. Indicate whether each of the following statements is true or false:

(a) Heat is the increase in the thermal energy of a system.

(b) $\Delta_{vap}S$ is always greater than zero.

(c) q is not a state function, but q_{rev} is a state function.

(d) In a reversible process, $\Delta S = 0$.

(e) For one cycle of a refrigerator, $\Delta S = 0$.

9. For each of the following processes, indicate whether ΔS is less than, equal to, or greater than zero, or whether there is insufficient information to decide (note that ΔS refers to the system):

(a) Melting of ice in a room at 25°C

(b) Reversible melting of ice at 0°C

(c) Reversible adiabatic expansion of an ideal gas

(d) Reversible adiabatic expansion of a van der Waals gas

(e) Reversible isothermal expansion of an ideal gas

(f) Joule expansion of an ideal gas

(g) Joule–Thomson expansion of an ideal gas

10. Rephrase each of the following statements in more exact thermodynamic language:

(a) The heat of a metallic rod is increased.

(b) The power company is building a new plant for producing energy.

(c) The country is suffering from an energy shortage.

11. Which of the following integrals must be zero when integrated over one cycle of a cyclic process (indicated by the symbol $\oint$):

(a) $\oint \dfrac{\delta q}{T}$ (e) $\oint U dT$

(b) $\oint \dfrac{\delta q_{rev}}{T}$ (f) $\oint (P\, dV + V\, dP)$

(c) $\oint P\, dV$ (g) $\oint U\, dT$ for an ideal gas

(d) $\oint dU$

Problems

1. Calculate ΔS for vaporization of water at 1.0 atm pressure, given that ΔH for this process is 40.6 kJ/mol.

2. Calculate ΔS for increasing the pressure on 1.0 mol of an ideal gas from 2 atm to 7 atm at 300 K. If the process is carried out reversibly, what is ΔS_{sur}?

3. Calculate ΔS for 1 mol of ideal gas expanding isothermally from a pressure of P_1 to a pressure of P_2 against a constant opposing pressure of P_2. What is ΔS_{sur}?

4. One gram of ice at $-10°C$ is placed in a large amount of liquid water at $+10°C$. What is the entropy change of the water originally in the ice as it reaches the temperature of the liquid? What is the entropy change of the rest of the water (ΔS_{sur})? What is ΔS_{univ} for this process? Take C_P for ice as 2.1 J/g K and for liquid water as 4.2 J/g K.

5. Using data from Table 1 in Chapter 2, find the change of entropy change as ammonia is heated from 25°C to 250°C.

6. Using Trouton's rule, estimate $\Delta_{vap}H$ of naphthalene, which has a normal boiling point of 211°C.

7.* Show that for heating at constant volume, with a single heat reservoir at the final temperature, $\Delta S_{univ} > 0$.

8. Did you ever wonder whether the light in your refrigerator stays on when you close the door and whether this is costing you much money? Assume that your refrigerator operates with a box temperature of 5°C in a room at 25°C and that electricity costs $0.20 per kilowatt-hour. What is the minimum cost per month if the 25-W bulb in the refrigerator stays on continuously? Why is this a "minimum"?

9. What is the minimum rate of heat deposition in a river by a 100-MW power plant operating with a boiler at 800°C and using the river water at 30°C to condense steam? Such heat deposition is known as *thermal pollution* and can contribute to making rivers unsuitable for many forms of aquatic life.

10. Considerations of heat losses from homes often show that they are proportional to the temperature difference between the home and its surroundings. Show that as a result of this, the work required by a heat pump will be roughly proportional to the square of the temperature difference between the home and its surroundings.

11.* A reversible heat pump can just keep a house at 25°C when the outside temperature is 10°C. What is the highest temperature that the same pump can maintain in the house when the outside temperature is −10°C? Assume that the heat loss from the house (which has to be provided by the pump) is proportional to the difference between the inside and outside temperature.

12. The average rate at which solar radiation is absorbed by the earth is 970 W/m² of surface. Because the Earth does not change appreciably over a short time (this is called the steady-state approximation), we can assume that the same rate of energy is radiated from Earth to space. The characteristic temperature at which the Earth absorbs radiation is, however, different from the temperature at which it radiates energy. Take an average temperature on the half of the Earth that is receiving solar radiation as 280 K and the temperature of the part of the atmosphere from which the earth radiates as 256 K. Because the steady-state approximation also requires that the entropy of the Earth is constant, what must be the rate of entropy generation of processes on and in the Earth. The radius of the Earth is 6400 km.

13.* A heat engine is proposed that operates with heat reservoirs at temperatures T_1 and T_2, using a monatomic ideal gas $[C_V = (3/2)R]$ as the working fluid. The engine cycle consists of the following three steps:

1. Expansion at constant pressure, P', from V_1 and T_1 to V_2 and T_2
2. Cooling at constant volume from T_2 to T_1
3. Reversible isothermal compression at T_1 from V_2 to V_1

(a) Draw the cycle on a work (P–V) diagram. Steps 1 and 2, not being reversible, cannot rigorously be drawn on the diagram. Use dashed lines to represent these steps.
(b) Calculate the heat and work for each step in the cycle.
(c) Find ΔS, ΔS_{sur}, and ΔS_{univ} for one cycle of operation of the engine.
(d) Derive an expression for the efficiency of the engine in terms of T_2 and T_1.
(e) For $T_2 = 600$ K and $T_1 = 300$ K, what is the efficiency of this engine? How does this compare with the efficiency of a Carnot cycle engine operating between reservoirs at the same two temperatures?

14. It has been proposed that temperature gradients in the oceans could be used as a source of power. Assuming that water is at 4°C at lower levels, where it has maximum density, and at upper levels, it has a temperature of 25°C, typical of the tropical ocean, what is the maximum efficiency with which heat could be extracted from the upper ocean to produce work?

15. Derive a formula for the entropy change for stretching an ideal rubber for which $f^* = a\varepsilon - b\varepsilon^2$.

16. What is the temperature of boiling water on a temperature scale defined by the efficiency or a reversible Carnot cycle engine and on which the reference temperature of an ice bath is taken as 100°.

17. A thermometer is based on a reversible heat engine, which operates between a boiling water bath and a heat reservoir at a lower temperature. The boiling water bath is defined to be at 373°. The temperature of the low temperature bath is determined from the efficiency with which the engine converts heat withdrawn from the boiling water bath to mechanical energy. Derive an explicit equation, $T_c = f(\varepsilon)$, from which the temperature of the low-temperature bath can be calculated.

18. Calculate the efficiency of a combined-cycle electricity generating plant. In this plant, the gaseous products of combustion at 1500°C are first directed into a gas turbine, from which the gas exits at 900°C and which operates at 90% of theoretical efficiency. Eighty-five percent of the energy of the effluent from the turbine is used to produce steam at 500°C. This steam is used in a steam electric generating plant which uses a river at 30°C as a cold reservoir and operates at 55% of theoretical efficiency.

Notes

1. Although these less useful forms of energy may still have some use, for example, in *cogeneration*, waste thermal energy from an electric generating plant is used for some other purpose, such as heating residences or greenhouses. In *combined-cycle* energy, generation electricity is produced by two different types of electric generator; for example, hot combustion gases are first directed into a gas turbine and then the effluent from the turbine is used to boil water for a steam electric generating plant.
2. Until Chapter 10, we will require that the system begins and ends the process in equilibrium states.
3. Other monotonically decreasing functions of T can be expanded in terms of the type T^{-m}.
4. $x^o = 1$.
5. We have adopted the standard procedure of not explicitly indicating that the temperature is reversible for a reversible process. Because heat is transferred reversibly, the system and surroundings are at thermal equilibrium and the temperature of the system must equal that of the external reservoir.
6. This is because such laws involve forces, which are related to acceleration. The acceleration, being a second derivative, is unchanged when t is replaced by $-t$.
7. Even in the absence of "giveaways," such as people walking backward and so on.
8. But not necessarily if this change was brought about by all the King's horses and all the King's men, whose entropy increase could compensate for Humpty's decrease in entropy.
9. HM Morris. Scientific Creationism. San Diego, CA: Creation Life Publishers, 1974, pp 37–46.
10. Of course, this does not rule out a higher force, a belief in which has been and is presently held by many scientists whose understanding of the second law cannot be questioned.

4

The Third Law and Free Energies

In a certain city the cold was so intense that words were congealed as soon as spoken.

Plutarch

Because a real engine operating with a cold reservoir at 0 K violates the second law, a third law, *it is impossible to reach 0 K in a finite number of steps*, is invoked. Thus, a real engine cannot employ such a reservoir. Practical methods of achieving very low temperatures are discussed. With the use of heat diagrams, the third law is then restated in the chemically useful form that *the entropies of all pure, perfectly crystalline materials approach zero as the temperature approaches 0 K*. This statement permits the calculation of absolute entropies. The definitions of the Helmholtz and Gibbs free energies allow thermodynamic criteria to be developed based only on properties of the system. Many important differential relations, including those of Maxwell, are developed from these functions, and the relations are utilized in considering heat capacities and entropies of mixing. Standard states for tabulating thermodynamic data are discussed. Finally, the thermodynamic equations are applied to the stretching of elastomeric materials.

4.1 Absolute Zero and the Third Law of Thermodynamics

Because the entropy increase of a heat reservoir is the heat added to the reservoir divided by its absolute temperature, as the temperature of the reservoir becomes very low, just a little heat added to it produces a very large entropy increase. A reversible heat engine operating with such a reservoir becomes more and more efficient ($\varepsilon = 1 - T_c/T_h$) until, with a cold reservoir at $0\,\mathrm{K}$, a heat engine could convert heat into work with 100% efficiency! Is this a contradiction of the second law, which requires net energy degradation in any real process? Once again we must look to the little word "real" to remove this difficulty. There must be something "unreal" about a heat engine operating with a reservoir at $0\,\mathrm{K}$ (in addition to its using reversible processes for maximum efficiency). The third law of thermodynamics provides us with a way out of this dilemma by asserting that it is infinitely difficult to reach $0\,\mathrm{K}$. A real engine thus cannot have one of its reservoirs at this temperature. We will state the third law as follows:

It is impossible to reach $0\,\mathrm{K}$ in a finite number of steps.[1]

In addition to being necessary for the consistency of the second law, the third law is a summary of our experience in attempting to achieve lower and lower temperatures. The third law also allows us to discuss *absolute entropies*, rather than just entropy changes. To see how this comes about, let us digress a bit and discuss how scientists actually attempt to reach very low temperatures.

We have already seen how gases below their Joule–Thomson inversion temperature (T_i) cool upon rapid expansion. By Joule–Thomson expansion, N_2 can be liquefied (77 K). Liquid N_2 can be used to cool H_2 below its T_i (195 K), and then further Joule–Thomson expansion can produce liquid H_2 (20.4 K), which can be used to cool He below its T_i (44.8 K). Joule–Thomson expansion of this cooled He can produce liquid He (4.2 K), and reducing the pressure above the liquid can conveniently produce temperatures as low as about 1 K.

To achieve temperatures lower than 1 K, the process of *adiabatic demagnetization* is usually employed. The best way to describe this process is by the *T–S* (reversible heat) diagram shown in Fig. 1a. In the presence of a magnetic field, it is energetically favorable for unpaired electrons of paramagnetic salts (usually of transition or lanthanide metals) to have their spins aligned with the field. Such a state also has a lower entropy than the state with randomly aligned spins in the absence of the field at the same temperature. The magnetization is carried out in contact with a thermal reservoir (liquid He, using gaseous He for heat transfer), which absorbs the energy released as the spins adopt their lower-energy orientation. Although it is not necessary in practice, we will assume that the isothermal magnetization is carried out reversibly, so that it can be represented by a horizontal line on the heat diagram. After magnetization, contact with the

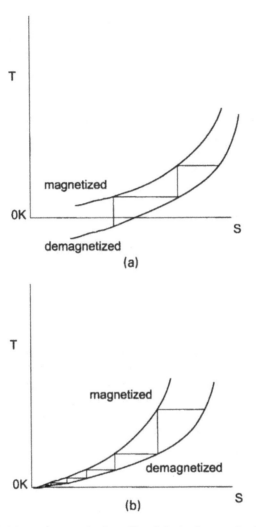

Figure 1 (a) Adiabatic demagnetization; (b) adiabatic demagnetization in agreement with the third law.

reservoir is removed by pumping away the gaseous He, thermally isolating the sample, and slowly reducing the magnetic field. Assuming that this step is reversible, it is isoentropic and is represented by a vertical line on the diagram. As the field is reduced, the spins randomly align, and the energy for this is taken from the thermal energy of the sample, with the resulting temperature drop shown on the diagram. The lower-temperature sample can now be used to reduce the temperature of the reservoir and the process is repeated, achieving an even lower temperature. Temperatures below 10^{-7} K have been achieved by this method, although it becomes necessary to use nuclear rather than electron spins, in order to obtain a less hindered reorientation at lower temperatures.[2]

It would appear that by a finite number of successive applications of the magnetization–demagnetization process shown in Fig. 1a, 0 K could be achieved. The only thing that could prevent this goal, as required by the third law, is if, as the temperature is reduced, the two curves approach each other, as shown in Fig. 1b. In this case, the temperature change for the isentropic demagnetization approaches zero as T approaches 0 K:

$$\left(\frac{\Delta T}{\Delta H}\right)_S \to 0 \quad \text{as } T \to 0 \tag{1}$$

where H is the magnetic field. In this situation, it takes an infinite number of steps to achieve 0 K. Also obvious in Fig. 1b is that the two curves also approach each other along the direction of the S axis, giving

$$\left(\frac{\Delta S}{\Delta H}\right)_T \to 0 \quad \text{as } T \to 0 \tag{2}$$

The choice of magnetic field variation as the process that distinguishes the two curves in Figs. 1a and 1b is of practical, but not of theoretical, significance. If we could find *any* reversible isothermal process whose entropy change remained finite as 0 K was approached, it would be described by a diagram similar to Fig. 1a and theoretically permit the attainment of 0 K in a finite number of steps. The third law therefore requires the following:

The entropy change of any reversible isothermal process must approach zero as T approaches 0 K.

$$\left(\frac{\Delta S}{\Delta \beta}\right)_T \to 0 \quad \text{as } T \to 0 \tag{3}$$

where β is the parameter whose variation determines the process.

One such process that we could consider is the chemical reaction $A + B \to C + D$, with $\beta = \xi$, the extent of the reaction, which will be introduced in Chapter 7. Equation (3) then indicates that the entropy change of *every* reversible[3] chemical reaction must approach zero at 0 K. This can be achieved if the entropy of every compound approaches the same value at 0 K. In Chapter 5, we will see that for conformity with the microscopic definition of entropy, we should call this common value of entropy zero.[4] The requirement that the considered reactions are all reversible demands that all reactants and products have sufficient time to achieve their equilibrium states, which, generally, will be the perfectly ordered crystalline state of lowest energy. A corollary of the third law is therefore as follows:

> The entropies of all pure, perfectly crystalline materials approach zero as temperature approaches 0 K.

As the temperature is reduced, the thermal energy available to overcome kinetic barriers to the lowest-energy state is reduced and, in some cases, *residual entropy* is difficult to remove at low temperatures. In other words, in the low-temperature cooling of these substances, reversibility cannot be approached. An example of a substance with residual entropy is solid CO. Carbon monoxide has a very small dipole moment, which indicates that there is a preferential orientation of molecules at low temperature. The magnitude of the dipole is so small, however, that at temperatures at which the preference becomes appreciable, there is insufficient thermal energy to overcome kinetic barriers for rotation of the molecules in the solid.

4.2 Absolute Entropies

The third law gives us a reference point for entropy and allows us to calculate absolute entropies as

$$S^\circ(T) = S^\circ(T) - S^\circ(0) = \int_0^T \left(\frac{\partial S^\circ}{\partial T}\right)_P dT \qquad (4)$$

The integral has contributions from phase transitions as well as from heating:

$$S^\circ(T) = \int_0^{T_m} \frac{C_{P,\,sol}^\circ}{T'} dT' + \frac{\Delta H_f^\circ}{T_m} + \int_{T_m}^{T_b} \frac{C_{P,\,liq}^\circ}{T'} dT' + \frac{\Delta H_v^\circ}{T_b}$$
$$+ \int_{T_b}^{T} \frac{C_{P,\,gas}^\circ}{T'} dT' \qquad (5)$$

The superscript $^\circ$ indicates that this equation is for a gas at the standard state of 1 bar. Standard states will be discussed in greater detail later in this chapter. For a solid, the calculation just includes the first integral (to T), and for a liquid, the first two terms plus the second integral (to T). If there are any solid-phase transformations at temperatures T_ϕ, with heats of transformation $\Delta_\phi H$, additional terms of the form $\Delta_\phi H/T_\phi$ must be included.

One problem that arises in calculating absolute entropies is that heat capacities of solids are not known to 0 K; thus, the first integral in Eq. (5) cannot be performed. The T' in the denominator of this integrand suggests that this contribution to the entropy might be very large. However, heat capacities also go to zero as $T \to 0$. The usual procedure is to make use of the Debye theory of solids, which predicts that for nonmetallic materials, heat capacities vary as T^3 at low temperatures. This theory has been verified by many measurements. (For metals, an additional term proportion to T must be included to account for the heat capacity of the electrons.)

Some values for standard entropies at 298 K are given in Table 1.

The following generalizations can be made concerning absolute entropies:

1. In a given phase, molar entropies are larger for molecules that are heavier and contain more atoms.
2. For molecules of similar mass and complexity, entropies are larger for gases than for liquids than for solids.

These trends are explained by the larger heat capacities of more complex molecules and heavier species (except for monatomic gases) and the large latent heat contributions introduced upon melting and vaporization.

4.3 Helmholtz and Gibbs Free Energies

The first and second laws of thermodynamics tell us what type of processes occur in nature, namely those that conserve energy but result in net energy degradation (entropy increase). For isolated systems, where there are no interactions between the system and its surroundings, the energy of the system must remain constant and its entropy must increase in any real process. For closed systems, however, the second law is somewhat inconvenient to use, because it forces us to consider entropy changes in both the system and its surroundings. In processes that occur at constant volume and constant temperature or at constant pressure and constant temperature, the surroundings interact with the system in very restricted ways. In this section, we will see that for such processes, it is possible to reformulate our

TABLE 1   Standard Entropies at 298.15 K (J/mol K)

Substance	S_m°	Substance	S_m°
C (graph)	5.74	N_2	191.61
PbO (solid)	68.70	Cl_2	223.08
Hg (liq)	76.03	Br_2 (liq)	152.21
He	126.15	CO_2	213.80
Ne	146.33	SO_2	248.21
Ar	154.84	NH_3	192.77
H_2	130.68	CH_4	186.25

Source: Data from NIST–JANAF Thermochemical Tables, 4th ed., Am Inst Physics, New York, 1998.

thermodynamics so that the criteria for real processes consider only properties of the system.

4.3.1 Constant-Volume and Temperature Processes: The Helmholtz Free Energy

In order to explore the tendency of a system to change in one direction or another, it will be sufficient to consider infinitesimal changes in these directions. If, for one of these changes, the energy remains constant and there is net entropy increase, the change can occur. At constant volume, expansion work is zero. Assuming that no other forms of work are done on the system, $dU = \delta q$. The net entropy change for a real process is then

$$dS_{univ} = dS - \frac{\delta q}{T} \stackrel{const. \ V}{=} dS - \frac{dU}{T} > 0 \qquad (6)$$

Multiplying by T and adding $S \ dT$ (which is zero for a constant temperature process) gives

$$T \ dS + S \ dT - dU = -d(U - TS) \stackrel{const. \ V, \ T}{>} 0 \qquad (7)$$

We define a new thermodynamic state function of the system:

$$A \equiv U - TS \qquad (8)$$

the *Helmholtz free energy.* (A is sometimes called the Helmholtz function, the Helmholtz energy, or the work function.) We see that in any real process at constant volume and constant temperature

$$-dA \stackrel{const. \ V, \ T}{>} 0 \quad \text{or} \quad dA \stackrel{const. \ V, \ T}{<} 0 \qquad (9)$$

A is a state function of the system because U, T, and S are all state functions of the system. We have thus remarkably transformed our criteria for real processes (at constant T and V) so that we do not have to consider the surroundings at all! We can say that a system with constant T and V will *spontaneously* undergo a process if it lowers the system's Helmholtz free energy.

A is sometimes called the *work function* because the decrease in A is the maximum work that the system can perform in any isothermal process. To see this, we calculate $-\delta w_{rev}$, which is the maximum work done by the system, because no energy is lost by friction in a reversible process:

$$-\delta w_{rev} = \delta q_{rev} - dU \stackrel{isoth.}{=} T \ dS + S \ dT - dU = -d(U - TS) = -dA \qquad (10)$$

Note that we can add a $S \ dT$ term because we are considering constant-temperature processes. Likewise, if A increases in a constant-temperature process,

dA is the minimum amount of work (δw_{rev}) that must be done on the system in the process.

If an analysis of a process at constant V and T shows that $dA > 0$, then that process will not occur. However, because A is a state function, the reverse process will have a negative dA and therefore be real. Only if $dA = 0$ will there be no tendency for a process to occur in either direction. We then say that the system is in *equilibrium* with respect to the process. Equilibrium corresponds to a reversible process, which has no driving force in either direction. Much of the interest in thermodynamics is in determining the conditions under which various processes are at equilibrium.

4.3.2   Constant Pressure and Temperature Processes: The Gibbs Free Energy

When dealing with condensed phases, it is more usual to employ systems exposed to the atmosphere, keeping pressure constant, than systems confined in vessels of fixed volume.[5] We are therefore particularly interested in transforming the equations dealing with constant P and T processes. Under such conditions, assuming that there is no work other than expansion work, $\delta q = dH$, and

$$dS_{univ} = dS - \frac{\delta q}{T} \overset{\text{const. } P}{=} dS - \frac{dH}{T} > 0 \tag{11}$$

Multiplying by T and adding $S\, dT$, which is zero at constant T, gives

$$T\, dS + S\, dT - dH = -d(H - TS) \overset{\text{const. } P,\, T}{>} 0 \tag{12}$$

We now define the state function,

$$G \equiv H - TS = U + PV - TS \tag{13}$$

the *Gibbs free energy* (sometimes just called the free energy). From Eq. (12),

$$dG \overset{\text{const. } P,\, T}{<} 0 \tag{14}$$

for any real process at constant T and P. G is a property only of the system; we no longer have to consider the surroundings, even though the system interacts with the surroundings. A process will spontaneously occur in a system at constant T and P, if it lowers the system's Gibbs free energy. If for a given process $dG > 0$, the reverse process will be spontaneous. Only if $dG = 0$ will a system held at constant T and P be at equilibrium for the considered change. We will make repeated use of this criterion for equilibrium in our applications of thermodynamics.

We can see another important use of the Gibbs free energy by writing the first law in the form

$$-\delta w_{rev} = P\,dV - \delta w_{oth,\,rev} = \delta q_{rev} - dU \tag{15}$$

At constant T and P, this becomes

$$-\delta w_{oth,\,rev} \overset{const.\,P,\,T}{=} -dU + T\,dS - P\,dV + S\,dT - V\,dP$$
$$= -d(U + PV - TS) = -dG \tag{16}$$

However, $-\delta w_{oth,\,rev}$ is the maximum (because reversible means that no energy is lost to friction) amount of useful work (because expansion against the atmosphere is not usually useful) that can be done by the system. This amount of work could be obtained from a chemical reaction, for example, if it is run in an electrochemical cell, with an opposing voltage that makes the reaction proceed infinitely slowly. This is a reversible (equilibrium) situation, because slight changes in the opposing voltage can make the reaction run in either direction. More commonly, the free energy decrease of one reaction drives another reaction for which $\Delta G > 0$. This commonly occurs in biochemical systems (which operate at constant T and P).

4.4 Partial Derivatives of Energy-like Quantities

As we have seen from our previous discussions of heat capacities, thermal expansion coefficients, and compressibilities, partial derivatives are the key to discussing changes in thermodynamic systems. In a single-component system of fixed size, the specification of two state variables completely determines the state of the system. Calling one of the molar energy quantities Z, we can write $Z = Z(X, Y)$, where X and Y are any two state variables, such as T and P, or T and V. Using the general mathematical properties of functions of two variables that are discussed in Appendix A,

$$dZ = \left(\frac{\partial Z}{\partial X}\right)_Y dX + \left(\frac{\partial Z}{\partial Y}\right)_X dY = M\,dX + N\,dY \tag{17}$$

where $M = (\partial Z/\partial X)_Y$ and $N = (\partial Z/\partial Y)_X$ are, in general, both functions of X and Y. The cross-derivative rule [Eq. (17) in Appendix A] gives

$$\left(\frac{\partial M}{\partial Y}\right)_X = \left(\frac{\partial N}{\partial X}\right)_Y \tag{18}$$

which is a useful route for deriving important partial derivative relationships.

The starting point for our derivation is the first law, which we will write in the following manner:

$$dU = \delta q + \delta w = \delta q_{rev} + \delta w_{rev} \tag{19}$$

Equation (19) acknowledges that U is a state function and, thus, the (differential) change in U between any two states can be calculated for any process that goes between the two states. In particular, we can choose a reversible process connecting the two states. However, an important restriction applies to the use of Eq. (19). In a reversible process, internal and external equilibria are maintained at all times. Because the symbol dU refers to specific states of the real process, these must be states that can also be initial and final states of the reversible process. Thus, Eq. (19) can only apply if the initial and final states are in internal equilibrium.

For a process in which the only work is expansion work, Eq. (19) becomes

$$dU \overset{\text{m.e.}}{=} T\, dS - P\, dV \tag{20}$$

Because Eq. (20) is of great importance and can be a source of considerable confusion, it will be discussed in detail. As we have stated, the initial and final states must be in internal equilibrium. Internal equilibrium requires mechanical equilibrium, thermal equilibrium, and material equilibrium. Because Eq. (20) contains P and T, these variables must be uniform throughout the system, ensuring internal mechanical and thermal equilibria. Because internal material equilibrium is not explicitly required by any of the variables in the final equation, it is indicated by "m.e." above the equals sign. Thus, if chemical reactions or phase transformations can occur in the system, they must be at equilibrium. The designation "m.e." is also appropriate for systems not in material equilibrium, if constraints result in transformation processes being negligibly slow. Thus, Eq. (20) could be used to treat the expansion of a mixture of H_2 and O_2, as long as no catalysts or initiators of the water-forming reaction are present. For single-component, single-phase systems, the designation "m.e." can be ignored.

For a single-component, single-phase system or a system at material equilibrium, the change of internal energy is completely determined by the change in two state variables. Thus, Eq. (20) is valid for any process that goes between the initial and final states of the infinitesimal process; its application is not limited to reversible processes. It would, for example, apply to a Joule–Thomson expansion, a distinctly nonreversible process.

From the definitions of H, A, and G, three more differential relationships are immediately obtained:

$$dH = d(U + PV) \stackrel{\text{m.e.}}{=} T\,dS - P\,dV + P\,dV + V\,dP$$
$$\stackrel{\text{m.e.}}{=} T\,dS + V\,dP \tag{21}$$

$$dA = d(U - TS) \stackrel{\text{m.e.}}{=} T\,dS - P\,dV - T\,dS - S\,dT$$
$$\stackrel{\text{m.e.}}{=} -P\,dV - S\,dT \tag{22}$$

$$dG = d(H - TS) \stackrel{\text{m.e.}}{=} T\,dS + V\,dP - T\,dS - S\,dT$$
$$\stackrel{\text{m.e.}}{=} V\,dP - S\,dT \tag{23}$$

To this can be added a differential relationship for S derived from Eq. (20):

$$dS \stackrel{\text{m.e.}}{=} \frac{1}{T}\,dU + \frac{P}{T}\,dV \tag{24}$$

Because of the simplicity of these relationships, we sometimes say that the *natural variables* of U are S and V, of H are S and P, of A are V and T, of G are P and T, and of S are U and V. It is noteworthy that the natural variables of U are both extensive variables and those of G are both intensive variables; the natural variables of H and A are mixed—one extensive and one intensive variable for each. Because Eqs. (20)–(24) only hold for systems at material equilibrium, they will become our criteria for material equilibrium under each set of conditions. For example, at constant T and P, for a system to be at material equilibrium for a process, dG must equal zero for the (infinitesimal) process.

Applying Eq. (18) to the first four differential relationships gives

$$\left(\frac{\partial T}{\partial V}\right)_S \stackrel{\text{m.e.}}{=} -\left(\frac{\partial P}{\partial S}\right)_V \quad \text{from Eq. (20)} \tag{25}$$

$$\left(\frac{\partial T}{\partial P}\right)_S \stackrel{\text{m.e.}}{=} \left(\frac{\partial V}{\partial S}\right)_P \quad \text{from Eq. (21)} \tag{26}$$

$$\left(\frac{\partial S}{\partial V}\right)_T \stackrel{\text{m.e.}}{=} \left(\frac{\partial P}{\partial T}\right)_V \quad \text{from Eq. (22)} \tag{27}$$

$$\left(\frac{\partial S}{\partial P}\right)_T \stackrel{\text{m.e.}}{=} -\left(\frac{\partial V}{\partial T}\right)_P \quad \text{from Eq. (23)} \tag{28}$$

Equations (25)–(28) are known as the Maxwell relations. We will find the latter two equations to be particularly useful. They give the surprising result that the volume and pressure dependence of entropy are determined by equations of state, from which the temperature derivatives of P and V can be calculated.

An additional useful relationship is

$$\left(\frac{\partial (T^{-1}G)}{\partial T^{-1}}\right)_P \stackrel{\text{m.e.}}{=} T^{-1}\left(\frac{\partial G}{\partial T}\right)_P\left(\frac{\partial T}{\partial T^{-1}}\right)_P + G \stackrel{\text{m.e.}}{=} H \tag{29}$$

which is known as the Gibbs–Helmholtz equation.

In considering changes in U, H, A, G, and S, all of which are extensive state functions with units of energy, it is usually convenient to divide each of these functions by the number of moles in the system, giving the intensive properties: molar energy, U_m; molar enthalpy, H_m; molar Helmholtz free energy, A_m; molar Gibbs free energy, G_m. All of the above equations hold for the corresponding molar quantities.

4.4.1   Internal Pressure

In Chapter 2, we called the quantity $(\partial U/\partial V)_T$ the internal pressure. An equation can be found for the internal pressure by writing Eq. (20) for a constant-temperature process and dividing by dV_T:

$$\left(\frac{\partial U}{\partial V}\right)_T \stackrel{\text{m.e.}}{=} T\left(\frac{\partial S}{\partial V}\right)_T - P \stackrel{\text{m.e.}}{=} T\left(\frac{\partial P}{\partial T}\right)_V - P \tag{30}$$

where the last step follows from Eq. (27). In a similar manner, we can write, from Eqs. (21) and (28),

$$\left(\frac{\partial H}{\partial P}\right)_T \stackrel{\text{m.e.}}{=} T\left(\frac{\partial S}{\partial P}\right)_T + V \stackrel{\text{m.e.}}{=} -T\left(\frac{\partial V}{\partial T}\right)_P + V = V(1 - \alpha T) \tag{31}$$

Because the right-hand sides of Eqs. (30) and (31) can be evaluated from equations of state, we see that such equations plus heat capacity data allow us to completely calculate changes of U and H. Equations (30) and (31) are known as the *thermodynamic equations of state*.

Example 1.   Find an expression for the internal pressure of an ideal gas.

Solution: For an ideal gas $P = nRT/V$; therefore, $(\partial P/\partial T)_V = nR/V$ and

$$\left(\frac{\partial U}{\partial V}\right)_T = \frac{nRT}{V} - P = 0$$

This shows that for an ideal gas, the energy does not depend on how far apart the molecules are, and the forces between molecules must be zero. This was previously taken as a postulate of the Bernoulli model of the ideal gas. Example 1 shows that the result does not depend on a particular model, but follows directly from the equation of state of the ideal gas. [In Problem 9, you show that in addition $(\partial H/\partial P)_T = 0$ for an ideal gas.]

For real gases, due to forces between molecules, the internal energy does depend on how far apart the molecules are. We define the difference between the internal energies of real and ideal gases at given volume and temperature as the *molecular interaction energy*, U_{int}. Because the internal energy of a real gas approaches that of an ideal gas as volume becomes infinite, we can write

$$U_{int}(V) = U(V) - U_{id}(V) = U(V) - U(\infty) = \int_{\infty}^{V} \left(\frac{\partial U}{\partial V}\right)_T dV \qquad (32)$$

Example 2. Find an expression for the internal pressure and the molecular interaction energy of a van der Waals gas. For He at 298 K and 1.0 atm pressure, compare the molecular interaction energy with the kinetic energy of the atoms.

Solution: For a van der Waals gas,

$$P = \frac{nRT}{V - nb} - \frac{an^2}{V^2}$$

therefore,

$$\left(\frac{\partial P}{\partial T}\right)_V = \frac{nR}{V - nb}$$

and the internal pressure is

$$\left(\frac{\partial U}{\partial V}\right)_T = \frac{an^2}{V^2}$$

$$U_{int}(V) = \int_{\infty}^{V} \frac{an^2}{V^2} dV' = -\frac{an^2}{V}$$

Note that for a van der Waals gas, only the attractive interactions contribute to the internal pressure. The molar interaction energy is

$$U_{int, m} = \frac{U_{int}}{n} = -\frac{an}{V} \approx -\frac{aP}{RT}$$

where we used the ideal gas approximation for the molar volume in the last step.

For He, $a = 0$, 0.00345 Pa $\cdot$ m^6/mol^2; therefore,

$$U_{int, m} = -\frac{0.00345 \text{ (Pa} \cdot \text{m}^6/\text{mol}^2) \times 10^5 \text{ Pa}}{(8.314 \text{ J/mol K})(298 \text{ K})} = -0.14 \frac{\text{J}}{\text{mol}}$$

This has to be compared with the total internal energy, which is very close to

$$\frac{3}{2}RT = 3716 \ \frac{J}{mol}$$

4.5 Heat Capacities

Because we have found the variation of U and H with V and P, we can do the same for the heat capacities, C_V and C_P:

$$\left(\frac{\partial C_V}{\partial V}\right)_T = \left(\frac{\partial}{\partial V}\left(\frac{\partial U}{\partial T}\right)_V\right)_T \tag{33}$$

Reversing the order of differentiation gives

$$\left(\frac{\partial C_V}{\partial V}\right)_T = \left(\frac{\partial}{\partial T}\left(\frac{\partial U}{\partial V}\right)_T\right)_V \tag{34}$$

Substituting for the internal pressure from Eq. (30),

$$\left(\frac{\partial C_V}{\partial V}\right)_T = \left(\frac{\partial}{\partial T}\left[T\left(\frac{\partial P}{\partial T}\right)_V - P\right]\right)_V = T\left(\frac{\partial^2 P}{\partial T^2}\right)_V \tag{35}$$

In a similar manner, it can be shown that (Problem 10)

$$\left(\frac{\partial C_P}{\partial P}\right)_T = -T\left(\frac{\partial^2 V}{\partial T^2}\right)_P \tag{36}$$

We can also write Eq. (32) of Chapter 2 for the difference in C_P and C_V as

$$C_P = C_V + T\left(\frac{\partial P}{\partial T}\right)_V \left(\frac{\partial V}{\partial T}\right)_P \tag{37}$$

4.6 Generalization to Additional Displacements

It will be useful to generalize these ideas to cases in which the system can have additional displacement coordinates, l_j. This will give rise to terms in δw_{oth} of the form $L_{i,\,ext}\,dl_i$, where $L_{i,\,ext}$ is the external driving force conjugate to the coordinate l_i. For example, l_i could be the length of a wire and $L_{i,\,ext}$ could be the force on the wire. Equation (20) then becomes

$$dU = \delta q_{rev} + \delta w_{rev} \stackrel{m.e.}{=} T\,dS - P\,dV + \sum_i L_i\,dl_i \tag{38}$$

As mentioned in Chapter 2, we choose to define enthalpy as

$$H = U + PV - \sum_i L_i l_i \tag{39}$$

The corresponding relation for the Gibbs free energy is

$$G = U - TS + PV - \sum_i L_i l_i \tag{40}$$

We then obtain

$$dH \stackrel{\text{m.e.}}{=} T\, dS + V\, dP - \sum_i l_i\, dL_i \tag{41}$$

$$dG \stackrel{\text{m.e.}}{=} -S\, dT + V\, dP - \sum_i l_i\, dL_i \tag{42}$$

We also have

$$dA \stackrel{\text{m.e.}}{=} -S\, dT - P\, dV + \sum_i L_i\, dl_i \tag{43}$$

$$dS \stackrel{\text{m.e.}}{=} \frac{1}{T}\, dU + \frac{P}{T}\, dV - \frac{1}{T} \sum_i L_i\, dl_i \tag{44}$$

These equations give criteria for equilibrium similar to those obtained from Eqs. (20)–(24). For example, at constant T, P, and generalized driving forces L_i, a process is at equilibrium if $dG = 0$ and proceeds spontaneously if $dG < 0$.

We will always choose G so that it is a criterion for spontaneity under conditions of constant *intensive* variables, T, P, and conjugate driving forces, L_i. It should be realized, however, that new displacement coordinates introduce the possibility of a number (see Question 9) of other new energy-like functions, and consideration of some of these may provide useful thermodynamic relationships. In particular, we will have some use for the function

$$F = U + PV - TS \tag{45}$$

(like G, but without the additional displacement variable terms). We will call this function the *F function*.

4.7 Standard States

Above all, thermodynamics is a useful subject. Its usefulness is largely dependent on the tabulation of thermodynamic quantities in an efficient and convenient form. Because it takes at least two variables to determine the state of a pure material, tables could get rather unwieldy. To avoid this, properties are tabulated at a standard (pressure) state and then converted to the pressure that is desired. For liquids and solids, the standard state is just that of the pure material at 1.0 bar pressure.

The properties of matter in so-called "ideal states" vary in ways that are mathematically much simpler than those of matter in real states. We have seen an example of this in Chapter 1, which dealt with ideal and real gases, and a similar result will be seen when we consider solutions in Chapters 8 and 9. Because of this mathematical simplicity, it will often be advantageous to use ideal states for tabulations. In particular, properties of gases are tabulated in the ideal gas state, where all intermolecular interactions are zero.

For the standard state of gaseous substances, we want to calculate thermodynamic properties in the 1.0-bar ideal gas state from measured values of properties at the measurement pressure, P. The calculation consists of the following three steps:

1. The thermodynamic property of the real gas is extrapolated to zero pressure.
2. At zero pressure, intermolecular interactions are removed, and the real gas becomes an ideal gas.
3. The ideal gas is then extrapolated to 1.0 bar pressure.

In general, thermodynamic properties are not changed by step 2, because at zero pressure, there are no intermolecular interactions.

For enthalpy, the calculation is

$$\Delta H = \Delta_a H + \Delta_c H = \int_P^0 \left(\frac{\partial H}{\partial P}\right)_T dP + \int_0^{1.0} \left(\frac{\partial H}{\partial P}\right)_{T,\,i.g.} dP \tag{46}$$

where the subscript i.g. represents ideal gas. The second integral is zero because enthalpies of ideal gases do not depend on pressure. For the first integral, we use Eq. (31), giving

$$\Delta H = \int_P^0 V(1 - \alpha T)\, dP \tag{47}$$

The conversion from a tabulated value at 1.0 bar pressure to a real gas value at pressure P is, of course, the reverse of process 1–2–3, and so its enthalpy change is the negative of that given in Eq. (47).

For entropy, the calculation is

$$\Delta S = \Delta_a S + \Delta_c S = \int_P^0 \left(\frac{\partial S}{\partial P}\right)_T dP + \int_0^{1.0} \left(\frac{\partial S}{\partial P}\right)_{T,\,i.g.} dP \tag{48}$$

$(\partial S / \partial P)_T$ is obtained from Eq. (28) and becomes $-nR/P$ for an ideal gas. Equation (48) then becomes

$$\Delta S = -\int_P^0 \left(\frac{\partial V}{\partial T}\right)_P dP - \int_0^{1.0} \frac{nR}{P} dP = \int_{1.0}^P \left(\frac{\partial V}{\partial T}\right)_P dP$$
$$+ \int_0^{1.0} \left[\left(\frac{\partial V}{\partial T}\right)_P - \frac{nR}{P}\right] dP \tag{49}$$

Note the method of dealing with two integrals, each of which is infinite. A difference of the two integrands is taken and this difference goes to zero as the pressure goes to zero. Standard Gibbs free energies can be calculated from standard enthalpies and entropies. Because real gases behave quite ideally up to 1.0 bar, Eqs. (47) and (49) show that negligible error is made by just correcting the real gas H and S to 1.0 bar.

4.8 Entropy of Mixing of Ideal Gases

So far, our discussion of entropy has been concerned with pure substances. In order to calculate the entropy of mixtures, we can first calculate the entropies of the pure components of the mixture and then add the entropy change of mixing to the sum of these. Entropy of mixing is most easily calculated for ideal gas mixtures. Because ideal gas molecules do not interact, their mixing is at constant temperature. Probably the simplest way of thinking about mixing two ideal gases is to have them in separate containers at the same pressure and then open a valve connecting the two containers. In this case, we are mixing the components from their partial molar volumes in the final mixture to the total volume of the mixture. Because this process involves no heat, we might think that it has no entropy change. However, this is not correct, because this "mixing at constant pressure" is not reversible, and the entropy change for this process cannot be directly calculated.

The same change of state can be achieved in the following manner: First, we expand each gas from its partial molar volume, V_j, to the final volume of the gas mixture, $V_f = V_j(n_{tot}/n_j)$. This gives for each gas, from Eq. (11) of Chapter 3, an entropy increase of expansion of

$$n_j R \ln\left(\frac{V_f}{V_j}\right) = n_j R \ln\left(\frac{n_{tot}}{n_j}\right) = -nRX_j \ln X_j$$

where X_j is the mole fraction of component j and we have used n for n_{tot}. The total entropy of expansion is then

$$\Delta_{exp}S = -nR \sum_j X_j \ln X_j \tag{50}$$

To the entropy of expansion, we add the entropy change on reversible mixing at constant temperature in a manner such that the final volume is equal to the initial volume of each of the component gases. In other words, we mix the gases from their partial pressures in the final mixture to the total pressure of the mixture. However, if the temperature is constant, so is the energy for an ideal gas, and Eq. (24) shows that there is no entropy change for each component of the mixture for this process. (This assumes that the process can be carried out with the system in material equilibrium.) The entropy change for this "mixing at constant volume" is thus zero. (In Question 6, a device for reversible mixing at constant volume is shown.) The entropy of mixing therefore is just

$$\Delta_{mix}S = -nR \sum_j X_j \ln X_j \tag{51}$$

Because we have achieved the same change of state as for "mixing at constant pressure" and entropy is a state function, this must be the entropy change for opening the valve between bulbs at the same pressure. Note that, as expected, entropy of mixing is always positive.

4.9 Thermodynamics of Stretching Rubbers

The change of internal energy upon stretching a rubber is

$$dU = \delta q_{rev} + \delta w_{rev} \overset{m.e.}{=} T\,dS + f\,dl - P\,dV \tag{52}$$

Material equilibrium is not an issue and this designation will be dropped. The change in the Helmholtz free energy is

$$dA = d(U - TS) = -S\,dT + f\,dl - P\,dV \tag{53}$$

from which a Maxwell relationship

$$\left(\frac{\partial f}{\partial T}\right)_{l,\,V} = -\left(\frac{\partial S}{\partial l}\right)_{T,\,V} \tag{54}$$

can be obtained. Also, from Eq. (53),

$$f = \left(\frac{\partial A}{\partial l}\right)_{T,\,V} = \left(\frac{\partial U}{\partial l}\right)_{T,\,V} - T\left(\frac{\partial S}{\partial l}\right)_{T,\,V} = \left(\frac{\partial U}{\partial l}\right)_{T,\,V} + T\left(\frac{\partial f}{\partial T}\right)_{l,\,V} \tag{55}$$

which gives

$$\left(\frac{\partial U}{\partial l}\right)_{T, V} = f - T\left(\frac{\partial f}{\partial T}\right)_{l, V} \tag{56}$$

Equations (54) and (56) suggest that measurements of the force required to keep a rubber at constant length as a function of temperature would determine the thermodynamic properties of the rubber. However, due to thermal expansion, very large changes in pressure would be required to keep the polymer volume constant as its temperature is varied. Thus, measurements of $(\partial f/\partial T)_l$ are usually made at constant pressure. Extending Eq. (10) of Appendix A to an additional variable gives

$$\left(\frac{\partial f}{\partial T}\right)_{l, V} = \left(\frac{\partial f}{\partial T}\right)_{l, P} + \left(\frac{\partial f}{\partial P}\right)_{l, T}\left(\frac{\partial P}{\partial T}\right)_{l, V} \tag{57}$$

Using the chain rule for partial derivatives [Eq. (7) of Appendix A], Eq. (57) can be written as

$$\left(\frac{\partial f}{\partial T}\right)_{l, V} = \left(\frac{\partial f}{\partial T}\right)_{l, P} - \left(\frac{\partial f}{\partial P}\right)_{l, T}\left(\frac{\partial V}{\partial T}\right)_{l, P}\left(\frac{\partial V}{\partial P}\right)_{l, T}^{-1} = \left(\frac{\partial f}{\partial T}\right)_{l, P} + \left(\frac{\partial f}{\partial P}\right)_{l, T}\frac{\alpha_l}{\kappa_l} \tag{58}$$

$(\partial f/\partial P)_{l, T}$ is small. However, the ratio of the thermal expansion coefficient to the isothermal compressibility in SI units of solid materials is so large (see Table 3 of Chapter 1) that $(\partial f/\partial T)_{l, V}$ is not well approximated by $(\partial f/\partial T)_{l, P}$. It can be shown,[6] however, that to a good approximation, $(\partial f/\partial T)_{l, V}$ is well approximated by $(\partial f/\partial T)_{l, \alpha}$

$$\left(\frac{\partial f}{\partial T}\right)_{l, V} \approx \left(\frac{\partial f}{\partial T}\right)_{l, \alpha} \tag{59}$$

where α, the *elongation* of the rubber, is defined by $\alpha \equiv l/l_0$. l_0 is the unstretched length of the polymer. Using Eq. (59) for estimating $(\partial f/\partial T)_{l, V}$ requires slight adjustments of l as the temperature is varied, to account for changes of l_0 with temperature due to bulk thermal expansion.

An ideal rubber is defined as one for which

$$\left(\frac{\partial U}{\partial l}\right)_{T, V} \overset{\text{id. rub.}}{\equiv} 0 \tag{60}$$

which, from Eq. (55), means that

$$f \overset{\text{id. rub.}}{\equiv} -T\left(\frac{\partial S}{\partial l}\right)_{T, V} = T\left(\frac{\partial f}{\partial T}\right)_{l, V} \tag{61}$$

The force resisting extension for an ideal rubber is completely entropically derived. Because it certainly requires a positive force to stretch a rubber,

$$\left(\frac{\partial f}{\partial T}\right)_{l, V} \overset{\text{id. rub.}}{>} 0 \tag{62}$$

This is in contrast with most other solid materials, which weaken as temperature is increased. In addition,

$$\left(\frac{\partial S}{\partial l}\right)_{T, V} \overset{\text{id. rub.}}{<} 0 \tag{63}$$

This result will be discussed in Chapter 5.

We can also write

$$\left(\frac{\partial T}{\partial l}\right)_{S, V} = -\left(\frac{\partial S}{\partial l}\right)_{T, V} \left(\frac{\partial S}{\partial T}\right)_{l, V}^{-1} \tag{64}$$

By analogy to other heat capacities, we identify $(\partial S/\partial T)_{l, V}$ as $C_{l, V}/T$, where $C_{l, V}$ is the heat capacity at constant extension and volume, definitely a positive quantity. We then have

$$\left(\frac{\partial T}{\partial l}\right)_{S, V} = -\frac{T}{C_{l, V}} \left(\frac{\partial S}{\partial l}\right)_{T, V} \tag{65}$$

With Eq. (63), we see that

$$\left(\frac{\partial T}{\partial l}\right)_{S, V} > 0 \tag{66}$$

(i.e., the rubber gets warm when it is stretched adiabatically and reversibly). (When stretching with different constraints, the magnitude, but not the sign, of the temperature change would vary.)

Questions

1. In order to make use of a heat reservoir at 0 K, we need to be able to add some heat to the reservoir while maintaining its temperature. Show that the second law precludes this, in that an infinite amount of work would be necessary to maintain a heat reservoir at 0 K against any heat addition.

2.* Consider how you might use a chemical reaction in a *gedanken* experiment to reduce temperature by a process described by Fig. 1. One suggestion is to use a an exothermic reaction, such as $H_2 + Cl_2 \rightarrow 2HCl$ ($\Delta S_{eq} < 0$), and vary the extent of the reaction by changing the amount of H_2 in contact with Cl_2. This could be done by using a piston through which H_2, but not Cl_2, could pass. The apparatus would be used in a reversible cycle, with alternating isothermal and adiabatic steps. Of course, the availability of an appropriate apparatus and reaction catalyst would have to be assumed.

3. Why are heat capacities at constant pressure used in Eq. (5)?

4. Is $-dA$ the maximum work that can be obtained in a constant-T process if volume is allowed to change in the process?

5. Show by both a mathematical and by a physical argument that if $dS = 0$ for a process at constant U and V with no "other work," the system is at equilibrium for that process.

6. A *gedanken* device is shown in the following diagram whereby two gases, each at volume V, could be mixed to give a mixture that also has volume V.

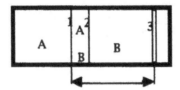

Movable partition 1 is permeable only to A, stationary partition 2 is permeable only to B, and partition 3 is impermeable and rigidly attached to partition 1 through the handle. Show that as the handle is moved to the left, the gases are mixed, and as it is moved to the right, they are unmixed. Moreover, show that the gases exert no net force on the device, so that in the absence of friction, it can be moved in either direction with negligible force, and thus reversibly.

7. In mixing ideal gases at constant pressure by opening a stopcock between bulbs containing them, T and U remain constant and $q = 0$. Why is not $\Delta S = 0$ for this process?

8. From the form of Eq. (28), why would a zero-pressure ideal gas state be inconvenient for tabulation of standard molar entropies?

9. Show that a system that has one additional displacement variable, l, in addition to V and S, can be described by a total of eight energylike functions (i.e., four in addition to U, H, A, and G).

10. Show that the first law of thermodynamics requires that heat be evolved when an ideal rubber is stretched.

11. At large elongation, attractive interaction between extended polymer chains in many rubbers occurs. These interactions are called strain-induced crystallization. What would be the effect of these interactions on $(\partial U/\partial l)_{T,V}$ and $(\partial S/\partial l)_{T,V}$?

12. Verify that each term in each of Eqs. (5), (29), (37), (43), and (54) has the same units.

13. Give all the Maxwell-type relationships that can be derived from the equation $dG = V\,dP - S\,dT + l\,dL$.

Problems

1. Show that in order to use the third law to calculate entropies, the heat capacity must approach zero as $T \to 0$ at least as quickly as does T.

2. Prove the equation for U analogous to the Gibbs–Helmholtz equation, Eq. (48), namely

$$\left(\frac{\partial(T^{-1}A)}{\partial T^{-1}}\right)_V = U$$

3. Show that for a nonmetallic crystalline substance at low temperature, in the range where the heat capacity varies as T^3, the entropy is given as $S = C_p/3$.

4.M Some heat capacity data for benzene is given in the following table. Estimate the entropy of liquid benzene at 298 K and compare your estimate with the literature value of 173 J/K mol.

T (K)	$C_{P,m}$ (J/K mol)	T (K)	$C_{P,m}$ (J/K mol)
13	2.87	160	67.9
15	4.16	180	75.4
20	8.37	200	83.7
25	13.16	220	93.4
30	18.0	240	104.1
40	26.5	260	116.1
50	33.0	278.7	128.7 (solid)
60	37.9	$\Delta_{fus}H^\circ = 9866$ J/mol at 278.7 K	
80	45.0	278.7	131.7 (liquid)
100	50.4	298	136
120	55.7		
140	61.5		

Source: Data from GD Oliver, M Eaton, HM Huffman. J Am Chem Soc 70:1502, 1948.

5. The heat capacity of a nonmetal at 7 K is 1.5 J/K mol. Assuming that the Debye theory holds up to this temperature, what is the molar entropy of the substance at 5 K?

6. Find an expression for the entropy change resulting from the expansion of a van der Waals gas from a molar volume of V_{m1} to a molar volume of V_{m2} at temperature T.

7. Find a formula for the change of Gibbs free energy of 1 mol of a gas that expands from a volume V_1 to a volume V_2 at constant temperature T. Use terms in the virial expansion up to and including the second virial coefficient to describe the equation of state of the gas.

8. Show that $\mu_{JT} = (V_m/C_{P,m})(\alpha T - 1)$.

9. Show that $(\partial H/\partial P)_T = 0$ for an ideal gas.

10. Derive Eq. (36).

11.* For a van der Waals gas find $(\partial C_V/\partial V)_T$ and $(\partial C_P/\partial P)_T$. You do not have to simplify your results.

12. For a Redlich–Kwong gas, find $(\partial C_V/\partial V)_T$.

13. Find an expression for the internal pressure of a Berthelot gas.

14. Show that for a two-component mixture, the entropy of mixing is a maximum when $X_1 = 0.5$.

15. Bromine has two isotopic forms of isotopic mass 79 and 81 and almost equal isotopic abundance. Calculate $\Delta_{mix}S^\circ$ (0 K) for forming 1 mol of the random mixture of Br_2 isotopomers from 0.5 mol of each of the isotopically pure Br_2 isotopomers.

16.* Show that the Gibbs free energy of 1 mol of a mixture of components A and B, with individual molar free energies $G_{m,\,A}$ and $G_{m,\,B}$ is a minimum when

$$\frac{x_A}{1-x_A} = \exp\left(\frac{-(G_{m,\,A}-G_{m,\,B})}{RT}\right)$$

Because molar free energies are finite, this minimum will occur at any real temperature when both A and B are present in the mixture.

17. Show that $(\partial S/\partial V)_P = C_P/TV\alpha$.

18. Calculate ΔG_f° of liquid methanol from enthalpies and entropies given in Appendix C. Compare your answer with the value given in this table.

19. Show that a rubber for which the force required to achieve a given elongation (at a particular pressure) is proportional to the temperature is an ideal rubber.

20. Show that from measurements of the stress versus strain curve and the force required to keep a rubber at constant extension as temperature is varied (and pressure is held constant) $(\partial H/\partial l)_{T,\,P}$, $(\partial S/\partial l)_{T,\,P}$, and $(\partial G/\partial l)_{T,\,P}$ can be obtained. (Hint: the F function is useful for this problem.)

21. Show that for a system at constant T and P and subject to constant driving forces L_i, processes proceed spontaneously if they lower the Gibbs free energy, with G defined as in Eq. (40).

22.* Show that the Joule coefficient can be obtained from the equation of state of a gas by

$$\mu_J = \frac{1}{C_P}\left(P - T\left(\frac{\partial P}{\partial T}\right)_V\right)$$

and that the Joule coefficient of an ideal gas is zero.

23.*   Show that the Joule–Thomson coefficient can be obtained from the equation of state of a gas by

$$\mu_{JT} = \frac{1}{C_P}\left(T\left(\frac{\partial V}{\partial T}\right)_P - V\right)$$

and that the Joule–Thomson coefficient of an ideal gas is zero.

24.   Find the change in entropy when 1 mol of a van der Waals gas is expanded from a volume of 5 L to a volume of 10 L at a temperature T. (Express your answer in terms of the van der Waals constants of the gas and the temperature.)

Notes

1. Some authors have argued that because this is required for the validity of the second law, it should not be considered a separate law of thermodynamics. For a discussion of this point, see AB Pippard, Elements of Classical Thermodynamics. Cambridge: Cambridge University Press, 1957, p 48.
2. Comparable low temperatures have also been obtained by trapping gaseous particles in magnetic fields and lowering their velocity by absorption and reemission of laser energy. Using these methods, a new state of matter, the Bose–Einstein condensate has been created.
3. We can think of chemical reactions occurring reversibly in electrochemical cells, where the driving force of the reaction is opposed by an electrical force. The difficulty of constructing electrochemical cells at 0 K is a practical, not a theoretical limitation.
4. If, at 0 K, entropies of compounds are the sum of nonzero values assigned to each atom, conservation of atoms would ensure that the entropy change for all chemical reactions would be zero. This way of satisfying the third law would not be in agreement with the microscopic interpretation of entropy given in Chapter 5.
5. Condensed phase systems can be studied at constant volume by using very strong cells, such as those made of diamonds, which can withstand the very high pressure buildup that occurs as these systems are heated.
6. PJ Flory, Principles of Polymer Chemistry. Ithaca, NY: Cornell University Press, 1953.

5

Statistical Mechanics

*Perhaps, the atomic hypothesis will be replaced by another some day—
perhaps, but not probably.*

Ludwig Boltzmann

In this chapter, the macroscopic world of thermodynamics is linked to the microscopic world of atoms and molecules through the idea of the probability of a macroscopic state. The basic assumption of statistical mechanics is that this probability is proportional to the number of microscopic arrangements that can give the macroscopic state. This number is calculated for distributions in space and among energy levels. The Boltzmann distribution is derived from the second law of thermodynamics. The partition function is used to obtain formulas to calculate thermodynamic functions from molecular properties. The development is applied to a system with only translational energy to give the partition function and thermodynamic properties of the monatomic ideal gas. The Maxwell velocity distribution and several of its averages from kinetic theory are calculated. Extension of the theory to polyatomic molecules in the classical limit gives the classical equipartition theorem. Foreshadowing the discussion of steady-state systems, thermal transpiration is treated by kinetic theory. Statistical theory is applied to the model of the freely jointed chain to discuss the mechanical properties of macromolecules.

5.1 The Microscopic World

Our discussion of thermodynamics has been thoroughly grounded on observations on the real macroscopic world. It is not necessary to be aware of the existence of atoms in order to discuss and use the first, second, and third laws of thermodynamics. The basis for the interpretation of thermodynamics in terms of the microscopic world of atoms, molecules, and ions was first given by Ludwig Boltzmann in 1896. However, at that time, there was so little direct evidence for the existence of these microscopic particles that Boltzmann's ideas were much disparaged and only slowly gained acceptance. The lack of appreciation of his theories may have been a contributing factor in Boltzmann's suicide in 1906. If Boltzmann had not killed himself, he might have lived long enough to see the general acceptance of his concepts and the increased utility of thermodynamics derived from the ability to calculate macroscopic thermodynamic functions from microscopic molecular properties. In this chapter, we will introduce some of the basic ideas behind the relation of thermodynamic properties to microscopic properties. This subject is known as statistical mechanics because it involves averaging over the properties of the microscopic entities in order to obtain the macroscopic properties.

One problem that arises is that although Newtonian mechanics is sufficient to describe the overall translational motion of particles of the size of atoms and molecules, quantum mechanics is required to describe their rotational and internal motion. Quantum mechanics is essential in dealing with the motion of particles as small as electrons. Because knowledge of quantum mechanics is not assumed of the reader of this book, we will be content to develop the framework into which quantum mechanical results can be later be inserted. We will apply this framework to two systems that can be treated classically, namely the monatomic ideal gas and polymer chains.

It is fairly simple to explain the first law of thermodynamics on the basis of atoms. (We will use the term "atom" to generically denote the microscopic particles: atoms, molecules, ions, and electrons.) Newtonian mechanics is known to conserve energy[1] and when we add huge numbers of interactions, each conserving energy, the energy of the sum should also be conserved. The second law, which tells us the direction of real processes, is much more difficult to explain by Newtonian mechanics. The direction of a Newtonian process depends on its initial conditions. If we reverse the velocities of all of the interacting particles, the process runs backward. The reason for this is that Newton's second law involves acceleration, which is a second derivative with respect to time and thus is unchanged upon reversing the "direction" of time. We shall see that Newtonian mechanics does have a forward direction in a statistical or averaged sense.

5.2   The Joule Process

In order to gain a sense of the statistical aspects of thermodynamic processes, let us look in detail at one simple process—the Joule expansion of an ideal gas. This spontaneous process involves the rapid (and thus adiabatic) expansion of a gas into vacuum (thus producing no work). It is described by the Joule coefficient, $\mu_J \equiv (\partial T/\partial V)_U$, which we have shown equals $-(1/C_V)(\partial U/\partial V)_T$ and is zero for an ideal gas. For an ideal gas, energy depends only on temperature and not on volume, and so the Joule process is isothermal as well. Because the Joule expansion is adiabatic and without work, the system is isolated and there are no changes in the surroundings. The driving force for this process must come from some change in the system. In Chapter 3, we called this driving force the entropy increase of the system. From a microscopic point of view, why does the gas spontaneously expand?

We will find it easier to consider the reverse Joule process, which is not spontaneous. We ask the following: If gas occupies an entire container, why at a later time do we *not* find it contracted into some fraction of this container? Choosing a halving of volume, let us look at this process from a microscopic point of view. We will first consider a system containing only four argon atoms ($N = 4$), which initially may be anywhere in the container of volume V. In the reverse Joule process, the atoms all move to the left half of the container. We could try to explain the nonoccurrence of this process using Newton's laws, by following the trajectories of the particles. However, this would require a computer and we would find that if there were any uncertainties in the initial conditions, after a short time (say 1 s) we could claim very little accuracy for our calculation.[2] This leads us to give up our attempt for a dynamical description of this process and think about how we might investigate it experimentally.

For an experimental investigation, we could make use of a "magic camera" that could photograph individual argon atoms and take a million photographs of the contents of the box. The camera would be performing a time average of the motion of the atoms in the box.[3] With a million photos and only four atoms in the box, it is likely that we would find some photos with all the atoms on the left side of the box. How would we interpret this result? A reasonable interpretation would be in terms of probability. We could count the pictures that showed all four atoms on the left side of the box (consistent with our definition of the final state of the reverse Joule process) and divide it by the total number of pictures, where the four atoms could be anywhere in the box (consistent with the initial state). This ratio would be taken as the fraction of the time that the system spent in the final state of the reverse Joule process, and we would assume that as the number of photos became infinite, the ratio would approach the *probability* of the spontaneous occurrence of the final state.

Unfortunately, we do not have a magic camera, and even if we did, the bill for developing the number of photos required would be prohibitive for any macroscopic number of particles. This leads us to attempt to figure out the result of this experiment without doing it. Because we have already given up on the possibility of following the microscopic behavior of the system over time, we are not able to calculate time averages. Instead, we adopt the procedure of calculation using ensemble averages. By an *ensemble*, we mean a large number of systems, each satisfying the macroscopic definition of the initial state of our reverse Joule process (fixed N, V, and U). An ensemble in which every system has the same N, V, and U is appropriate for isolated systems and is known as a *microcanonical ensemble*. Every microscopically distinguishable arrangement of the atoms is included once in the ensemble. We then make the basic postulate of statistical mechanics:

> The probability of a macroscopic state is proportional to the number of distinguishable microscopic configurations that are consistent with the definition of that state.

We can justify that the configurations must be distinguishable, because quantum mechanics asserts that there can be no real difference between configurations that cannot be distinguished. Of course, Boltzmann predated quantum mechanics, but he realized that distinguishable configurations were necessary for agreement with observation.

The key word for the calculation in the basic postulate is "number." What we are claiming is that there is no difference in probability between any distinguishable microscopic configuration of the system. All we have to do is count the possible microscopic configurations corresponding to a state. Therefore, what the microscopic description of thermodynamics comes down to in the end is just counting!

Substituting an ensemble average for the time average rests on a number of assumptions. It is necessary that all configurations represented in the ensemble be accessible during the motion over time of the system. In addition, no distinguishable configuration can be favored over other configurations in the motion over time. It is difficult to see why some configurations would not be accessible in the motion over time and what criteria could be used to favor certain configurations in a system defined only in terms of N, V, and U. Although there has been considerable discussion of these assumptions in the literature, we will just accept the above basic postulate.[4]

In counting distinct configurations, we must realize that coordinate space is continuous, and because positions that differ infinitesimally can be distinguished, theoretically we can enumerate an infinite number of configurations. Ratios of infinities are not much use to us for calculating probabilities! The way to get around this is to divide the container into a large number of microscopic compartments, say M for each half of the box. The size of the compartments should be large compared to that of an atom but small enough that there are many more compartments than atoms and little probability that two atoms are in the same compartment.[5] This will make our counting much easier. A configuration is then defined by which of the M compartments is occupied (by the center of the atom). Note that even though the atoms are indistinguishable, these configurations would give distinguishable photographs with our magic camera.

For the reverse Joule process, initially the atoms can occupy $2M$ compartments. Counting configurations for four atoms in the initial system is equivalent to asking how many ways can four indistinguishable atoms occupy $2M$ compartments, with no more than one atom in a compartment? This is a fundamental problem in probability theory. The way it is solved is by first taking all possible permutations (orderings) of the $2M$ compartments [there are $(2M!) = 2M(2M - 1)(2M - 2) \cdots 1$ of these] and then dividing by the possible permutations of the filled (4!) and empty [$(2M - 4)!$] compartments, both of which are indistinguishable sets. We will call the number of configurations consistent with a macroscopic state Ω. We thus have for the initial state

$$\Omega_i = \frac{(2M)!}{4!(2M - 4)!} = \frac{2M(2M - 1)(2M - 2)(2M - 3)}{4!} \approx \frac{(2M)^4}{4!} \tag{1}$$

where the final approximation results from M being much larger than 4. Likewise, the number of configurations consistent with the final state (four atoms in M compartments) is

$$\Omega_f = \frac{(M)!}{4!(M - 4)!} \approx \frac{M^4}{4!} \tag{2}$$

This gives

$$\frac{\Omega_f}{\Omega_i} = \frac{1}{2^4} = \frac{1}{16} \tag{3}$$

as the ratio of the number of configurations of the final state to that of the initial state. Now let us say that instead of four atoms in the box, there are N atoms. Redoing our calculation, we would get

$$\frac{\Omega_f}{\Omega_i} = \frac{1}{2^N} \tag{4}$$

With N corresponding to any macroscopic amount of matter (say 1 nmol $= 6 \times 10^{14}$ atoms), this is such a small number that the probability of all of the atoms being in the left-hand side of the box is totally negligible. This is a dramatic example of the microscopic meaning of the second law. Although there is a *possibility*, in a macroscopic amount of a gas, of all the atoms suddenly finding themselves in one-half of a container, the *probability* of this happening is so small that we can completely neglect it. The reason that the reverse Joule process does not spontaneously occur is therefore a statistical one. It is very, very unlikely that it would ever occur. Conversely, the Joule process, the spontaneous expansion of a gas, is statistically very likely.

For systems containing macroscopic numbers of particles, the probability of finding even 51% of the atoms (or any measurable difference) in one-half of the box is also negligible. We can state the following:

> For macroscopic numbers of molecules, states whose properties differ measurably from the state with the maximum number of configurations have negligible probability. The properties of the system are therefore just the properties of the state with the maximum number of configurations.

Note that from our calculation on the four-atom case, we can see that for a small number of particles, the probability of appreciable percentage fluctuations from the uniform distribution is more probable.

Because both entropy and the number of configurations indicate the direction of spontaneous processes, these must be related. What is the relationship between these quantities? Although both Ω and S tend to maximum values in real processes, S cannot be equal (or proportional) to Ω because Ω is not an extensive function. For example, if in a given state of the container considered here, there are Ω_l configurations for the left side of the container and Ω_r configurations for the right side of the container, then there are $\Omega_l\Omega_r$ configurations for the entire container, because each left-side configuration can go with any right-side configuration. In order to convert this product into the sum needed for an extensive property, we set S proportional to $\ln \Omega$,[6] giving Boltzmann's famous relationship[7]

$$S = k \ln \Omega \tag{5}$$

To find k, the Boltzmann constant, we calculate ΔS for the halving of the volume of an ideal gas, considered here in the reverse Joule process. From Eq. (4), if $\Omega_i = C$, then $\Omega_f = 2^{-N}C$, and

$$\Delta S = k \ln (2^{-N}C) - k \ln C = k \ln 2^{-N} = -Nk \ln 2 \tag{6}$$

However, we have seen in Chapter 3 that the entropy change for the isothermal (and constant energy) halving of a volume of n moles of an ideal gas is $-nR \ln 2$. If our statistical result is to agree with our thermodynamic result, then $Nk = nR$ or $k = R(N/n)^{-1} = R/N_A$. This is the same constant that arose in calculating the average energy of a molecule in the Bernoulli model of the ideal gas in Section 1.4 of Chapter 1. In fact, the Boltzmann constant will arise whenever we calculate macroscopic properties from a microscopic model of a system. The value of Boltzmann's constant is 1.38×10^{-23} J/K.

We have discussed Eq. (5) in connection with a system at constant N, V, and U (a microcanonical ensemble). There is no reason, however, why the same relationship between Ω and S should not hold under other constraints—in particular, at constant N, V, and T (called a *canonical ensemble*).

Equation (5) also gives a microscopic explanation of the third law of thermodynamics. In most perfectly crystalline materials, there is a unique arrangement that has the lowest energy. As the temperature is lowered and energy is removed from the material, it tends to this lowest-energy state. With a unique lowest-energy state, $\Omega = 1$, and $S = 0$ is approached as $T \to 0$. Some species (such as CO, discussed in Chapter 4) have residual entropy, due to an alternative arrangement that is very close in energy to the lowest-energy configuration. In this case, it may be that at temperatures at which the lowest energy configuration becomes favored by the Boltzmann distribution, there is no longer sufficient thermal energy to overcome the kinetic activation barrier to make a transition to the lowest energy state. The random distribution between the two low-energy arrangements then becomes "frozen in" as $T \to 0$.

Example 1. Calculate the residual entropy at $T = 0$ for carbon monoxide.

Solution: Carbon monoxide has a small electric dipole moment (approx 0.1 Debye), which gives the molecules an energetically preferred orientation as $T \to 0$. However, this dipole moment is so small that the preference is not appreciable until very low temperatures, and the random orientation of the molecules (the dipole has equal probability of pointing in one direction or its opposite) remains as the temperature is lowered. For a mole of CO, each molecule can point in either of two directions and there are 2^{N_A} configurations that are about equally probable. This model predicts a residual entropy of

$$S(0) = k \ln 2^{N_A} = R \ln 2 = 5.76 \text{ J/mol K}$$

which is close to the measured value of 5.0 J/mol K for CO.

Note that in solid CO, a configuration is determined by *which* CO molecules have the less favorable orientation, not by just how many molecules have this orientation. This is because, unlike in a gas, in a solid, molecules can be identified by their position in the lattice. Thus, configurations in which the same number, but different, molecules are less favorably oriented can be distinguished.

Using Eq. (5), we can write

$$\frac{\Omega_i}{\Omega_{max}} = \exp\left(\frac{S_i - S_{max}}{k}\right) = \frac{P_i}{P_{max}} \tag{7}$$

where "max" refers to the state with the maximum number of configurations (i.e., the equilibrium state). Because numbers of configurations are proportional to the probability of states, Eq. (7) also gives the probability of a spontaneous fluctuation of the system away from the equilibrium state with maximum entropy to a state of lower entropy. Entropy differences for macroscopic processes are usually of the order of R. Thus the probability of a spontaneous fluctuation of a system by a macroscopically measurable amount from the equilibrium state is of the order of $\exp(-N_A)$, a very, very small number.

5.3 Distribution Among Energy States

So far, we have considered how molecules are distributed among compartments of equal volume. These compartments can be thought of as molecular states of equal energy, and we postulated that there is equal probability of occupying such states. How do molecules occupy states when they are not at equal energy? We will consider a system comprised of noninteracting particles in contact with a heat reservoir at temperature T. Because the particles are noninteracting, each particle has energy levels that are independent of the other particles. Energy being continuous in classical mechanics,[8] we will envision small ranges of energy, which we designate as ε_j. We seek the *occupation numbers*, N_j of these ranges, such that the sum of the N_j equals the total number of molecules in the system, $N = \sum_j N_j$. We will investigate a process in which we change the occupation numbers.

We will use the microscopic definition of entropy, Eq. (5), to calculate the entropy change of the system in this process. In order to do this, we must calculate the number of distinguishable microscopic configurations corresponding to the initial and final distributions of the molecules over their energy levels. Although the molecules are indistinguishable, we will assume that they can be numbered, perhaps by their position in the container. This way of calculating the number of configurations is called *Boltzmann statistics*. With N numbered molecules, the number of ways of assigning a set of occupation numbers $\{N_0, N_1, \ldots, N_j, \ldots\}$ to the molecules is

$$\Omega = \frac{N!}{N_0! N_1! \cdots N_i! \cdots} = \frac{N!}{\prod_i N_i!} \tag{8}$$

The correctness of this very important formula is illustrated for the example $N = 3$, $N_0 = 2$, $N_1 = 1$. There are obviously three configurations for this distribution, corresponding to the three choices of which molecule we place in state 1. The general formula, Eq. (8), is arrived at by including a partition in the ordered listing of the molecules as follows: 1 2|3. All 3! permutations of the three molecules will now place different molecules in different positions, but these will include configurations such as 2 1|3, which differs from 1 2|3 only by an indistinguishable permutation of identical molecules in the same state. Therefore, we must divide 3! by the 2! possible permutations of the identical molecules in state 0 to get our final result.

Quantum mechanics tells us that particles in the gas phase cannot be distinguished by their position; thus, the above-considered example cannot correspond to more than a single configuration. In fact, in quantum systems, there are two different ways of counting the number of configurations, each appropriate for different types of particles.[9] Gibbs first suggested that for indistinguishable particles, the result of Eq. (8) be divided by $N!$ to give *corrected Boltzmann statistics*. Although this procedure cannot be justified in any rigorous manner, it can be shown to approach the correct results for both types of quantum particles as the temperature is increased. Dividing the result of Eq. (8) by $N!$ is seen to have no effect on the results derived in this section.

From Eq. (8), the minimum number of configurations is one, which occurs when all the molecules are in the same state; the maximum number of configurations is $N!$, which occurs when each N_i is either 0 or 1 (because $1! = 0! = 1$). In other words, a broad distribution of molecules over energy states has a higher entropy than a narrow distribution. Why then, in order to satisfy the second law, does the system not achieve a state where there is no more than a single molecule in each state? The reason for this is that such a distribution would require that the system have a large amount of energy, so that high-energy levels could be occupied. For an isolated system, the energy is limited and the system will achieve the broadest distribution consistent with that energy. For a closed system in contact with a thermal reservoir at temperature T, the system can increase the breadth of its distribution and its entropy by extracting energy from the reservoir. However, this will lower the entropy of the surroundings, and because $\Delta S_{sur} = -q/T$, ΔS_{sur} will be lowered further, the lower the temperature of the reservoir and the more energy that is transferred to the system.

To find the equilibrium distribution, we consider a system of fixed volume[10] and number of particles in contact with a thermal reservoir at temperature T. By the second law, at equilibrium there must be a zero entropy change of the universe for all infinitesmal changes in the system, because S_{univ} is a maximum at equilibrium. In particular, for the change involved in increasing the number of molecules in the jth state (with $\varepsilon = \varepsilon_j$) of the system by 1, while decreasing the number of molecules in the lowest state of the system (with $\varepsilon = 0$)

by 1, thermal energy of the amount ε_j must be transferred to the system. This produces an entropy decrease in the surroundings of

$$\Delta S_{sur} = -\frac{\varepsilon_j}{T} \tag{9}$$

Because the system volume is constant, no energy can be transferred to the system by compression work; we assume that no other work is done on the system.) Using Eqs. (5) and (8), the entropy change of the system becomes

$$\Delta S = S_f - S_i = k \ln \Omega_f - k \ln \Omega_i = k \ln \left(\frac{\prod\limits_{j,i} N_{j,i}!}{\prod\limits_{j,f} N_{j,f}!} \right) \tag{10}$$

Note that for the process that we are considering, one molecule going from the lowest ($\varepsilon = 0$) state to the jth state, all terms in the numerator and denominator cancel except the $N_0!$ and $N_j!$ terms. ΔS then becomes

$$\Delta S = k \ln \left(\frac{N_{0,i}! N_{j,i}!}{(N_{0,i} - 1)!(N_{j,i} + 1)!} \right) = k \ln \left(\frac{N_{0,i}}{(N_{j,i} + 1)} \right) \approx k \ln \left(\frac{N_0}{N_j} \right) \tag{11}$$

In the last step, we have assumed that the range of energy chosen for each level is large enough that the N's are very large numbers, giving $N_{j,i} + 1 = N_{j,f} \approx N_{j,i} \equiv N_j$.

If the system is at equilibrium, the entropy change (of the universe) for this process must be zero:

$$\Delta S_{univ} = 0 = -\frac{\varepsilon_j}{T} + k \ln \left(\frac{N_0}{N_j} \right) \tag{12}$$

or

$$\frac{N_j}{N_0} = \exp\left(-\frac{\varepsilon_j}{kT} \right) \tag{13}$$

When using this distribution function, we usually know N, the total number of molecules, rather than N_0. N is just the sum over molecules occupying all states

$$N = \sum_{states} N_j = N_0 \sum_{states} \exp\left(-\frac{\varepsilon_j}{kT} \right) = N_0 q \tag{14}$$

where q is known as the *molecular partition function*:

$$q \equiv \sum_{states} \exp\left(-\frac{\varepsilon_j}{kT} \right) \tag{15}$$

The result gives the *Boltzmann distribution*:

$$f_j = \frac{N_j}{N} = \frac{1}{q} \exp\left(-\frac{\varepsilon_j}{kT}\right) \tag{16}$$

where f_j is the fraction of the molecules that are in state j.

The Boltzmann distribution is *normalized*, in that the sum of f_j, the fraction of the molecules that are in state j, over all states is unity:

$$\sum_{states} f_j = 1 \tag{17}$$

There is an alternative way of writing the Boltzmann distribution. Recognizing that there may be two or more states with the same energy (e.g., the six $2p$ states or ten $3d$ states of an atom), we introduce the quantity g_j, the *degeneracy* of states at energy ε_j. To find the fraction of molecules that have energy ε_j, we just multiply the fraction of molecules that are in one state at the energy [Eq. (16)] by the number of states, g_j, at the energy

$$f_j = \frac{g_j}{q} \exp\left(-\frac{\varepsilon_j}{kT}\right) \tag{18}$$

If the states span a continuous, rather than a discrete, energy range, we can introduce a continuous *distribution function*, $f(\varepsilon)\, d\varepsilon$, the fraction of molecules that have energy in a range $d\varepsilon$ around ε:

$$f(\varepsilon)\, d\varepsilon = \frac{g(\varepsilon)}{q} \exp\left(-\frac{\varepsilon}{kT}\right) d\varepsilon \tag{19}$$

where $g(\varepsilon)$ is the number of states per unit energy at ε and

$$q = \int_0^\infty g(\varepsilon) \exp\left(-\frac{\varepsilon}{kT}\right) d\varepsilon \tag{20}$$

In this case, normalization becomes

$$\int_0^\infty f(\varepsilon)\, d\varepsilon = 1 \tag{21}$$

A variable (or variables) other than energy can be used to represent a continuous range of states. Calling this or these variables h, we have

$$f(h)\, dh = \frac{g(h)}{q} \exp\left(-\frac{\varepsilon(h)}{kT}\right) dh \tag{22}$$

and

$$q = \int g(h) \exp\left(-\frac{\varepsilon(h)}{kT}\right) dh \tag{23}$$

One problem that arises in dealing with continuous distributions is that because states are infinitesimally close together, the number of states in any finite energy range is infinite. Such a distribution cannot be normalized unless an arbitrary decision is made on what range of energy (or other variables) corresponds to the definition of a "state." We will return to this question later in this chapter.

From Eq. (13), the Boltzmann factor, $\exp(-\varepsilon_j/kT)$, is the probability of a state at energy ε_j being occupied compared to the probability of a zero-energy state being occupied. At $T = 0$, the only states that can be occupied are the states at zero energy. At higher temperatures, we must add to these g_0 states the sum of the g_j states at energy ε_j, weighted by the probability of states at this energy being occupied (i.e., the Boltzmann factor) for all energy levels. Performing this addition is just a calculation of the partition function. It follows that the partition function is a measure of the number of states thermally available at temperature T.

In addition to being a function of T, the partition function is also a function of V, on which the quantum description of matter tells us that the molecular energy levels, ε_j, depend. Because, for single-component systems, all intensive state variables can be written as functions of two state variables, we can think of $q(T, V)$ as a state function of the system. The partition function can be used as one of the independent variables to describe a single-component system, and with one other state function, such as T, it will completely define the system. All other properties of the system (in particular, the thermodynamic functions U, H, S, A, and G) can then be obtained from q and one other state function.

5.4 Thermodynamic Functions from the Partition Function

Unlike the case for entropy, there is no absolute zero for energy. We measure energy with respect to a chosen value, U_0, per mole of each substance at $0\,\text{K}$ (corresponding to each molecule in its lowest-energy state), to which we add the excitation energy of the molecules:

$$U_m = U_0 + N_A\langle\varepsilon\rangle = U_0 + N_A \sum_{\text{states}} f_j\varepsilon_j = U_0 + \frac{N_A}{q} \sum_{\text{states}} \varepsilon_j e^{-\varepsilon_j/kT} \qquad (24)$$

where $\langle\varepsilon\rangle$ is the average energy per molecule. Noting that

$$\left(\frac{\partial \ln q}{\partial T}\right)_V = \frac{1}{q}\left(\frac{\partial q}{\partial T}\right)_V = \frac{1}{q}\left(\frac{\partial \sum\limits_{\text{states}} e^{-\varepsilon_j/kT}}{\partial T}\right)_V$$

$$= \frac{1}{qkT^2} \sum_{\text{states}} \varepsilon_j e^{-\varepsilon_j/kT} \qquad (25)$$

we get, with $kN_A = R$,

$$U_m = U_0 + RT^2 \left(\frac{\partial \ln q}{\partial T}\right)_V \tag{26}$$

We hold V constant in these derivatives because quantum mechanics tells us that the molecular energy levels are functions of V.

In order to calculate the entropy, we use its fundamental definition [Eq. (5)]. For distinguishable particles (Boltzmann statistics), we use formula (8) for the number of configurations:

$$S_m \overset{\text{dis}}{=} k \ln N_A! - \sum_i k \ln N_i! \tag{27}$$

with $\sum_i N_i = N_A$. *Stirling's approximation*, for the logarithm of the factorial of a large number,[11] is

$$\ln A! = A \ln A - A \tag{28}$$

giving

$$S_m \overset{\text{dis}}{=} kN_A \ln N_A - kN_A - k \sum_i N_i \ln N_i + k \sum_i N_i$$

$$= kN_A \ln N_A - k \sum_i N_i \ln N_i \tag{29}$$

From the Boltzmann distribution function, for $N = N_A$

$$\ln N_j = \ln N_A - \ln q - \frac{\varepsilon_j}{kT} \tag{30}$$

Substituting into Eq. (29),

$$S_m \overset{\text{dis}}{=} kN_A \ln N_A - k \sum_i N_i \ln N_A + k \sum_i N_i \ln q + \frac{1}{T} \sum_i N_i \varepsilon_i \tag{31}$$

The first two terms in this equation cancel, and the last term is just $(U_m - U_0)/T$ per mole, yielding

$$S_m \overset{\text{dis}}{=} R \ln q + \frac{U_m - U_0}{T} \tag{32}$$

Remembering that we count configurations for indistinguishable particles by dividing the number for distinguishable particles by $N!$, we get

$$S_m \overset{\text{indis}}{=} R \ln q + \frac{U_m - U_0}{T} - k \ln N_A! = R \ln q + \frac{U_m - U_0}{T}$$

$$- k(N_A \ln N_A - N_A) \tag{33}$$

$$S_m \overset{\text{indis}}{=} R \ln \left(\frac{q}{N_A}\right) + R + \frac{U_m - U_0}{T} \tag{34}$$

With these formulas for U_m and S_m, equations can be directly derived for the other thermodynamic functions: H_m, A_m, and G_m. For condensed phases $H_m = U_m$ to a very good approximation. For gases, assumed ideal, $H_0 = U_0$ and $H_m = U_m + RT$, giving

$$H_m \overset{i.g.}{=} U_0 + RT^2 \left(\frac{\partial \ln q}{\partial T}\right)_V + RT \tag{35}$$

For the Helmholtz free energy, $A_m = U_m - TS_m$, giving

$$A_m \overset{dis}{=} U_0 - RT \ln q \tag{36}$$

and

$$A_m \overset{indis}{=} U_0 - RT \ln\left(\frac{q}{N_A}\right) - RT \tag{37}$$

For condensed phases $G_m \approx A_m$; whereas for an ideal gas (indistinguishable), $G_m = A_m + RT$; thus,

$$G_m \overset{i.g.}{=} \mu = U_0 - RT \ln\left(\frac{q}{N_A}\right) \tag{38}$$

In calculations of thermodynamic functions, q is calculated at the appropriate temperature and volume. For example, for standard thermodynamic functions at 298 K, q is calculated at 298 K and 24.7 L.

5.5   System Partition Functions

We have derived a formula for the molecular partition function by considering a system containing many molecules at equilibrium with a heat bath. We can generalize our statistical mechanics by a gedanken experiment of considering a large number of identical *systems*, each with volume V and number of particles N at equilibrium with the heat bath at temperature T. Such a "supersystem" is called a *canonical ensemble*. Our derivation is the same; the fraction of systems that are in a state with energy E_i is

$$f_i = \frac{1}{Q} \exp\left(-\frac{E_i}{kT}\right) \tag{39}$$

where Q is the *system partition function*, given by

$$Q = \sum_{states} \exp\left(-\frac{E_i}{kT}\right) \tag{40}$$

and the sum is over system states. Alternatively, we can sum over energies rather than states:

$$Q = \sum_{\text{energies}} g(E_i) \exp\left(-\frac{E_i}{kT}\right) \tag{41}$$

where $g(E_i)$ is the degeneracy, the number of system states at energy E_i. Usually, for systems, the number of states is so high that they can be considered continuous, and Eq. (41) is written as

$$Q = \int_0^\infty g(E) \exp\left(-\frac{E}{kT}\right) dE \tag{42}$$

where $g(E)$ is the state density (the number of states between E and $E + dE$).

Because for systems, unlike particles, there are no requirements for them to be indistinguishable, we use the thermodynamic formulas analogous to those for distinguishable particles. For systems containing N particles and having volume V,[12]

$$U = kT^2 \left(\frac{\partial \ln Q}{\partial T}\right)_{V,N} \tag{43}$$

$$S = \frac{U}{T} + k \ln Q \tag{44}$$

$$A = -kT \ln Q \tag{45}$$

$$\mu = \left(\frac{\partial A}{\partial n}\right)_{T,V} = N_A \left(\frac{\partial A}{\partial N}\right)_{T,V} = -RT \left(\frac{\partial \ln Q}{\partial N}\right)_{T,V} \tag{46}$$

It is difficult to evaluate Q in cases in which there are interactions between the particles of the system, because, in this case, E_i is truly an energy level of the entire macroscopic system. In the absence of such interactions, E_i is just a sum over the energies of the individual particles (atoms, molecules, etc.) comprising the system:

$$E_i = \sum_{k=1}^{N} \varepsilon_{ik} \tag{47}$$

where ε_{ik} is the energy of the kth particle when the system is in the ith state. If the particles are distinguishable, each particle can be in any of its states, independent of the other particles, and

$$Q \overset{\text{dis}}{=} q_1 q_2 \cdots q_N \tag{48}$$

(Problem 12). If the noninteracting particles comprising the system are identical, then they have the same molecular partition function and

$$Q \overset{\text{dis}}{\underset{\text{iden}}{=}} q^N \tag{49}$$

If the particles are indistinguishable and identical, this formula can just be divided by $N!$, as long as there are many more particle states than particles in the system.

5.6 Velocity Distributions

The *kinetic theory of gases* deals with the translational motions of gaseous molecules. Translational motion of molecules can be considered separately from rotational and internal motions and has energy

$$\varepsilon_{tr} = \tfrac{1}{2}mc^2 = \tfrac{1}{2}m(v_x^2 + v_y^2 + v_z^2) \tag{50}$$

where c is the molecular speed. Because classical mechanics is appropriate for treating translational energy in macroscopic containers, all values of velocity are possible and we should use Eq. (19), the distribution function for continuous-energy levels.

$$
\begin{aligned}
f(\varepsilon)d\varepsilon &= \frac{g(\varepsilon)}{q} \exp\left(-\frac{\varepsilon}{kT}\right) d\varepsilon \\
&= \frac{g(\varepsilon)}{q} \exp\left(-\frac{mc^2}{2kT}\right) d\varepsilon \\
&= \frac{g(\varepsilon)}{q} \exp\left(-\frac{mv_x^2}{2kT}\right) \exp\left(-\frac{mv_y^2}{2kT}\right) \exp\left(-\frac{mv_z^2}{2kT}\right) d\varepsilon
\end{aligned}
\tag{51}
$$

The exponential terms tell us how the molecules are distributed in velocity space, and the degeneracy, $g(\varepsilon)\,d\varepsilon$, is proportional to the volume in velocity space that gives rise to energies between ε and $\varepsilon + d\varepsilon$:

$$g(\varepsilon)\,d\varepsilon = C4\pi c^2\,dc \tag{52}$$

C is a normalization factor determined by integrating over all of velocity space:

$$\frac{C}{q}4\pi \int_0^\infty c^2 \exp\left(-\frac{mc^2}{2kT}\right) dc = 1 \tag{53}$$

Two formulas that may be found in most tables of integrals and are very useful in kinetic theory are

$$\int_0^\infty x^{2n} \exp(-\beta x^2)\,dx = \frac{1}{2}\sqrt{\pi}\,\frac{(2n)!\beta^{-(n+1/2)}}{2^{2n}n!} \tag{54}$$

and

$$\int_0^\infty x^{2n+1} \exp(-\beta x^2)\,dx = \tfrac{1}{2}n!\beta^{-(n+1)} \tag{55}$$

where n is an integer. From Eq. (53), this gives

$$\frac{C}{q} = \left(\frac{m}{2\pi kT}\right)^{3/2} \tag{56}$$

and

$$f(\varepsilon)\,d\varepsilon = 4\pi\left(\frac{m}{2\pi kT}\right)^{3/2} c^2 \, \exp\left(-\frac{mc^2}{2kT}\right) dc = F(c)\,dc \tag{57}$$

$F(c)$, the distribution over molecular speeds, is known as the *Maxwell–Boltzmann velocity distribution*. In Fig. 1a, the distribution are shown for N_2 and H_2, both at 300 K. Note that because the distributions are normalized, the areas under both curves are the same. The most probable speed, average speed, and root-mean-

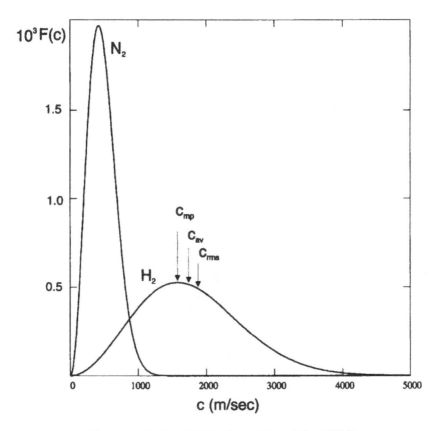

Figure 1   (a) Speed distribution of N_2 and H_2 at 300 K.

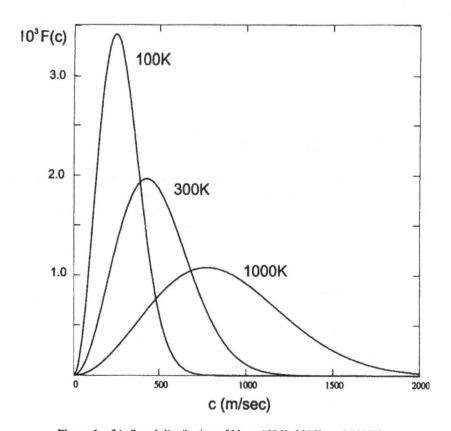

Figure 1 (b), Speed distribution of N_2 at 100 K, 300 K, and 1000 K.

square speed are marked on the H_2 distribution curve. In Fig. 1b, distributions are shown for N_2 at 100 K, 300 K, and 1000 K.

Equation (57) is the product of three *one-dimensional velocity distributions*,

$$f(v_i) = \left(\frac{m}{2\pi kT}\right)^{1/2} \exp\left(-\frac{mv_i^2}{2kT}\right) dv_i \qquad (58)$$

(where $i = x$, y, or z), integrated over the spherical shell in velocity space, which encompasses all speeds between c and $c + dc$.

The velocity distribution functions allow us to evaluate the average value of any function of v_i or c by integrating over the appropriate distribution. For example, the average molecular speed, $\langle c \rangle$, is, using Eqs. (55) and (57),

$$\langle c \rangle = \int_0^\infty c4\pi\left(\frac{m}{2\pi kT}\right)^{3/2} c^2 \exp\left(-\frac{mc^2}{2kT}\right) dc = \sqrt{\frac{8kT}{\pi m}} \qquad (59)$$

Another useful quantity is the rate at which molecules hit a unit area of a wall (or of a hole). Placing the x axis perpendicular to the wall, $(N/V)v_x$ molecules with the x-component of velocity v_x will hit a unit area of the wall in unit time (all of the molecules in the rectangular parallelepiped of height v_x and unit cross-sectional area). This quantity has to be averaged over the distribution of v_x for positive values of v_x (only molecules moving in the positive direction will hit the wall):

$$\text{Rate}_{\text{wall}} = \int_0^\infty \left(\frac{N}{V}\right) v_x \left(\frac{m}{2\pi kT}\right)^{1/2} \exp\left(-\frac{mv_x^2}{2kT}\right) dv_x = \left(\frac{N}{V}\right)\frac{\langle c \rangle}{4} \tag{60}$$

The x component of kinetic energy transported by these molecules is

$$\text{KE}_x = \int_0^\infty \left(\frac{N}{V}\right) v_x \left(\frac{mv_x^2}{2}\right)\left(\frac{m}{2\pi kT}\right)^{1/2} \exp\left(-\frac{mv_x^2}{2kT}\right) dv_x = \left(\frac{N}{V}\right)\frac{\langle c \rangle}{4}kT \tag{61}$$

[Equation (55) is used to evaluate both of these integrals.] Equation (61) means that kT of x-component of kinetic energy is transported by each molecule passing through the hole, twice the x component of kinetic energy that the molecule has in the bulk gas. Because the y and z components of kinetic energy that the molecules transport are just the values that they have in the bulk, the average total kinetic energy transported is

$$\langle\text{KE}_{\text{trans}}\rangle = 2kT \tag{62}$$

We will make use of this result in Chapter 12.

5.7   A Steady-State Example

Consider two chambers separated by a thin wall in which there is a small hole. We will assume that effusive flow holds for molecules passing through the hole in both directions (i.e., that molecules hitting the hole pass through the wall without collisions with other molecules or with the wall). At thermal equilibrium, both chambers are at the same temperature and, therefore, by Eq. (59), the average speeds of the molecules in the two containers are equal. For mechanical equilibrium, the rate of flow in both directions through the hole must be equal. As a result, from Eq. (60), the density and the pressure in both chambers are equal.

What happens if, by some external means, we maintain the two chambers at different temperatures, say T_1 and T_2? In order to do this, we must continually inject heat into the higher-temperature chamber and remove it from the lower-temperature chamber, to counterbalance the tendency of the temperature of the chambers to equalize.[13] In this case as well, equality of flow in both directions through the hole will quickly be attained. The system will then be unchanging in time, but because the unchanging condition relies upon interactions with the surroundings (the heat transfers), the system is at steady state, rather than equilibrium.

Using Eq. (60) to equate the flows in the two directions at steady state,

$$\left(\frac{N}{V}\right)_1 \langle c_1 \rangle = \left(\frac{N}{V}\right)_2 \langle c_2 \rangle \tag{63}$$

$$\left(\frac{N}{V}\right)_1 \left(\frac{N}{V}\right)_2^{-1} = \frac{P_1 T_2}{P_2 T_1} = \frac{\langle c_2 \rangle}{\langle c_1 \rangle} = \sqrt{\frac{T_2}{T_1}} \tag{64}$$

or

$$\frac{P_1}{P_2} = \sqrt{\frac{T_1}{T_2}} \tag{65}$$

We see that establishing a temperature difference between the chambers has resulted in a density and pressure difference! The movement of material due to a temperature differential is known as *thermal diffusion*, and in the special case where the flow is molecular, it is known as *thermal transpiration*.

5.8 Thermodynamic Functions of the Monatomic Ideal Gas

Excitation of the internal energy of monatomic ideal gases requires so much energy that it can be neglected at ordinary laboratory temperatures. The only type of energy that must be considered for such species is that of their translational motion, which we can write as

$$\varepsilon_{tr} = \frac{1}{2} m v^2 = \frac{p^2}{2m} = \frac{p_x^2}{2m} + \frac{p_y^2}{2m} + \frac{p_z^2}{2m} \tag{66}$$

where $p = mv$ is the momentum, which, like velocity, is a vector quantity.

Because the motion of a classical point particle is completely determined by specifying its initial position and momentum, we choose these variables to represent the state of the particle. The six-dimensional space of coordinates and momenta, in which the state of each particle is represented by a point, is called *phase space*. Equation (23) then becomes

$$q = \int g(\mathbf{r}, \mathbf{p}) \exp\left(-\frac{p^2}{2mkT}\right) d\mathbf{r}\, d\mathbf{p} \tag{67}$$

In using Eq. (67), we are faced with the problem of determining $g(\mathbf{r}, \mathbf{p})$, the number of states corresponding to each volume in phase space. Classical mechanics does not quantize variables, so there is no obvious way of doing this. How can we make the connection between classical phase space and quantized states?[14] Having no reason to favor one region of phase space over another, we assume the following:

> Equal volumes of phase space correspond to the same number of states, regardless of the specific values of **r** and **p** to which they apply.

We note that volumes in phase space have units of kg m^2/s = J s for each Cartesian coordinate. In order to have a dimensionless degeneracy, g, we divide volumes in phase space, $dr\,dp$, by a constant, h^3, for which h has units of J s. In other words, $g(r, p) = 1h^{-3}$, independent of position and momentum:

$$q = \frac{1}{h^3} \int \exp\left(-\frac{p^2}{2mkT}\right) dr\, dp \tag{68}$$

Because the three Cartesian directions are indistinguishable, Eq. (55) becomes

$$q = \frac{1}{h^3} \int \exp\left[-\frac{(p_x^2 + p_y^2 + p_z^2)}{2mkT}\right] dx\, dp_x\, dg\, dp_y\, dz\, dp_z = q_x^3 \tag{69}$$

where

$$q_x \equiv \frac{1}{h} \int_0^L dx \int_{-\infty}^{\infty} \exp\left(-\frac{p_x^2}{2mkT}\right) dp_x \tag{70}$$

From Eq. (54), the second integral is $\sqrt{2\pi mkT}$, giving[15]

$$q_x = \frac{L}{h}\sqrt{2\pi mkT} \tag{71}$$

or

$$q = \frac{V}{h^3}(2\pi mkT)^{3/2} \tag{72}$$

Using Eq. (26) to calculate the energy (realizing that the only term in $\ln q$ that depends on T is $\ln(T^{3/2})$:

$$U_m = U_0 + \tfrac{3}{2}RT \tag{73}$$

For $C_{V,m}$, this gives

$$C_{V,m} = \left(\frac{\partial U}{\partial T}\right)_V = \frac{3}{2}R \tag{74}$$

just the value that we obtained in Chapter 2.

Quantum mechanics clearly denies the possibility of distinguishing between particles in translational motion.[16] For the ideal gas entropy, we must therefore use Eq. (34) for indistinguishable particles:

$$S_m = R \, \ln\left(\frac{V(2\pi mkT)^{3/2}}{N_A h^3}\right) + R + \frac{3}{2}R \tag{75}$$

Writing this in terms of the pressure for 1 mol gives

$$S_m = \frac{5}{2}R + R \, \ln\left(\frac{(2\pi M)^{3/2} R^{5/2} T^{5/2}}{N_A^4 h^3 P}\right) \tag{76}$$

which is called the *Sackur–Tetrode equation*. Note that Eq. (76) gives the proper dependence for the entropy of an ideal gas on pressure at constant temperature [Eq. (28) of Chapter 4] and on temperature at constant pressure [Eq. (17) of Chapter 3]. In order to use Eq. (76) to calculate absolute entropies, a numerical value for h is required. This can be obtained by choosing h to fit the experimental absolute entropy for 1 mol of monatomic ideal gas (Problem 14). The result, $h = 6.6 \times 10^{-34}$ J s, is called Plank's constant. It is the fundamental constant in the quantum description of matter. With the use of this constant, the Sackur–Tetrode equation gives good agreement with measured absolute entropies for monatomic gases at room temperature.

5.9 Energy of Polyatomic Ideal Gases

A polyatomic molecule contains more than one atom[17] and can undergo rotational and vibrational motions in addition to translational. Overall, rotational motion is unhindered and therefore consists only of a kinetic energy term of the form

$$\varepsilon_{rot} = \frac{p_\theta^2}{2I} \tag{77}$$

where p_θ is a rotational angular momentum and I is a moment of inertia. A vibrational motion in the harmonic approximation is of the form

$$\varepsilon_{vib} = \frac{p_d^2}{2\mu} + \frac{kd^2}{2} \tag{78}$$

where d is the displacement of the vibrating oscillator from its equilibrium position and p_d is the momentum associated with vibration. (k and μ are the force constant and mass, respectively, appropriate for the oscillator.) Of importance is that in the harmonic approximation, the internal energy of a polyatomic molecule can be written as

$$\varepsilon_{\text{int}} = \sum_i C_i l_i^2 \tag{79}$$

where l_i is a coordinate *or* a momentum, and the sum goes over all of the coordinates of the rotations and the coordinates and momenta of the vibrations of the molecule.

Although vibrational and some rotational motions certainly require quantum mechanics for their accurate consideration, we will treat these motions in their classical limit. Using Eq, (23) and integrating over a $2M$-dimension phase space for a total of M rotations and vibrations, we get, using a normalization factor of h^{-3M},

$$
\begin{aligned}
q &= \frac{1}{h^{3M}} \int \exp\left(-\frac{1}{kT} \sum_i C_i l_i^2\right) d\mathbf{l}_j \\
&= \frac{1}{h^{3M}} \int d\mathbf{l}_{\text{rot}} \prod_i \int_{-\infty}^{\infty} \exp\left(-\frac{C_i}{kT} l_i^2\right) dl_i
\end{aligned}
\tag{80}
$$

$\mathbf{l}_j$ includes a coordinate and momentum for each rotation and vibration, whereas l_i includes a coordinate and momentum for each vibration, but only a momentum for each rotation. Using Eq. (54) for each of the exponential integrals gives

$$q = \frac{V_\theta}{h^{3M}} \prod_i \sqrt{\frac{\pi kT}{C_i}} \tag{81}$$

where V_θ is a factor of 4π or $8\pi^2$ obtained by integrating over the rotational coordinates. The final result does not depend on the value of this term. Inserting Eq. (81) into Eq. (26) gives

$$U_m = U_0 + RT^2 \left\{ \frac{\partial}{\partial T} \left[\ln\left(\frac{V_\theta}{h^{3M}}\right) + \sum_i \ln\left(\frac{\pi k}{C_i}\right)^{1/2} + \sum_i \ln T^{1/2} \right]_V \right\} \tag{82}$$

Only the last term has a nonzero temperature derivative; therefore,

$$U_m = U_0 + \sum_i \frac{RT}{2} \tag{83}$$

This is the *classical equipartition theorem*. It states that each rotation (which only contributes one term to the sum) adds $RT/2$ to the energy, whereas each vibration (which contributes two terms) adds RT to the energy. From Eq. (73), each of the three translational motions of a molecule also contributes $RT/2$ to the energy. In terms of heat capacities,

$$C_{V,m} = \left(\frac{\partial U_m}{\partial T}\right)_V = \frac{3}{2}R + \sum_{rot} \frac{1}{2}R + \sum_{vib} R \tag{84}$$

with an additional factor of R to calculate $C_{P,m}$. In calculating energies and heat capacities, one must enumerate the number of rotational and vibrational degrees of freedom of the molecule of interest. The key to this task is realizing that the same number of coordinates is required to localize the constituent atoms of the molecule, whether these atoms are bonded or not! If the molecule contains N atoms, $3N$ coordinates (degrees of freedom) are required to localize these atoms in Cartesian space. When the atoms are bonded, we choose to use three coordinates to localize the center of mass of the ensemble of atoms. We call these the three translational degrees of freedom of the molecule. Likewise, the orientation of an axis of a linear molecule in space can be specified by two degrees of freedom, and if the molecule is nonlinear, an additional degree of freedom is required to specify the orientation of the molecule around this axis. The remaining $3N - 6$ degrees of freedom ($3N - 5$ in the case of linear molecules) must correspond to internal motions of the molecule, which we call vibrations.[18]

Example 2. Calculate the classical constant-pressure heat capacity of CH_4. Compare this with the literature value of 35.46 J/mol K at 15°C.

Solution: CH_4 has 5 atoms, therefore $N = 5$ and there are a total of $3N = 15$ degrees of freedom. Three of these are translations, and because the molecule is nonlinear, three are rotations. There are, thus, $15 - 6 = 9$ vibrational degrees of freedom. The internal energy is

$$U_m = \tfrac{3}{2}RT(\text{trans}) + \tfrac{3}{2}RT(\text{rot}) + 9RT(\text{vib}) = 12RT$$

and the heat capacities are $C_{V,m} = (\partial U/\partial t)_V = 12R$ and

$$C_{P,m} = C_{V,m} + R = 13R = 108 \ \frac{J}{\text{mol K}}$$

This is over three times the literature value.

Real molecules follow quantum mechanics. However, the quantum results approach those of classical mechanics at high temperatures. "High temperature" is much lower for rotations and low-frequency vibrations than for high-frequency vibrations. For the latter, the molecule may dissociate before the classical approximation becomes appropriate. Comparing Eq. (84) for heat capacity with the data presented in Fig. 4 of Chapter 2, we note the following:

1. For monatomic gases, we have the complete translational contribution at all temperatures.
2. For diatomic gases, we have the translational and rotational (two degrees of freedom) contributions at room temperature, whereas vibrations only begin to contribute at higher temperature.
3. For polyatomic gases, some vibrational degrees of freedom start to contribute at room temperature, but the complete vibrational contribution is not even achieved at quite high temperature.

Observations that real energies and heat capacities are lower than those predicted by classical mechanics were one of a number of unsettling mysteries of physics that were cleared up by the development of quantum mechanics.

5.10 Configurations of a Polymer Chain

When we talk about a linear polymer chain, such as non-cross-linked polyethylene, we do not mean that all of the atoms in the chain are on a straight line or even that spatial structure repeats in some regular manner. Rather, the bond angles at the atoms in the chain are tetrahedral, and different conformations are obtained by rotation around the single bonds. A complete set of angles at all of the bonds gives a *configuration* of the chain, which can range from compact to extended. One particular configuration of a short polymer chain is shown in Fig. 2. We seek the components of the displacement vector, r, between the ends of the chain. The magnitude of this vector, r, is called the *displacement length* and is a measure of the extension of the polymer chain.

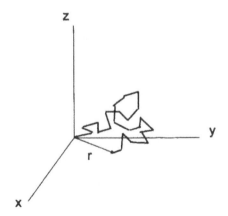

Figure 2 Configuration of a polymer chain.

For ease of calculation, we make a number of simplifying assumptions. These are relaxed in advanced treatments of the subject. First, rather than requiring tetrahedral bonds at each vertex of the chain, we allow all bond angles and assume that these are randomly distributed. Second, we ignore any excluded volumes or interactions between the segments of the chain. In this sense, our calculation is similar to the Bernoulli model of the ideal gas, which neglects intermolecular interactions. Our approximation is called the *freely jointed chain* model.

We first consider a single bond in the chain. We want a measure of the projection of the length of this bond, l, in a direction, say x. The projection is shown in Fig. 3, with θ varying from 0 to π radians. Because the average value of the projection is zero, we take its root-mean-square value as its measure. For a bond of length l at an angle of θ with respect to the x axis, the projection is $l \cos \theta$. The root-mean-square (rms) projection is calculated by squaring this value, averaging it over all solid angles ($d\Omega = 2\pi \sin \theta \, d\theta$), and taking the square root of the result:

$$x_{rms} = \left(\frac{l^2}{4\pi} \int_0^\pi 2\pi \cos^2 \theta \sin \theta \, d\theta \right)^{1/2} = \frac{l}{\sqrt{3}} \qquad (85)$$

Whereas x_{rms} is a measure of the magnitude of the projection of a single bond in the chain on the x axis, the projection can be in either the positive or negative direction. The actual length of the projection of the displacement vector on the x axis is proportional to the *difference* between the number of projections in the positive and negative x directions. Positive and negative projections are random and each has probability of $\frac{1}{2}$. For a chain having N segments, each of length l, the probability that the x projection of the extension vector has a value mx_{rms} is calculated in the following manner:

1.   The probability of any particular ordering of positive and negative projections of the N bonds is $(1/2)^N$.
2.   In order to have the difference between the number of positive projections and the number of negative projections equal m, with the

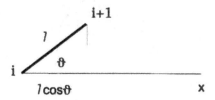

Figure 3   Projection of a bond on the total extension vector.

total number of projections equal N, there must be $(N + m)/2$ positive projections and $(N - m)/2$ negative projections.

3. The number of ways of getting a difference of m out of a total of N projections is calculated by writing down all $N!$ permutations of the segments and then dividing by the permutations of the positive and negative projections,

$$\left(\frac{N+m}{2}\right)!\left(\frac{N-m}{2}\right)!$$

because the order of these does not make a difference.

The probability is then

$$P(m, N) = \frac{N!}{[(N+m)/2]![(N-m)/2]!}\left(\frac{1}{2}\right)^N \tag{86}$$

This probability distribution is called the (equal probability) *binomial distribution*, and is the same distribution that is obtained for tossing an unbiased coin. In books on statistics,[19] it is shown that for large values of N, the binomial distribution approaches the continuous *normal distribution*:

$$P(m, N)dm \approx \frac{1}{\sqrt{2\pi N}}\exp\left(-\frac{m^2}{2N}\right)dm \propto \Omega(m, N) \tag{87}$$

which is said to have the *Gaussian* form. We have indicated that the probability is proportional to the number of configurations that give rise to a chain with m and N. Substituting $m = x/x_{rms} = x\sqrt{3}/l$ gives

$$P(x, N)\,dx = \sqrt{\frac{3}{2\pi N l^2}}\exp\left(-\frac{3x^2}{2Nl^2}\right)dx \propto \Omega(x, N)\,dx \tag{88}$$

Similar distributions result for the projections along the y and z axes, and because these are independent,

$$P(x, y, z, N)\,dx\,dy\,dz = \left(\frac{3}{2\pi N l^2}\right)^{3/2}\exp\left(-\frac{3(x^2 + y^2 + z^2)}{2Nl^2}\right)dx\,dy\,dz$$
$$\propto \Omega(x, y, z, N)\,dx\,dy\,dz$$
$$\tag{89}$$

To find the probability of a displacement length between r and $r + dr$, independent of direction, we integrate over the corresponding volume of space, much as we did in deriving the Maxwell–Boltzmann distribution:

$$P(r, N)\,dr = 4\pi\left(\frac{3}{2\pi N l^2}\right)^{3/2}r^2\exp\left(-\frac{3r^2}{2Nl^2}\right)dr \propto \Omega(r, N)\,dr \tag{90}$$

A single polymer chain is not a macroscopic system. However, for N of the order of 100 or more, the number of configurations is very large. We shall not hesitate to use our statistical thermodynamic equations in the mesoscopic range of single polymer chains.

5.11 Theory of Ideal Rubber Elasticity

Theories of different levels of complexity have been used to explain the elastic behavior of macromolecules (Fig. 12 of Chapter 1). The simplest theory, which nevertheless captures the essentials of the phenomenon, will be described here.[20] In this theory, a rubberlike polymer is pictured as a three-dimensional network of identical linear chains, each chain consisting of N bonds of length l. The chains are joined together at junctions, from which more than two chains extend. The junctions may be atoms chemically bonding more than two chains or tiny crystalline regions. In the language of polymer science, the chains are cross-linked and the cross-links prevent permanent deformation. The chains between the junctions are long and are assumed to adopt a random configuration by means of rotations around bonds. In the simplest theory, all rotational configurations are equally available; there are no excluded volume effects or intrachain interactions.[21] This is the model of the freely jointed chain, discussed in the previous section. All interactions between different chains are also neglected. The elasticity is completely intrachain and entropically produced.

Our object is to calculate a stress–strain relationship for comparison with Fig. 12 of Chapter 1. To do this, we calculate the force along the x direction necessary to distort a polymer sample, initially of dimensions X_0, Y_0, and Z_0 to dimensions X, Y, and Z. We introduce distortion ratios, α_x, α_y, and α_z such that

$$X = \alpha_x X_0, \qquad Y = \alpha_y Y_0, \qquad Z = \alpha_z Z_0 \tag{91}$$

(In Chapter 4, α_x was called α, the elongation.) For stretching at constant volume, we have $\alpha_x \alpha_y \alpha_z = 1$, which, with $\alpha_z = \alpha_y$, gives

$$\alpha_y = \alpha_z = \frac{1}{\sqrt{\alpha_x}} \tag{92}$$

From Eq. (55) of Chapter 4 with X as the length of the polymer sample,

$$f = \left(\frac{\partial A}{\partial X}\right)_{T,V} = \left(\frac{\partial A}{\partial \alpha_x}\right)_{T,V} \frac{d\alpha_x}{dX} = \frac{1}{X_0}\left(\frac{\partial A}{\partial \alpha_x}\right)_{T,V} \tag{93}$$

Because A, the Helmholtz free energy, is an extensive property, it is given by

$$A = \kappa a \tag{94}$$

where a is the Helmholtz free energy of a single chain and κ is the number of chains in the sample:

$$f = \frac{\kappa}{X_0}\left(\frac{\partial a}{\partial \alpha_x}\right)_{T,V} = -\frac{\kappa T}{X_0}\left(\frac{\partial s}{\partial \alpha_x}\right)_{T,V} \tag{95}$$

where s is the entropy of a single chain.

The last equality follows for an ideal rubber, in which, by definition, the internal energy does not depend on the distortion.

We assume that the distortion ratios for the individual polymer chains are the same as those for the sample (affine approximation):

$$x = \alpha_x x_0, \qquad y = \alpha_y y_0, \qquad z = \alpha_z z_0 \tag{96}$$

(Lowercase symbols refer to the dimensions of individual polymer chains.) Also, in the absence of a force, the individual chains are isotropic:

$$x_0 = y_0 = z_0 \tag{97}$$

Using $s = k \ln \Omega$ with Eq. (89),

$$s = -\frac{3k(\alpha_x^2 x_0^2 + \alpha_y^2 y_0^2 + \alpha_z^2 z_0^2)}{2Nl^2} \tag{98}$$

(Distortion-independent terms have been omitted.) Substituting Eqs. (92) and (97) into Eq. (98) gives

$$s = -\frac{3kx_0^2(\alpha_x^2 + 2/\alpha_x)}{2Nl^2} \tag{99}$$

Substituting into Eq. (95),

$$f = \frac{3kT\kappa x_0^2}{2Nl^2 X_0}(2\alpha_x - 2/\alpha_x^2). \tag{100}$$

Dividing by the area of the sample $Y_0 Z_0$ and defining $\kappa/X_0 Y_0 Z_0 \equiv \nu$, the number of network chains per unit volume gives, for the stress,

$$f^* = \frac{3kT x_0^2 \nu}{Nl^2}(\alpha_x - 1/\alpha_x^2) \tag{101}$$

Because $3x_0^2 = r_0^2$, and r_0, the displacement length of a polymer chain is close to $\sqrt{N}l$ (see Problem 19), Eq. (101) can be written

$$f^* \approx kT\nu(\alpha_x - 1/\alpha_x^2) \tag{102}$$

The simple theory presented accounts for the fall-off of the modulus of polymers with increased extension. The stiffening of the polymer at very large distortion results from interactions between the extended polymer chains, which is called *strain-induced crystallization*.

Questions

1. Show that dividing Eq. (8) by $N!$ produces no change in Eq. (10), and, thus, corrected Boltzmann statistics gives the same results as Boltzmann statistics.

2. Why is $\int_{-\infty}^{\infty} x^{2n+1} \exp(-\beta x^2)\, dx = 0$?

3. Show that if the energy of a molecule can be written as the sum of terms for translational, rotational, and vibrational energies, the partition function for the molecule is the product of translational, rotational, and vibrational partition functions.

4. Why is the heat capacity of monatomic gases, such as He and Ne, practically independent of temperature, whereas molecular heat capacities increase with temperature?

5. Use Eq. (5) to discuss why coherent laser light has lower entropy than noncoherent monochromatic light and why a troop of marching soldiers has lower entropy than that of noonday pedestrians.

6. Give an example of a system containing distinguishable particles and one containing indistinguishable particles.

7. In Section 5.2, the probability of an observable spontaneous fluctuation away from the equilibrium state is given as $1/\exp(N_A)$. To get an idea of the magnitude of this quantity, compare $\exp(N_A)$ with the number of atoms in the universe. You may make any assumptions that you like in calculating the latter quantity; you will see that they will make little difference in the comparison.

8. At a given temperature, which is larger: the molecular partition function, q or the system partition function, Q?

9. What would happen if the two chambers maintained at different temperatures in Section 5.7 had small openings connecting them to the atmosphere and were therefore at the same pressure? Would the flow in both directions be the same and, if not, in which direction would there be a greater flow?

10. Compare Eq. (102) with the ideal gas equation. With reasonable values of v and α could a stress of 1.0 atm be achieved?

11. Explain why the protein in egg white coagulates (becomes insoluble) when it is heated in boiling water.

Problems

1. How many ways can you choose 3 students from a group of 10 students for the following.

    (a) If the order in which you pick the students is important (corresponding to distinguishable molecules)

    (b) If the order in which you pick the students is not important (corresponding to indistinguishable molecules)

2. How many configurations are there for the distribution of 10 distinguishable items in 3 states with occupation numbers $N_0 = 5$, $N_1 = 3$, and $N_2 = 2$?

3. How many configurations are there for four distinguishable molecules distributed over two states, with two molecules in each state? Write down these configurations.

4. Show that Eq. (4) for the reverse Joule process is unchanged if the particles are distinguishable.

5.* Calculate the energy lost from a large container of monatomic gas if 1 mol of gas effuses out of a small hole in the wall of the container. Show that when some of the gas effuses out of the container, the remainder of the gas cools.

6. How accurate is Stirling's approximation, Eq. (28), for $A = 3$ and $A = 10$, and $A = 20$?

7. How many ways can you place 40 indistinguishable atoms in 1000 compartments?

8. Using Eq. (54), show that

$$\int_0^\infty \exp(-\beta x^2)\, dx = \tfrac{1}{2}\sqrt{\pi}\beta^{-1/2}$$

Note that $0! = 1$.

9. Verify Eq. (71).

10. Use the Maxwell–Boltzmann distribution to calculate c_{rms} and compare the result with that obtained from the Bernoulli model in Chapter 1.

11. A single particle with energy states at 0, 1, and 2 has three states with energy below 2.5. How many states does a system consisting of three such particles have with energy below 2.5, for the following:

    (a) The particles are distinguishable.
    (b) The particles are indistinguishable.

12. Show that the Sackur–Tetrode equation [Eq. (76)] is in agreement with Eq. (17) of Chapter 3 and Eq. (28) of Chapter 4 for an ideal gas.

13. Use the Sackur–Tetrode equation [Eq. (76)] to calculate the standard molar entropy of neon considered an ideal gas at 298 K.

14.* Use the Sackur–Tetrode equation [Eq. (76)] to evaluate h from the values of S_m° of He and Ar at 298 K. (See Table 1 of Chapter 4.)

15. Calculate the classical $C_{P,m}$ for Ar, Cl_2, and CO_2 and compare these with the experimental values at 298 K for these gases, which are 20.93, 34.13, and 36.62 J/mol K, respectively.

16.* Ice has a structure in which oxygen atoms are tetrahedrally arranged, and between every two oxygen atoms, there is a hydrogen atom. However, hydrogen atoms may be either covalently bonded or hydrogen-bonded to a given oxygen atom. Calculate the number of possible configurations of the hydrogen atoms in 1 mol of ice in the following manner:

    (a)   There are $2N_A$ hydrogen atoms, each of which can be in two positions. How many configurations does this give?

    (b)   However, not every configuration listed in part (a) is possible, because around each oxygen, there must be two covalently bonded and two hydrogen-bonded hydrogen atoms. For each of the N_A oxygen atoms, how many of the $2^4 = 16$ arrangements of the hydrogen atoms give acceptable configurations?

Use your results from parts (a) and (b) to calculate the residual entropy of ice. The measured value is 3.4 J/mol K.

17.   Derive the barometric distribution, $P = P_0 \exp(-Mgh/RT)$, discussed at the end of Chapter 1, starting from the Boltzmann distribution law [Eq. (16)].

18.   Write down all of the possible results of four tosses of an unbiased coin. By counting, show that the probability for different numbers of "excess" heads is given by Eq. (86).

19.   Find the most probable displacement length of a linear polymer chain, having 1000 bonds, each of length 0.14 nm.

20.   Using Eqs. (55) and (90), show that the average displacement length of a freely jointed polymer chain with N bonds each of length d is within 10% of $\sqrt{N}d$.

21.   The strain, ε, on a rubber is $\alpha_x - 1$. Using this and Eq. (102), find expressions for a and b in Eq. (54) of Chapter 1. (Hint: You may want to expand a term in a power series in ε.)

22.^{M} Calculate the fraction of molecules that have speeds greater than twice the average molecular speed.

23.   Find a formula for the most probable molecular speed, c_{mp}. Sketch the Maxwell–Boltzmann velocity distribution and show the relative positions of $\langle c \rangle$, c_{mp}, and c_{rms} on your sketch.

24.   Find a formula for the root-mean-cube velocity ($c_{rmc} \equiv \sqrt[3]{\langle c^3 \rangle}$). Explain the trend of $\langle c \rangle$, c_{rms}, and c_{rmc}.

Notes

1.   Quantum mechanics, the modern description of the microscopic world, also conserves energy.
2.   In fact, quantum mechanics tells us that there must be uncertainty in the initial conditions and we can only discuss this process in a probabilistic manner.
3.   We will assume that our camera could take the pictures without influencing the motion of the atoms. A real camera would observe light photons scattered off the atoms and these photons would influence the motion of the atoms.
4.   See, for example, JE Mayer, MG Mayer. Statistical Mechanics. New York: Wiley, 1959, pp 53–63.

5. Quantum mechanics denies the possibility of exactly locating an atom or, more rigorously, of simultaneously determining both its position and momentum. In a combined position–momentum space (generally called phase space) quantum mechanics provides a "natural" size for the volume in which an atom can be located.
6. Logarithms to other bases, (e.g., $\log_{10}$) would also be an appropriate definition, and they are related to ln by a constant multiplier.
7. This relationship, written as $S = k \ln W$ is carved on Boltzmann's gravestone in Vienna.
8. As a result of translational energy, energy is also often nearly continuous in quantum systems.
9. These types of particle are called fermions (electrons, protons, neutrons, and odd numbers of these) and bosons (photons, and even numbers of electrons, protons, and neutrons).
10. Quantum mechanical considerations indicate that it is necessary to keep the volume of the system constant in order for the energy levels of the system to be unchanged in the process.
11. This is the simplified form of Stirling's approximation; there are more accurate (more complicated) forms.
12. We take $U_0 = 0$ for the system.
13. We cannot avoid this by using a *gedanken* adiabatic wall between the two chambers, because molecules passing through the hole will transfer energy.
14. In quantum mechanics, *Heisenberg's uncertainty principle* states that there is a limit to which we can know the *product* of the uncertainties in a coordinate and its corresponding momentum, $\Delta x \Delta p_x$. Thus, even in quantum mechanics, there is a minimum volume in phase space in which we can localize a particle.
15. We have assumed the gas to be confined in a cubic box with side L.
16. We can neither mark them (they are too small) nor watch them (the light bouncing off them will make this impossible).
17. Sometimes, diatomic molecules are distinguished from polyatomic molecules, which are defined to contain three or more atoms.
18. A molecule may also have internal rotations, which, if unhindered, should be counted as rotational motions, as no potential energy is involved in their motion.
19. For example, AJ Thomasian. The Structure of Probability Theory with Applications. New York: McGraw-Hill, 1969, p 253.
20. For a more exact development, see DA McQuarrie. Statistical Thermodynamics. New York: Harper and Row, 1973, pp 280–284.
21. More exactly, excluded volume effects and intrachain interactions are assumed not to change as the rubber is stretched.

6

Phase Transformations in Single-Component Systems

Suppose water always froze as ice-one on Earth because it never had a seed to form... ice-nine—a crystal as hard as this desk—with... a melting point of one (hundred and)-thirty degrees.

Kurt Vonnegut, *Cat's Cradle*

Phase transformations are treated by the thermodynamics of open systems. Using the chemical potential, the equations for the (differential) changes of U, H, A, G, and S are generalized to the case of a variable amount of material. At equilibrium, the chemical potential is equal in coexisting phases. The variation of pressure with temperature required to maintain this equality is given by the Clapeyron or Clausius–Clapeyron equations. First-order, second-order, and lambda transitions are discussed. Activity and fugacity are used to deal with nonideal substances. The presence of an inert gas affects vapor pressure. Condensed phase equilibria, including triple points, are considered. Phase diagrams of CO_2 and H_2O are compared. Liquid crystals, glasses and some polymeric materials illustrate mesomorphic behavior.

6.1 Thermodynamics of Open Systems

In order to discuss phase transformations in this chapter and chemical reactions in the next chapter, we will need to develop the thermodynamics of open systems. In open systems, the number of moles of the various components of the system can change and the thermodynamic functions depend on the numbers of moles of these components, as well as on thermodynamic variables. For example, the natural variables for U become $U(S, V, n_i)$, where the index i ranges over the components of the system.

Mathematically, we can write for U, which is a state function of these variables,

$$dU = \left(\frac{\partial U}{\partial S}\right)_{V,n_i} dS + \left(\frac{\partial U}{\partial V}\right)_{S,n_i} dV + \sum_i \left(\frac{\partial U}{\partial n_i}\right)_{S,V,n_{j\neq i}} dn_i \tag{1}$$

In the derivatives with respect to the number of moles of each component of the system, the numbers of moles of all other components ($j \neq i$) are held constant. To simplify the notation while keeping it useful for treating phase transformation processes, the same species in different phases will be designated by different subscripts. We first apply this equation to a system with a constant number of moles (all $dn_i = 0$). With the number of moles of all components (including the same component in different phases) constant, material equilibrium is not an issue and we can use Eq. (20) of Chapter 4:

$$\left(\frac{\partial U}{\partial S}\right)_{V,n_i} = T \quad \text{and} \quad \left(\frac{\partial U}{\partial V}\right)_{S,n_i} = -P \tag{2}$$

We define μ_i, the *chemical potential* of component i, as

$$\mu_i \equiv \left(\frac{\partial U}{\partial n_i}\right)_{S,V,n_{j\neq i}} \tag{3}$$

giving

$$dU = T\,dS - P\,dV + \sum_i \mu_i\,dn_i \tag{4}$$

The use of T, P, and μ_i in Eq. (1) implies that these three quantities are uniform throughout the system and, therefore, that it is at thermal, mechanical and material equilibrium. Compared to Eq. (20) of Chapter 4, it is no longer necessary to explicitly indicate material equilibrium when writing the equation. Using the definitions of the other thermodynamic function, we can write

$$dH = d(U + PV) = T\,dS + V\,dP + \sum_i \mu_i\,dn_i \tag{5}$$

$$dA = d(U - TS) = -S\,dT - P\,dV + \sum_i \mu_i\,dn_i \tag{6}$$

$$dG = d(H - TS) = -S\,dT + V\,dP + \sum_i \mu_i\,dn_i \tag{7}$$

Equations (4)–(7), yield four equivalent definitions for the chemical potential of component i:

$$\mu_i \equiv \left(\frac{\partial U}{\partial n_i}\right)_{S,V,n_{j\neq i}} = \left(\frac{\partial H}{\partial n_i}\right)_{S,P,n_{j\neq i}} = \left(\frac{\partial A}{\partial n_i}\right)_{T,V,n_{j\neq i}} = \left(\frac{\partial G}{\partial n_i}\right)_{T,P,n_{j\neq i}} \tag{8}$$

From its form, the chemical potential is obviously an intensive property and therefore must be a function only of other intensive variables and independent of the size of the system. We can write it as $\mu_i(T, P, c_j)$, where c_j is some measure of concentration of component j.

Consider the following process: Starting with a small amount of a system with concentrations c_j at T and P, we add components to the system, at constant T and P, in the proper ratio to keep the concentrations constant. We can integrate Eq. (7) for the Gibbs free-energy change for this process:

$$G_2 - G_1 \overset{\substack{\text{const}\\T,P}}{=} \int_1^2 \sum_i \mu_i\,dn_i \overset{c_j}{=} \sum_i \mu_i \int_1^2 dn_i = \sum_i \mu_i(n_{2,i} - n_{1,i}) \tag{9}$$

If the initial number of moles is vanishingly small, $n_{1,i} \to 0$, $G_1 = 0$. With the final number of moles equal to that in the system, $n_{2,i} = n_i$,

$$G = \sum_i \mu_i n_i \tag{10}$$

For a single-component system, this becomes

$$G = \mu n, \qquad \mu = \frac{G}{n} = G_m \tag{11}$$

showing that, for single-component systems, the chemical potential is just the molar Gibbs free energy. Note that of the five definitions of μ in Eq. (8), only in the one in terms of G are intensive variables kept constant as the size of the system is increased; thus, only this relationship can be integrated with constant μ. Our extended definition of enthalpy and Gibbs free energy [Eq. (24) of Chapter 2] allows this to hold even in cases in which there are work terms in addition to the work of expansion.

6.2 Entropy Change for Open Systems

From Eqs. (4) and (5), the following equations for dS are obtained:

$$dS = \left(\frac{1}{T}\right) dU + \left(\frac{P}{T}\right) dV - \left(\frac{1}{T}\right) \sum_i \mu_i \, dn_i \tag{12}$$

$$dS = \left(\frac{1}{T}\right) dH - \left(\frac{V}{T}\right) dP - \left(\frac{1}{T}\right) \sum_i \mu_i \, dn_i \tag{13}$$

These give

$$\mu_i = -T\left(\frac{\partial S}{\partial n_i}\right)_{U,V,n_{j\neq i}} = -T\left(\frac{\partial S}{\partial n_i}\right)_{H,P,n_{j\neq i}} \tag{14}$$

Following Eqs. (33) and (36) of Chapter 4, if there are other work terms of the form $L_i \, dl_i$, Eqs. (12) and (13) become

$$dS = \left(\frac{1}{T}\right) dU + \left(\frac{P}{T}\right) dV - \left(\frac{1}{T}\right) \sum_i \mu_i \, dn_i - \left(\frac{1}{T}\right) \sum_i L_i \, dl_i \tag{15}$$

$$dS = \left(\frac{1}{T}\right) dH - \left(\frac{V}{T}\right) dP - \left(\frac{1}{T}\right) \sum_i \mu_i \, dn_i + \left(\frac{1}{T}\right) \sum_i l_i \, dL_i \tag{16}$$

In addition to Eqs. (12) and (13), which give entropy changes in terms of changes of other state functions of the system, we can also calculate entropy changes by considering the energy and material transported to the system during the process. The approach of Chapter 3 must be modified to include the entropy content of the matter transported between the system and the surroundings. The surroundings can be idealized as a reservoir, which maintains constant intensive properties (T, P, and μ_i) as heat and matter is withdrawn. The change of entropy of the surroundings as heat δq and moles δn_i of the various components are transported to the system is

$$dS_{sur} = -\left(\frac{\delta q}{T} + \sum_i S_{m,i} \, \delta n_i\right) \tag{17}$$

where $S_{m,i}$ is the molar entropy of component i. We will see in Chapter 8 that when the components of the system interact, *partial* molar entropies must be used in place of molar entropies. Note that δn_i is not the total differential of n_i, because components may also be created or destroyed by chemical reactions *within* the system. If the system is at equilibrium with the surroundings, the transfers are reversible and there can be no net entropy change of the universe. The entropy transport to the system is therefore

$$\partial S = \left(\frac{\delta q}{T} + \sum_i S_{m,i} \, \delta n_i\right)_{rev} \tag{18}$$

Because

$$S_m = \frac{H_m - G_m}{T} = \frac{H_m - \mu}{T}$$

we can write Eq. (18) as

$$\partial S = \left(\frac{\delta q}{T} + \frac{1}{T}\sum_i H_{m,i}\delta n_i\right)_{rev} - \left(\frac{1}{T}\sum_i \mu_i\delta n_i\right)_{rev} = \left(\frac{\delta q_{tot}}{T}\right)_{rev}$$
$$- \left(\frac{1}{T}\sum_i \mu_i\delta n_i\right)_{rev} \qquad (19)$$

where δq_{tot}, the total heat added to the system, includes the enthalpy of the added material. Because the reservoir and the system maintain constant pressure during the reversible transfer and δq_{tot} at constant pressure is the change of enthalpy, Eq. (19) is identical to Eq. (13). *Reversible* in these equations indicates that in using them, the surroundings must be at thermal and material equilibrium with the system.

6.3 Phases and Phase Transformations

A phase is a homogeneous region of matter (i.e., a region with uniform properties). The region does not have to be connected; droplets of water on a cold surface are a single phase. Phases of single substances are usually gases, liquids, or a particular arrangement of molecules in a solid. (Among the other types of region to which the term *phase* is sometimes applied are: *plasmas*, highly ionized *gases*, and *interfaces*, where properties vary from those of one adjoining phase to the other.) Phases that have properties between those of liquids and solids are called *mesomorphic* and will be discussed later in this chapter. In this chapter, we consider only systems in which all phases are composed of the same substance. The phase transitions that we will initially discuss are vaporization (liquid to gas, sometimes called evaporation), melting (solid to liquid, often called *fusion*), sublimation (solid to gas) and their reverse—condensation, freezing, and deposition, respectively.

The change of a property in a phase change will be written as (for the volume change of vaporization as an example), $\Delta_{vap}V$, or, in the general case, $\Delta_\phi X$. It will be assumed that these quantities are on a molar basis. Thus, for water, $\Delta_{fus}V$ is the volume of 1 mol of liquid water minus the volume of 1 mol of ice. $\Delta_{fus}V$ of water is negative, as anyone who has had a water-filled auto radiator freeze can testify. $\Delta_{fus}V$ is positive for most materials, including ethylene glycol, which is currently used in most auto radiators. Because molar volume is a state function, $\Delta_{freezing}V = -\Delta_{fus}V$.

Of particular interest are changes in enthalpy, such as the heat of fusion, $\Delta_{fus}H$, and the heat of vaporization, $\Delta_{vap}H$, as well as the corresponding changes of entropy. A useful observation, discussed in Chapter 3, is Trouton's rule, that entropies of vaporization are often ~ 85 J K mol.

6.4 General Criterion for Equilibrium in a Multiphase System

Equilibrium in a multiphase system implies thermal, mechanical, and material equilibrium. Thermal equilibrium requires uniformity of temperature throughout the system, and mechanical equilibrium requires uniformity of pressure. To find the criterion for material equilibrium, we treat a two-phase system and consider a transfer of dn moles from phase β to phase α. First, we regard each phase as a separate system. Because material enters or leaves these phases, they are open systems and we must use Eq. (4) to write their change in internal energy:

$$dU_{\alpha} = T\, dS_{\alpha} - P\, dV_{\alpha} + \mu_{\alpha}\, dn \tag{20}$$
$$dU_{\beta} = T\, dS_{\beta} - P\, dV_{\beta} - \mu_{\beta}\, dn \tag{21}$$

Adding these two equations and realizing that $dS_{\alpha} + dS_{\beta} = dS$ and $dV_{\alpha} + dV_{\beta} = dV$ (extensive quantities without subscripts refer to the combined system) gives

$$dU = T\, dS - P\, dV + (\mu_{\alpha} - \mu_{\beta})\, dn \tag{22}$$

We may also consider as our system, however, the combined system consisting of the two phases, which is closed. If the phase transformation is at equilibrium, material equilibrium holds and Eq. (20) of Chapter 4 is applicable:

$$dU \overset{m.e.}{=} T\, dS - P\, dV \tag{23}$$

In order for Eqs. (22) and (23) to be compatible, at material equilibrium (phase equilibrium) we must have

$$\mu_{\alpha} = \mu_{\beta} \tag{24}$$

which is a general principle for phase equilibrium in single-component systems at thermal and mechanical equilibrium.

6.4.1 Phase Transfer at Constant Temperature and Pressure

Consider a one-component, two-phase system held at constant temperature and pressure by means of a piston as in Fig. 1 of Chapter 2. The change in Gibbs free energy for the transfer of dn moles from phase β to phase α is, from Eq. (7),

$$dG = dG_\alpha + dG_\beta = \mu_\alpha\, dn - \mu_\beta\, dn = (\mu_\alpha - \mu_\beta)\, dn \tag{25}$$

Because the overall system is closed, the transfer will occur spontaneously if dG is negative (i.e., from phase β to phase α if $\mu_\beta > \mu_\alpha$). If the chemical potentials of the two phases are equal, the system will be at equilibrium for phase transfer (as is known from the general principle derived in the last section). Chemical substances are driven from regions of high chemical potential to regions of low chemical potential, much like charge is driven from regions of high electrical potential to regions of low electrical potential. The chemical potential is one measure of the *escaping tendency* of a substance.

The chemical potential of a homogeneous material (a phase) is a function of two intensive variables, usually chosen as temperature and pressure. We say that such a material has two *degrees of freedom* (i.e., we are free to set two intensive variables). (Note that only intensive variables count as degrees of freedom.) In addition to being able to specify a number of intensive variables equal to the number of degrees of freedom of a system, we are also at liberty to specify the size of the phase with one extensive variable. The chemical potential can be represented as a *surface* on a plot of μ versus P and T. The condition for equilibrium between phase α and phase β is, according to Eq. (24),

$$\mu_\alpha(T, P) = \mu_\beta(T, P) \quad \text{or} \quad G_{m,\alpha}(T, P) = G_{m,\beta}(T, P) \tag{26}$$

Equation (26) represents the intersection of two surfaces in $\mu(P, T)$ space. The intersection of two surfaces is a curve in the three-dimensional space. The projection of this curve on the PT plane is given by $P(T)$. Because P is a function of T, at equilibrium between two phases, the system has been reduced to one degree of freedom by the requirement of Eq. (26). If one of the phases is a gas, $P(T)$ is the *vapor pressure* curve of the condensed phase. If both phases are condensed, P is the externally applied pressure. Alternatively, we could consider $T(P)$, which gives the temperature at which two phases are at equilibrium as a function of pressure.

6.4.2 Phase Transfer at Constant Temperature and Volume

Consider two phases of a single component in a container of fixed volume and temperature. Using Eq. (6) and noting that $dT = 0$, $dn_\alpha = -dn_\beta = dn$, and

$dV_\alpha = -dV_\beta$, we can write the change of Helmholtz free energy for transfer of dn_α moles from phase β to phase α as

$$dA = dA_\alpha + dA_\beta = -PdV_\alpha + \mu_\alpha dn + PdV_\alpha - \mu_\beta dn = (\mu_\alpha - \mu_\beta)dn. \quad (27)$$

Because processes in closed systems at constant T and V are spontaneous if the Helmholtz free energy decreases, phase transfer from β to α is spontaneous if $\mu_\beta > \mu_\alpha$. If $\mu_\beta(T, P) = \mu_\alpha(T, P)$, the system is at equilibrium. This determines $P(T)$. However, because the chemical potential of a phase is not a function of its size, we can set *both* T and V for a system containing two phases at equilibrium. The system still has only a single degree of freedom, because V, being an extensive variable, does not count as a degree of freedom. This can be seen most clearly in cases in which one of the phases is a gas. Temperature will determine the vapor pressure of the condensed phase and the vapor will expand to fill the volume of the container, which can be varied maintaining phase equilibrium, as long as some liquid phase remains.

6.4.3 Regimes of Phase Stability

At 0 K, entropy does not contribute to chemical potential because

$$\mu = G_m = H_m - TS_m \quad (28)$$

The ordering of chemical potentials of phases at 0 K is the same as that of their H_m (or of U_m), namely solid < liquid < gas. The molar free energy of a phase varies as

$$d\mu = dG_m = V_m \, dP - S_m \, dT \quad (29)$$

and, therefore,

$$\left(\frac{\partial \mu}{\partial T}\right)_P = -S_m \quad (30)$$

From the third law, entropies are always positive, requiring that the chemical potential of all phases decrease with temperature. However, because entropy is a measure of randomness, $S_{m,gas} > S_{m,liq} > S_{m,sol}$, and the chemical potential falls most rapidly with temperature for the gas phase and least rapidly for the solid phase. In Fig. 1a (drawn for a particular value of pressure), as the temperature is increased, $\mu_{liq}(T)$ crosses $\mu_{sol}(T)$ at the melting point, and the liquid remains the most stable phase until $\mu_{gas}(T)$ crosses $\mu_{liq}(T)$ at the boiling point. In Fig. 1b (drawn for a different substance or at a different pressure), $\mu_{gas}(T)$ falls so rapidly with temperature that it crosses $\mu_{sol}(T)$ before $\mu_{liq}(T)$ does. As a result, liquid is never the most stable phase and, at the given pressure, the solid *sublimates* directly to gas.

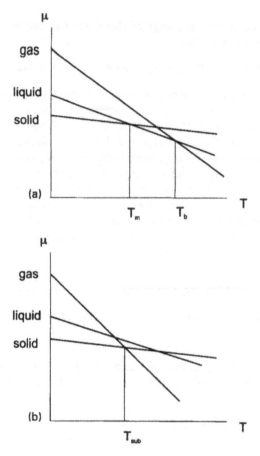

Figure 1 (a) Chemical potentials showing melting and boiling; (b) chemical potentials showing sublimation.

To discuss pressure regimes of phase stability, we use

$$\left(\frac{\partial \mu}{\partial P}\right)_T = V_m \tag{31}$$

For condensed phases, we can take the molar volume as independent of pressure,

$$\mu \stackrel{cp}{=} \mu° + V_m(P - P°) \tag{32}$$

where $\mu°$ is the chemical potential at the standard pressure, $P°$, chosen as $1.0\,\text{bar} \approx 1.0\,\text{atm}$. Because the molar volume of solids and liquids are quite small, the pressure correction to their chemical potential is small and can sometimes be neglected.

For the gas phase, we can usually use the ideal gas expression, $V_m = RT/P$, giving

$$d\mu \overset{\text{i.g.}}{=} RT \frac{dP}{P} = RT \, d \, \ln P \tag{33}$$

$$\mu \overset{\text{i.g.}}{=} \mu° + RT \, \ln\left(\frac{P}{P°}\right) \tag{34}$$

The chemical potential varies from $-\infty$ to $+\infty$ as the pressure of the gas varies from 0 to $+\infty$. As a result, there is always some gas pressure that will be at equilibrium with a condensed phase; this pressure is called the *vapor pressure* of the material at the given temperature. If all of the condensed phase does not evaporate (or sublimate) before its vapor pressure is reached, equilibrium will be attained.

In Fig. 2, the chemical potential curves of Fig. 1a are shown for two different pressures. Because the molar volume of a gas is greater than that of condensed phases, the chemical potential of the gas is increased much more than those of liquid or solid by increasing pressure. The boiling point and sublimation point therefore increase with pressure. The molar volume of the liquid and solid are comparable, and either one may be larger. As a result, the melting point may either increase or decrease with pressure.

The phase transitions illustrated in Fig. 1 are examples of *first-order phase transitions*. These are characterized by discontinuities in the derivatives of the

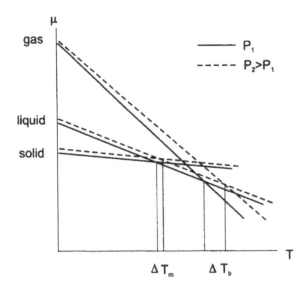

Figure 2 Chemical potentials at two pressures.

chemical potential at the phase transition. From Eq. (31), a discontinuity in $(\partial\mu/\partial P)_T$ requires a volume change for the transition. From Eq. (30), a discontinuity in $(\partial\mu/\partial T)_P$ results from an entropy change and an enthalpy change for the transition. In addition, because $C_P = (\partial H/\partial T)_P$, the heat capacity becomes infinite at a first-order phase transition. These changes are diagrammed in Fig. 3.

Second-order phase transitions, which are quite rare, involve a discontinuity in second derivatives at the transition point. They occur with zero enthalpy, entropy, and volume change. The transitions of certain metals (e.g., Hg and Sn) to a state in which they have no electrical resistance (superconductivity) are second-order phase transitions.

Another type of phase transition is called a lambda transition, because a graph of heat capacity versus temperature for this type of transition resembles the Greek letter λ, as shown in Fig. 4. This type of transition is usually associated with a change from an ordered state to a state with some disorder (order–disorder

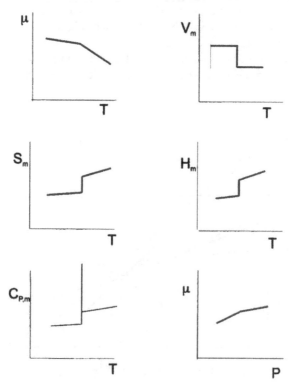

Figure 3   Changes at a first-order phase transition (drawn for the melting of ice).

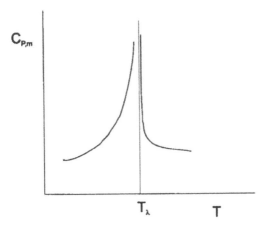

Figure 4 A lambda transition.

transition) as the temperature is raised. The transition occurs over a range of temperature, because it is cooperative—the first bits of disorder ease the way for additional disorder. At the transition temperature, all residual order disappears and the heat capacity goes to infinity. Throughout the transition, only a single phase is present. An example of a lambda transition occurs with an equimolar mixture of Cu and Zn (β brass) at 739 K. Below the lambda transition, each Cu atom is surrounded by four equidistant Zn atoms, whereas above the transition, Cu and Zn atoms are randomly arranged in the crystal.

6.5 Phase Equilibrium Conditions

The solution of Eq. (26) for the phase equilibrium condition gives $P(T)$ (e.g., vapor pressure as a function of temperature) or $T(P)$ (e.g., melting point as a function of pressure). A single point on one of these curves can be obtained by measurement or calculation.[1] We will now show how thermodynamics can be used to obtain the slope of these curves. Equation (26), $\mu_\alpha = \mu_\beta$, holds at phase equilibrium. If we change either P or T and the system remains at equilibrium, Eq. (26) must still hold:

$$\mu'_\alpha = \mu_\alpha + d\mu_\alpha \stackrel{\text{equil}}{=} \mu'_\beta = \mu_\beta + d\mu_\beta \tag{35}$$

Substituting Eq. (26) gives

$$d\mu_\alpha = dG_{m,\alpha} = d\mu_\beta = dG_{m,\beta} \tag{36}$$

Using Eq. (29),

$$dG_{m,\alpha} = V_{m,\alpha}\, dP - S_{m,\alpha}\, dT = dG_{m,\beta} = V_{m,\beta}\, dP - S_{m,\beta}\, dT \tag{37}$$

or

$$\frac{dP}{dT} = \frac{S_{m,\beta} - S_{m,\alpha}}{V_{m,\beta} - V_{m,\alpha}} = \frac{\Delta_\phi S}{\Delta_\phi V} \tag{38}$$

where $\Delta_\phi S$ and $\Delta_\phi V$ are the entropy and the volume change, respectively, for the phase change $\alpha \to \beta$. Because we are dealing with a system at equilibrium,

$$\Delta_\phi S = \frac{\Delta_\phi H}{T} \tag{39}$$

giving

$$\frac{dP}{dT} = \frac{\Delta_\phi H}{T\Delta_\phi V} \tag{40}$$

This is the *Clapeyron equation*, which applies to any type of phase equilibrium.

For equilibrium between a gas and a condensed phase, we can write (using vaporization as an example)

$$\mu_g = \mu_g^\circ + RT\, \ln\!\left(\frac{P}{P^\circ}\right) = \mu_l \approx \mu_l^\circ \tag{41}$$

(The chemical potentials of condensed phases are relatively pressure independent, due to their small molar volumes.) Rewriting,

$$\mu_g^\circ - \mu_l^\circ = \Delta_{vap} G^\circ = -RT\, \ln\!\left(\frac{P}{P^\circ}\right) \tag{42}$$

where $\Delta_{vap} G^\circ$ is the Gibbs free energy of vaporization when both liquid and vapor are in their standard (1.0 bar) states. Using Eq. (29) of Chapter 4, the Gibbs–Helmholtz equation, we obtain

$$\Delta_{vap} H^\circ = \left(\frac{\partial (T^{-1}\Delta_{vap} G^\circ)}{\partial T^{-1}}\right)_P = \frac{d(T^{-1}\Delta_{vap} G^\circ)}{dT^{-1}} = -R\, \frac{d\, \ln(P/P^\circ)}{dT^{-1}} \tag{43}$$

The partial derivative can be converted into an ordinary derivative because $\Delta_{vap} G^\circ$ does not depend on pressure. Because P° does not depend on temperature, this becomes

$$\frac{d\, \ln P}{dT^{-1}} = -\frac{\Delta_{vap} H^\circ}{R} \tag{44}$$

which is known as the *Clausius–Clapeyron equation*. The same formula can be obtained from Eq. (40) by neglecting the molar volume of the condensed phase and using the ideal gas law. The standard enthalpy change of vaporization (usually called the *heat of vaporization*) is given by

$$\Delta_{vap}H° = H°_{m,g} - H°_{m,l} \tag{45}$$

The standard state of condensed phases, such as liquids, are chosen as the pure substance at 1.0 bar pressure at the temperature of interest. The standard state of an ideal gas is also at 1.0 bar. Over a moderate range of temperature, heats of vaporization and sublimation usually do not vary greatly and, as shown by Eq. (44), can be obtained from plots of the vapor pressure versus $1/T$. An example of this for ethyl acetate is shown in Fig. 5. Note the slight deviation of Fig. 5b from a straight line.

The integrated form of Eq. (44), assuming temperature-independent $\Delta_\phi H°$,

$$\ln\left(\frac{P_2}{P_1}\right) = \frac{\Delta_\phi H°}{R}\left(\frac{1}{T_1} - \frac{1}{T_2}\right) \tag{46}$$

is useful for interpolating and extrapolation vapor pressure data.

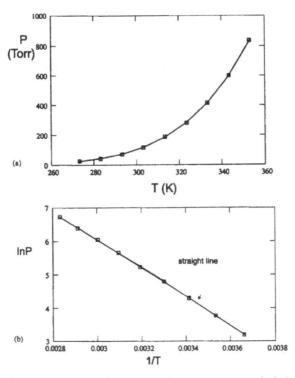

Figure 5 (a) Vapor pressure of ethyl acetate; (b) vapor pressure of ethyl acetate plotted according to Eq. (44).

Example 1. The normal boiling point of nonane is 150.8°C and its vapor pressure is 400 mm Hg at 128.2°C. What is the $\Delta_{vap}H°$ of nonane? Compare $\Delta_{vap}S°$ with the Trouton's law value, 85 J/K mol.

Solution:

$$\Delta_{vap}H° = \frac{R \ln(P_2/P_1)}{(1/T_1 - 1/T_2)}$$

The temperature must be in Kelvin.

$$\Delta_{vap}H° = 8.314 \ \frac{J}{K \ mol} \ \ln\left(\frac{760}{400}\right)\left[\left(\frac{1}{401.3} - \frac{1}{423.9}\right)\frac{1}{K}\right]^{-1}$$

$$= 40.1 \ mol/kJ$$

$$\Delta_{vap}S° = \frac{40,100 \ J}{mol(423.9 \ K)} = 94.6 \ K \ mol/J$$

   This example illustrates how thermodynamic quantities, the determination of which we might expect would require calorimetric measurements, can often be much more easily obtained from calculations based on equilibrium conditions.

6.6 Vapor Pressure When a Gas Is Not Ideal

The form of Eq. (34) is so useful that we generalize it by defining for any substance, an *activity, a*, so that

$$\mu = \mu° + RT \ \ln a \tag{47}$$

$\mu°$ is the chemical potential in the standard state, which must be defined for the substance. (Standard states were discussed in Section 4.7 of Chapter 4.) From Eq. (47), the activity is given by

$$a = \exp\left(\frac{\mu - \mu°}{RT}\right) \tag{48}$$

and depends on the choice of standard state. It is a dimensionless quantity. The activity is always 1.0 at the standard state. For condensed phases, the standard state is chosen to occur at 1.0 bar pressure, and the activity is 1.0 at this pressure. Because the molar volumes of condensed phases are very small, their activities remain close to 1.0 for moderate pressure deviations from 1.0 bar. (See Problem 3.)

   From Eq. (34), for an ideal gas, the activity is $P/P°$, or just the numerical value of the pressure in bars. For an ideal gas, pressure is therefore

$$P \stackrel{i.g.}{=} aP° \tag{49}$$

Because we would like to discuss the properties of real gases in terms of a variable analogous to pressure, we define the fugacity of a real gas as

$$f = aP^\circ \tag{50}$$

(For an ideal gas, $f = P$.) μ, a, and f are alternative measures of escaping tendency.

Using Eq. (50) in Eq. (47) gives

$$\mu = \mu^\circ + RT \ \ln\left(\frac{f}{P^\circ}\right) \tag{51}$$

We define the standard state of a real gas so that Eq. (51) is general (i.e., so that it also applies to ideal gases). For ideal gases, the standard state is at 1.0 bar pressure. For real gases, we also use a 1.0-bar ideal gas as the standard state. We find the standard state by the two-step process shown in Fig. 6. First we extrapolate the real gas to very low pressure, where $f \to P$ and the gas becomes ideal (Step I). We then convert the ideal gas to 1.0 bar (step II). The convenience of an ideal gas standard state is that it allows temperature conversions to be made with ideal gas heat capacities (which are pressure independent). Conversion to the real gas state is then made at the temperature of interest.

Some properties of an ideal gas, such as molar enthalpy, are independent of pressure, so that Step II is not necessary. The standard values for such properties

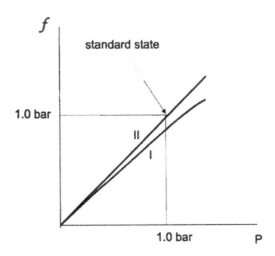

Figure 6 Correction of a real gas to its standard state.

are just obtained by extrapolating the measured values to zero pressure. Thus, for real gases, we get, instead of Eq. (44),

$$\frac{d \ln f}{dT^{-1}} = -\frac{\Delta_{vap}H^\circ}{R} \tag{52}$$

where $\Delta_{vap}H^\circ$ is the extrapolation of the heat of vaporization to zero pressure.

Although fugacity is theoretically and computationally significant, it is the pressure that we need for practical applications. To find the relation between f and P, we write, from Eqs. (31) and (51),

$$d\mu_T = V_m \, dP = RT \, d \ln f \tag{53}$$

Integrating this equation from $P = 0$, where f and P are equal,

$$\int_0^{P'} V_m \, dP = RT \int_0^{P'} d \ln f \tag{54}$$

Both of these integrands are infinite at their lower limit, and the integrals cannot be evaluated. To circumvent this difficulty, we define the *fugacity coefficient*, ϕ, so that

$$f = \phi P \tag{55}$$

with $\phi \to 1$ as $P \to 0$. Equation (54) then becomes

$$\int_0^{P'} V_m \, dP = RT \left(\int_0^{P'} d \ln \phi + \int_0^{P'} d \ln P \right) \tag{56}$$

or

$$\ln \phi(P') = \frac{1}{RT} \int_0^{P'} \left(V_m - \frac{RT}{P} \right) dP \tag{57}$$

The integrand of this integral goes to zero at the lower limit.

6.7 Equation of State for the Two-Phase Region

In Chapter 1, we remarked on the oscillatory behavior of two-parameter equations of state, such as those of van der Waals, Berthelot, and Redlich-Kwong in the two-phase region. Usually, the parameters for these equations are determined from the critical point; therefore, the equations should have some validity slightly below the critical point in the two-phase region. Because these equations of states are cubic equations, with three real roots below the critical point, they give three values of V_m for any combination of P and T. Referring to Fig. 7, V_{m1} and V_{m3} correspond to the liquid and gas, respectively, whereas V_{m2}, where

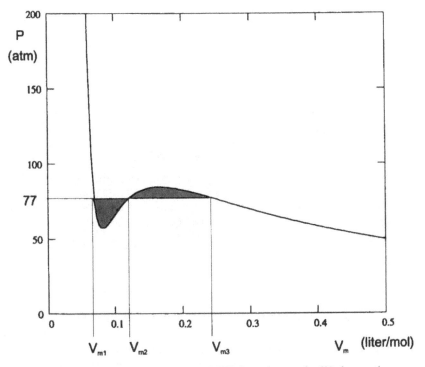

Figure 7 Vapor pressure of NH_3 at 370 K from the van der Waals equation.

$(\partial V_m / \partial P)_T > 0$ does not correspond to a real substance. However, we know that P and T are not independent in the two-phase region and we seek a criterion for finding $P(T)$ (i.e., the vapor pressure of the substance).

Because liquid and gas are at equilibrium in the two-phase region, the chemical potentials of the phases with V_{m1} and V_{m3} must be equal. Points 1 and 3 lie on the same isotherm; therefore, we can write

$$\mu_1 - \mu_3 = \int_1^3 V_m \, dP = 0 \qquad (58)$$

Because we prefer to deal with an integral over V_m, integration by parts is used to give

$$V_{m,3}P - V_{m,1}P - \int_{V_{m,1}}^{V_{m,3}} P \, dV_m = 0 \qquad (59)$$

which can be written as

$$\int_{V_{m,1}}^{V_{m,3}} P \, dV = P(V_{m,3} - V_{m,1}). \tag{60}$$

In other words, the pressure is chosen so that the area under the oscillating curve is the same as that under the straight line joining the end points. Estimates can be performed by shifting a transparent horizontal straight edge until the two shaded areas in Fig. 7 are visually equal, which occurs for the van der Waals equation for ammonia at 77 atm for 370 K. The measured vapor pressure is about 57 atm. Quantitative equalization of the areas can be done by trial and error, but is usually not worthwhile, because the major part of the inaccuracy is due to the poor fit of the equation in the two-phase regions. The Redlich–Kwong equation provides a better fit, and when used in a graphical procedure similar to that employed in Fig. 7, it gives a vapor pressure of 61 atm at 370 K.

6.8 Effect of Inert Gas on Phase Equilibria

In our consideration of phase equilibria at constant pressure, we imagined the overall system confined by a piston. However, when we talk about constant-pressure systems, we usually mean that the pressure is maintained by an inert gas (e.g., the atmosphere).[2] In some situations, much higher constant pressures of inert gases are applied to systems. If we take a gas-condensed phase equilibrium and apply an inert gas pressure to both phases, we get, from Eqs. (31) and (33),

$$d\mu_{cond} = V_m \, dP_{inert} = d\mu_{gas} = RT \, d \ln P \tag{61}$$

Note that μ of the gas depends only on the pressure of the gas at equilibrium, as we have assumed that it behaves ideally. On the other hand, in calculating change of μ of the condensed phase, we consider only the inert gas pressure. This pressure is much larger than the change in the vapor pressure of the active gas, and to the condensed phase, all types of pressure are the same. Integrating Eq. (61) from an inert gas pressure of zero at which the vapor pressure is P_0 to an inert gas pressure of P_{inert} gives

$$\ln\left(\frac{P}{P_0}\right) = \frac{V_m P_{inert}}{RT} \tag{62}$$

Example 2.   The vapor pressure of water is 23.76 torr at 25°C. What is its vapor pressure in the presence of 1.0 atm of air considered an inert gas?... in the presence of 100 atm of air?

Solution: Since the density of water is 1.0 g/mL, its molar volume is 18 mL/mol.

At 1.0 atm;

$$\ln\left(\frac{P}{23.76}\right) = \frac{(18 \times 10^{-3} \text{ L})(1.0 \text{ atm})}{(\text{mol})(0.082 \text{ L atm/mol K})(298 \text{ K})} = 0.00074$$

$$\frac{P}{23.76} = \exp(0.00074) = 1.00074, \qquad P = 23.78 \text{ torr}$$

At 100 atm;

$$\frac{P}{23.76} = \exp(0.074) = 1.0768, \qquad P = 25.58 \text{ torr}$$

The example indicates that the inert gas effect would be difficult to measure at 1 atm, but becomes appreciable at very high pressures.

6.9 Condensed-Phase Equilibria

Condensed-phase equilibria are treated by Eq. (40), the Clapeyron equation. The most important type of condensed-phase equilibrium is that between solid and liquid. For melting, $\Delta_\phi H$ is always positive, because the solid is the lowest-energy (and enthalpy) arrangement of molecules. The direction of change of the melting temperature with pressure,

$$\left.\frac{dT}{dP}\right|_{\text{melt}} = \frac{T(\Delta_{\text{melt}}V)}{\Delta_{\text{melt}}H} = \frac{T(V_{m,\text{liq}} - V_{m,\text{sol}})}{\Delta_{\text{melt}}H} \tag{63}$$

is thus determined by the sign of $\Delta_{\text{melt}}V$. Usually, molar volumes of liquids are greater than those of solids and $\Delta_{\text{melt}}V$ is positive. In this case, the melting temperature increases with pressure. In water, however, the solid has a hydrogen-bonded configuration containing cavities and a larger molar volume than the liquid. As a result, freezing water expands (as in an automobile radiator unprotected by antifreeze) and the resulting ice is less dense and floats on the liquid. Equation (63) then tells us that the melting temperature of water decreases with increasing pressure. This is often offered as an explanation for the speed that can be achieved in ice skating. The narrow blades concentrate the skater's weight and produce a large pressure, at which the melting temperature of water is lowered below ambient temperature. The resulting film of liquid water provides low friction, which enhances the skater's speed. Because volume changes on melting are usually quite small, melting temperatures do not change very much with pressure.

Example 3. For water, the freezing point at 1 atm pressure is $0°C = 273.15$ K. The density of liquid water is 1.00 g/cm^3 and that of ice is 0.92 g/cm^3. The heat of fusion of water is 6.0 kJ/mol. What is the freezing point of water at 100 atm pressure?

Solution: Separating variables in Eq. (63) and integrating,

$$\ln\left(\frac{T_2}{T_1}\right) = \frac{\Delta V_{melt}}{\Delta H_{melt}}(P_2 - P_1)$$

$$\Delta V_{melt} = \left(\frac{cm^3}{1.0\ g} - \frac{cm^3}{0.92\ g}\right)\frac{18\ g}{mol} = -1.6\ \frac{cm^3}{mol}$$

$$\ln\left(\frac{T_2}{273.15}\right) = -\left(1.6\ \frac{cm^3}{mol}\frac{L}{10^3\ cm^3}\right)$$

$$\times \left(6.0 \times 10^3\ \frac{J}{mol}\frac{0.082\ L\ atm}{8.314\ J}\right)^{-1}(100-1)\ atm$$

$$= -0.0027$$

Two different values for the gas constant were used to convert from Joule to liter atm:

$$T_2 = 273.15\ \exp(-0.0027) = 272.41\ K = -0.75°C$$

6.10 Equilibrium Between Three Phases

If a mixture of ice and water at 1 atm pressure and 0°C is placed in an insulated container and all of the air is pumped away and the container sealed, what will happen? As was shown in the previous section, at pressures lower than 1 atm, the melting point of ice is above 0°C. Water will, therefore, be solid at 0°C and reduced pressure. However, when some liquid water freezes, its latent heat is released and the temperature of the system is slightly increased. Equilibrium is reestablished at the higher temperature and reduced pressure. The pressure in the system is the vapor pressure of both liquid and solid water slightly above 0 K. The new equilibrium point of the system is called the *triple point* of water and is at 0.0098°C and 611 Pa. Three phases—solid, liquid, and gas—coexist at the triple point, and the chemical potential of water in each of the phases must be equal:

$$\mu_{solid} = \mu_{liq} = \mu_{vap} \tag{64}$$

This introduces two independent equations that must be satisfied and reduces the number of degrees of freedom of the system to zero. The triple point of a single-component system is invariant and provides a convenient temperature reference. Because vapor pressure, as well as chemical potential, is a measure of escaping tendency from a condensed phase, the vapor pressures of the solid and liquid are equal at the triple point.

6.11   Phase Diagrams

Regions of phase stability and equilibrium between phases are often presented by means of *phase diagrams*, which represent the properties of substances as a function of pressure and temperature. For a one-component system, on such a diagram, regions where a single phase exists are represented as areas (two degrees of freedom), regions where two phases exist at equilibrium are represented as curves (one degree of freedom), and where three phases exist at equilibrium as points (the triple point). In Fig. 8, the phase diagram of water is shown at low pressures. (As can be seen, the diagram is not drawn to scale.) The solid phase is represented by the area at high pressure and low temperature and the vapor phase by the area at low pressure. The liquid phase is represented by an intermediate area, with the curve separating the liquid and vapor ending at the critical point (220 atm and 374°C). Above the critical temperature, there is no observable phase transition between liquid and vapor and the material is called a *supercritical fluid*.

Water, of course, makes transitions from solid ice to liquid water (at 0°C) to water vapor (at 100°C) as it is heated at 1.0 atm pressure. Because at the pressure of the triple-point, the transformation is directly from solid to gas, the triple point pressure must be below 1.0 atm. Because $\Delta_{melt}V$ for water is negative, by Eq. (38), the solid–liquid equilibrium line has negative slope, and the triple point must be above 0°C. Considering the steepness of the line, the difference is very small and the measured triple point of water is 0.0098°C and 611 Pa (∼4.6 torr).

A substance that can exist in more than one crystalline form is said to exhibit *allotropy*, and the different forms are called *allotropes*. Figure 9 is the high-pressure part of the phase diagram of water and shows that water has a number of allotropes. The crystalline forms of water in the allotropes that melt are

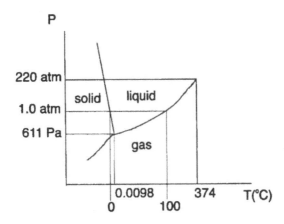

Figure 8   Phase diagram for water (not to scale).

denser than liquid water, as can be deduced from the positive slopes of their liquid–solid equilibrium lines. The allotropes of water result in a number of triple points, where three solid phases, or two solid phases and liquid water are at equilibrium. The phases are numbered in order of their discovery, from I through IX. There is no ice IV, because what was originally given that designation was later found to be *metastable*, converting to other phases over time. Conversion of one solid phase to another is often very slow. Assuring that equilibrium has been achieved is a major difficulty in studying high-pressure phase diagrams.

The ice IX shown in Fig. 9 is not that postulated in Vonnegut's science fiction book, *Cat's Cradle*, that is quoted at the beginning of this chapter. That ice-nine is the stable form of water at ambient conditions, never previously discovered, due to lack of a nucleation crystal. When the mad scientist in Vonnegut's book synthesizes ice-nine and a crystal finds its way into the ocean, great calamities ensue.

The low-pressure phase diagram of carbon dioxide, shown in Fig. 10, is different from that of water in a number of respects. Carbon dioxide sublimates at 1.0 atm, leading to solid CO_2 being called "dry ice." The triple point is at 5.1 atm

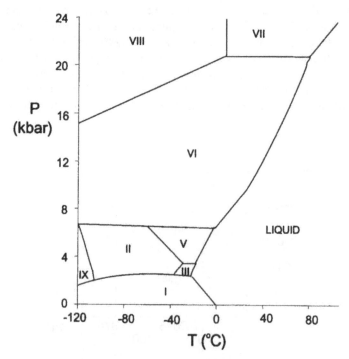

Figure 9 High-pressure phase diagram of water.

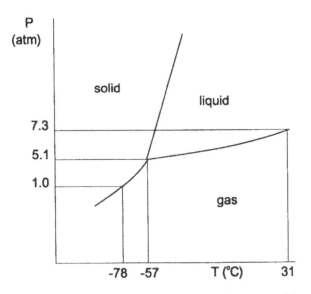

Figure 10   Phase diagram of carbon dioxide (not to scale).

and −57°C, and the solid is denser than the liquid (by far, the more common occurrence), leading to a positive slope of the solid–liquid equilibrium line. The critical point is at 73 atm and 31°C, conditions that are easily accessible with commercial high-pressure pumping equipment. As a result, supercritical carbon dioxide has found extensive use as a solvent in industrial processes. One of these is a method of removing caffeine from coffee that avoids using chlorinated solvents. Like many supercritical fluids, supercritical carbon dioxide is an excellent solvent. After the fluid CO_2 is passed through the coffee, dissolving the caffeine, the external pressure is released, and the carbon dioxide becomes gaseous and releases the caffeine. The carbon dioxide can be reused, by recompressing it, and no solvent residue is left in the coffee, because carbon dioxide is gaseous at ambient conditions.

6.12   Mesomorphic Phases

Mesomorphic phases have properties between those of solids and liquids. For example, rod-shaped or disk-shaped molecules often melt to form *liquid crystals*, which are fluid, but contain some order. One form of liquid crystals formed from rod-shaped molecules, called a *nematic* liquid crystal, is shown in Fig. 11. The molecules in the nematic phase are partially ordered in one dimension, with an orientation angle that can vary from one domain to another in the fluid. The forces

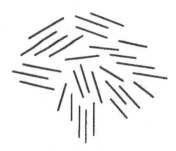

Figure 11   A nematic liquid crystal.

orienting the molecules are quite weak and the heat required to melt this type of liquid crystal is usually quite small. Small electric fields can be used to orient domains in liquid crystals, changing the way they scatter light. This is the basis for many of the displays used in digital watches, calculators, and laptop computers.

Glasses, which are rigid but noncrystalline (*amorphous*), are also meso-morphic. They undergo a *glass transition* at temperature T_g, above which they show fluid or elastomeric behavior. The glass transition occurs over a range of temperatures, rather than at a single, well-defined temperature characteristic of the melting of a crystalline material. Even at temperatures considerably below T_g, glasses are not totally rigid. This can be verified by accurately comparing the thickness of the top and the bottom of a pane of very old window glass. Some flow under the long-time influence of gravity can usually be noted. T_g depends on the heating rate, indicating that the glass transition is a kinetic as well as an equilibrium phenomenon.

Polymeric materials are often composed of crystalline regions imbedded in an amorphous matrix. As shown in Fig. 12, such a material undergoes several

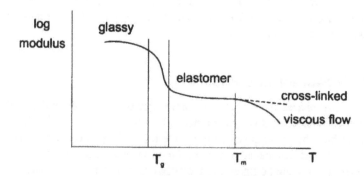

Figure 12   Phase transformations of a polymer.

changes of state as it is heated. At very low temperatures, thermal motion is insufficient to overcome attractions in both the crystalline and amorphous regions. Motions are limited to vibrations and the modulus is very high. The polymer is very brittle in this region and is easily fractured, as indicated by what happens when a rubber band is cooled to liquid-nitrogen temperature. At the glass transition temperature, T_g, the amorphous regions of the polymer become fluid, but the crystalline regions serve as cross-linking junctions, which produce the elastomeric behavior discussed in Chapter 5. Actually, the glass transition occurs over a temperature range (the transition is not first order), as thermal motion becomes sufficient to overcome the disparate attractive interaction in various amorphous regions. Elastomeric behavior persists up to a considerably higher temperature, T_m (the melting temperature). At T_m, the stronger interactions in the crystalline regions become disturbed and (highly viscous) flow of the polymer becomes possible. For polyethylene, T_g is around $-125°C$ and T_m is $140°C$. For polymers in which chains are held together by chemical cross-links, viscous flow is not possible and no melting temperature is observed.

Questions

1. Comment on the sign of each of the following: $\Delta_{vap}H$, $\Delta_{cond}H$, $\Delta_{fus}S$, $\Delta_{fus}V$, and $\Delta_{vap}G$ (for $T > T_b$).

2. An ice cube is dropped into a glass of room-temperature water. An instant after the last of the ice melts, is the system a single phase?

3. Does Eq. (8) imply that for a single-component system, the chemical potential is the molar internal energy as well as the molar Gibbs free energy?

4. What is the difference between δn_i and dn_i?

5. On Fig. 1b, draw $\mu(T)$ for a gas and solid at a higher pressure and show how the sublimation temperature is affected by the pressure increase.

6. It is desired to determine the heat of vaporization of a substance. However, the normal boiling point of the substance is inconveniently high for laboratory measurement. How could the heat of vaporization of the substance be determined at a lower temperature?

7. Why do we not choose as the standard state for a real gas the state where $f \rightarrow P \rightarrow 0$. (Hint: What would be the values of the chemical potentials that we would tabulate at the standard state?)

8. At the triple point of water, is the system's volume invariant? Explain your answer.

9. In Fig. 2, why is the separation for the curves at two different pressures relatively independent of temperature for the liquid and solid phases while it approaches zero for the gaseous phase as $T \rightarrow 0$?

10.   According to our analysis, does the effect of an inert gas on the vapor pressure of a liquid result from a change in the escaping tendency from the liquid or from the gas, or from both. What is the direction of the change(s) in escaping tendency? Explain your answer.

11.   What is wrong with the following phase diagram (drawn for an arbitrary substance)?

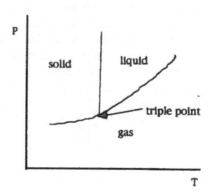

12.   What would happen if a crystal of Vonnegut's ice IX was thrown into a glass of water. Include discussion of energy flows as well as phase changes.

13.   Sketch a diagram analogous to Figs. 1a and 1b at the pressure at which a gas–liquid–solid triple point occurs.

14.   Redraw Fig. 2 for water.

15.   Redraw Fig. 3 for the boiling of water.

16.   Why is solid CO_2 called "dry ice"?

17.   The Earth's core is thought to be mainly iron, and seismic data indicate that the inner core is solid and the outer core is liquid. The pressure at the center of the Earth is 3.6×10^{11} Pa, and at this pressure, iron melts at 6350 K. From this information, what can you infer about the solid–liquid equilibrium boundary in the iron high-pressure phase diagram? (Pressure and temperature both increase toward the Earth's center.)

Problems

1.   How much heat must be added to 1.0 kg of water at 25°C to convert it into steam at 300°C? The C_P of liquid may be taken as 1 cal/gK = 4.18 J/gK (this is the definition of the calorie.). The heat of vaporization may be obtained from Table 1 of Chapter 3 and gaseous heat capacity data from Table 1 of Chapter 2.

2.   The enthalpy of vaporization of water is 40.66 kJ/mol; what is the boiling point of water in Denver, Colorado at an elevation of 1.0 mile? (See Chapter 5, Problem 17.)

3. The standard state of a condensed phase is taken at 1.0 bar pressure. Show that the activity of incompressible condensed matter at pressure P is from Eq. (47), $a = \exp[V_m(P-1)/RT]$. Apply this formula to find the activity of liquid octane, with density $0.7\,g/cm$ at $25°C$ and $100\,atm$ pressure. How much does the chemical potential of liquid octane change between 1 and $100\,atm$?

4. Show that for a gas that is described by the first two terms in the virial expansion given by Eq. (27) in Chapter 1, the fugacity is given by $f = P\,\exp(B'P/RT)$. Show that for a van der Waals gas, the fugacity equals $P\,\exp(bP/RT)\,\exp(-aP/R^2T^2)$. (See Example 4 in Chapter 1.)

5. Using the results of Problem 4, for NH_3, at what pressure at $250\,K$ is there a 10% difference between pressure and fugacity?

6.* What is the vapor pressure of water at $500°C$? Use Eq. (52), with $\Delta H^\circ_{vap}(298) = 40.66\,kJ/mol$ and average heat capacities $C_P(gas) = 34\,J\,K\,mol$ and $C_P(liq) = 75\,J\,K\,mol$. The results of Problem 4 can be used to convert fugacity to pressure.

7. The vapor pressure of water is $634\,mm\,Hg$ at $95°C$ and $1074\,mm\,Hg$ at $110°C$. Estimate the standard heat of vaporization of water using the Clapeyron equation and the Clausius–Clapeyron equation.

8. Determine the heat of vaporization of water from its triple point and normal boiling point.

9.^{M} Use graphical equalization of areas with the van der Waals and Redlich–Kwong equations to estimate the vapor pressure of water at $600\,K$. Compare these with the literature value of $122\,atm$.

10. Using expression (38) of Chapter 5 for its partition function and $U_0 = 0$, find a formula for the chemical potential of an ideal monatomic gas. Show that $(\partial\mu/\partial P)_T = V_m$.

11. Show that the requirement that second derivatives of μ are discontinuous at a second-order phase transition also requires that C_P be discontinuous at the transition.

12. Give a mathematical statement of the first law appropriate for open systems.

13. Derive the Clausius–Clapeyron equation [Eq. (44)] from Eq. (40) by neglecting the volume of the condensed phase and using the ideal gas law for the vapor.

14. Dry ice has a vapor pressure of $1.0\,atm$ at $-72.2°C$ and $2.0\,atm$ at $-69.1°C$. Calculate the $\Delta_{sub}H^\circ$ of CO_2 and the vapor pressure of dry ice at $-50°C$.

15. The density of liquid CO_2 is greater than that of liquid water. It has been suggested that to ameliorate the greenhouse effect caused by atmospheric CO_2, CO_2 from power plants could be pumped deep enough under the ocean so that liquid CO_2, being heavier than the surrounding water, would sink to the ocean bottom. Using the data in Fig. 10, estimate the depth in the ocean to which CO_2 would have to be pumped so that it is a liquid.

16. From the data in Appendix B, graphite is seen to be the stable form of carbon at 298 K and atmospheric pressure. The densities of graphite and diamond are 2.25 g/cm^3 and 3.52 g/cm^3, respectively. Calculate the pressure to which graphite must be raised at 298 K in order for a diamond to become the stable form of carbon. Is this a viable method for converting graphite to diamond? Explain your answer.

17. Calculate the vapor pressure of water at 298 K, given that $\Delta G_f(298)$ is −237.129 kJ/mol for liquid water and −228.572 kJ/mol for gaseous water.

18. The normal boiling point of CS_2 is 46.3°C. Use Trouton's rule to estimate the vapor pressure of CS_2 at 25°C.

19. An auto radiator has a vent valve that releases at 3 atm pressure. If the radiator is filled with water in summer, what is the maximum temperature that the water can reach before the valve vents? The $\Delta_{vap}H°$ of water is 40.6 kJ/mol at the boiling point.

Notes

1. *Ab initio* (from first principles) calculations are difficult to do with reasonable accuracy, due to the difficulty in calculating the energy of liquids.
2. The atmosphere is not completely inert. N_2 and O_2 have some solubility in most solutions. We will see in Chapter 8 that dissolving gases in liquids lowers the vapor pressure of the liquid. Also, in a system open to the atmosphere, complete equilibrium will not be established because material will continually be transported to the ambient environment. In practice, however, a small opening will be sufficient to maintain the pressure of the atmosphere while allowing neglect of the loss of gaseous material by diffusion or convection.

7

Chemical Reactions

The changes of thermodynamic quantities in chemical reactions are calculated from tabulated values of absolute entropies and heat capacities. However, enthalpies and free energies of formation reactions are used. Such quantities can be measured in calorimetric experiments, such as those using the bomb calorimeter, which is used for combustion reactions. Alternatively, generic bond dissociation energies or group additivity relations may be employed to estimate thermodynamic quantities. At equilibrium, the Gibbs free energy change of a reaction is zero, allowing equilibrium constants to be calculated from standard free-energy changes. Because the free energies of condensed phases are approximately pressure independent, these species do not enter into equilibrium-constant expressions, except at high pressure. The Gibbs phase rule specifies the number of degrees of freedom of a system at equilibrium. Calculations of equilibrium concentrations are most systematically performed using a single variable, the extent of reaction, to express the concentrations of reactants and products at equilibrium. The enthalpy change of a reaction is used to convert its equilibrium constant from one temperature to another.

7.1 Nomenclature

We write a general chemical reaction as

$$\text{Reactants} \rightarrow \text{Products} \tag{1}$$

Therefore, a chemical reaction is a process in which the initial state is the reactants and the final state is the products of the reaction. Reactions may involve changes in temperature and pressure. However, for simplicity of calculation, we usually consider the reactants and products to be at the same T and P, and correct to the actual T and P in a separate step, using the energy released in the reaction. As discussed in Chapter 3, the standard pressure chosen for compilation of thermodynamic data is $1.0\,\text{bar} \equiv 10^5\,\text{Pa}$,[1] and it is designated by a degree sign, so we can write a standard reaction as

$$\text{reactants}^\circ(T) \rightarrow \text{products}^\circ(T) \tag{2}$$

In terms of individual reagents, a reaction may be written as

$$aA + bB + \cdots \rightarrow cC + dD + \cdots \tag{3}$$

where a, b, c, and d are the reagent numbers of the reaction, usually written as integer numbers. (Sometimes, fractional values are used.) Because, in regard to conservation of mass, atomic species, and energy, a chemical reaction implies equality of reactants and products, we choose to convert the arrow ("is converted to") in this reaction into an equal sign. The reactants are then brought over to the product side of the equality to give

$$0 = \sum_i \nu_i A_i \tag{4}$$

where i ranges over reactants and products and ν_i is the *stoichiometric coefficient*, positive for products and negative for reactants.

During a reaction, the change in the amount of any reactant or product is proportional to its stoichiometric coefficient, and a single variable is sufficient to specify the production of products and consumption of reactants. This variable, ξ, is the number of moles of product with unit stoichiometric coefficient that is formed and is called the *extent of the reaction*. The change in the number of moles of any reactant or product is given by

$$\Delta n_i = n_i - n_i(t = 0) = \nu_i \xi \tag{5}$$

The change of a property in a reaction is that property of the products minus that property of the reactants. The symbol Δ_{rxn} indicates the change in the property when molar amounts of reactants indicated by their stoichiometric coefficients form molar amounts of products indicated by their stoichiometric coefficients. It is important to always give the balanced equation of the reaction being considered, including the phase of each reactant and product if these are not evident. If the reagents and products are noninteracting (e.g., ideal gases), the change of any extensive property can be calculated from molar properties of the individual reactants and products. For example, for the volume change on reaction

$$\Delta_{rxn}V = cV_{m,c} + dV_{m,d} + \cdots - (aV_{m,a} + bV_{m,b} + \cdots) = \sum_i \nu_i V_{m,i} \qquad (6)$$

At temperature T and the standard pressure, we have

$$\Delta_{rxn}V^\circ(T) = \sum_i \nu_i V^\circ_{m,i}(T) \qquad (7)$$

which is called the standard volume change of the reaction (the volume change at the standard pressure).

7.2 Thermochemistry

Using the procedure of Eq. (7), the entropy change of chemical reaction (2) can be written as

$$\Delta_{rxn}S(T) = \sum_i \nu_i S_{m,i}(T) \qquad (8)$$

This is a useful equation because the third law sets the zero of entropy for every pure substance and thus permits calculations of absolute entropies at temperature T by Eq. (5) of Chapter 4.

Equations similar to Eq. (8) can also be written for U, H, A and G. For example,

$$\Delta_{rxn}H(T) = \sum_i \nu_i H_{m,i}(T) \qquad (9)$$

and

$$\Delta_{rxn}G(T) = \sum_i \nu_i G_{m,i}(T) \qquad (10)$$

An alternative form for Eq. (10), which is also applicable for systems in which there exist interactions between species, is from Eq. (11) of Chapter 6:

$$\Delta_{rxn}G(T) = \sum_i \nu_i \mu_i(T) \qquad (11)$$

Equations like Eqs. (9)–(11) are of little value for calculations, however, because we have no way of determining absolute values for thermodynamic functions that have an energy component. Even though absolute values of energy can be defined by using Einstein's $E = mc^2$ to convert mass into energy, there are both theoretical and practical difficulties in dealing with absolute energy-like quantities: theoretical because the definitions of potential and kinetic energies require the choice of zero-energy position and velocity reference; practical because the Einstein energy or even the total electrostatic energies of molecules are so huge that their uncertainties mask chemically interesting energies. We can circumvent this difficulty by arbitrarily choosing a reference state and measuring changes in thermodynamic properties with respect to this reference state. Taking the enthalpy change as a example, we have, according to Fig. 1,

$$\Delta_{rxn} H = \Delta H_1 + \Delta H_2 \tag{12}$$

Equation (12) follows from enthalpy being a state function, which means that its change is independent of the path taken in going from reactants to products. The enthalpy change of a reaction is known as the *heat of the reaction*, and Eq. (12) is an example of *Hess' law*, which is the procedure of calculating the change of a thermodynamic property of a process by using an alternative path to go from the initial to the final state of the process. At temperature T and the standard pressure, Eq. (12) becomes

$$\Delta_{rxn} H^\circ(T) = \Delta H_1^\circ(T) + \Delta H_2^\circ(T) \tag{13}$$

where $\Delta_{rxn} H^\circ(T)$ is the standard enthalpy change or standard heat of the reaction at temperature T.

In choosing a reference state, we are allowed to make a choice for each element, because elements cannot be transformed into each other by chemical means. The choice usually made for the reference state of an element is the form in which it is stable at temperature T and the standard pressure $= 1.0$ bar. For example, at most temperatures, for O_2 this would be gaseous diatomic molecules; for iron, it would be the solid metal, and for bromine, it would be the diatomic in the liquid state below 59°C and in the gaseous state above 59°C. We call the standard enthalpy change of the reaction in which 1 mol of compound i is formed from its component elements in their reference states the *heat of formation* of compound i, $\Delta_f H_i^\circ(T)$. The heat of reaction is related to heats of formation as

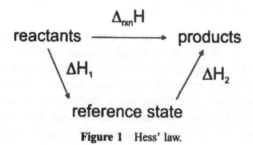

Figure 1   Hess' law.

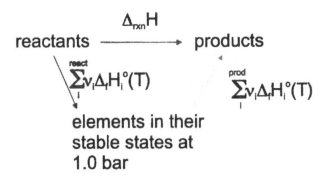

Figure 2 Using heats of formation.

shown in Fig. 2. (Note that v_i for the reactants are negative numbers.) Because enthalpy is a state function,

$$\Delta_{rxn}H°(T) = \sum_i v_i \Delta_f H_i°(T) \tag{14}$$

where the sum extends over both reactants and products. In Appendix B, heats of formation are tabulated for a variety of compounds at 298 K, which allows the heat of reaction to be calculated for many reactions at 298 K by Eq. (14). The entries for the heats of formations of elements in this table are, of course, zero, because the reaction to form them from the elements are null reactions (reactions in which nothing happens). Note the efficiency of tabulating thermodynamic data in this form. For n compounds, each of which can react with any other, roughly n^2 heats of reaction can be calculated. Care must be taken in using the data that values are chosen for compounds in the state of interest. This is illustrated in Example 1.

Example 1. Calculate $\Delta_{rxn}H°(298.15\,K)$ of the reaction

$$CH_4 + 2O_2 \rightarrow CO_2 + 2H_2O$$

Solution: The solution depends on whether the water is formed in the liquid or gaseous state. Gaseous water will result if water is formed with a partial pressure less than its vapor pressure of 23.7 torr at 298.15 K. In this case,

$$CH_4(g) + 2O_2(g) \rightarrow CO_2(g) + 2H_2O(g)$$

$$\Delta_{rxn}H°(298.15) = 2(-241.8) + (-393.5) - [(-74.8) + 2(0)]$$
$$= -802.3 \text{ kJ/mol}$$

"mol" in the result implies that it is for the reaction *as written* in molar quantities. If the water is formed in the liquid state,[2] we have

$$CH_2(g) + 2O_2(g) \rightarrow CO_2(g) + 2H_2O(l)$$

$$\Delta_{rxn}H°(298.15) = 2(-285.8) + (-393.5) - [(-74.8) + 2(0)]$$
$$= -890.3 \text{ kJ/mol}$$

Appendix B also includes values for $S°(298.15)$ and $\Delta_f G°(298.15)$, from which entropies of reaction and Gibbs free energies of reactions can be calculated at 298.15 K from

$$\Delta_{rxn}S°(T) = \sum_i v_i S_i°(T) \tag{15}$$

and

$$\Delta_{rxn}G°(T) = \sum_i v_i \Delta_f G°(T) \tag{16}$$

Because absolute entropies, rather than entropies of formation, are tabulated, the entries for entropies of elements are not zero and must not be neglected in using Eq. (15).

In most tabulations of thermochemical data, such as that of Appendix B, values for the thermodynamic functions are given only at a single temperature, usually 298.15K.[3] In order to convert standard heats of reaction to other temperatures, temperature variations at constant pressure must be considered. From Eq. (9),

$$\left(\frac{\partial \Delta_{rxn}H°}{\partial T}\right)_P = \sum_i v_i \left(\frac{\partial H_{m,i}°}{\partial T}\right)_P = \sum_i v_i C_{P,m,i}° \equiv \Delta_{rxn}C_p° \tag{17}$$

where $\Delta_{rxn}C_p°$ is the standard heat capacity change for the reaction. Integration gives

$$\Delta_{rxn}H°(T) = \Delta_{rxn}H°(298.15) + \int_{298.15}^{T} \Delta_{rxn}C_p° \, dT \tag{18}$$

Temperature corrections are usually not too large, due to partial cancellation of the heat capacities of the reactants and products. If the temperature of interest is not too different from 298.15 K, it is usually sufficient to use temperature-independent heat capacities, such as those obtained from Appendix B, and remove $\Delta_{rxn}C_p°$ from under the integral sign.

Example 2. Calculate the heat of the reaction at 50°C for

$$CH_4(g) + 2O_2(g) \rightarrow CO_2(g) + 2H_2O(g)$$

Solution: Since we only have to make a temperature correction over a range of 25 K, we can consider the heat capacities to be constant at the values given in Appendix B:

$$\Delta_{rxn}C_p^\circ = 37.11 + 2(33.58) - [35.31 + 2(29.36)] = 10.44 \text{ J/K mol}$$

Using the value of $\Delta_{rxn}H^\circ(298.15)$ from Example 1,

$$\Delta_{rxn}H^\circ(323.15) = \Delta_{rxn}H^\circ(298.15) + (25 \text{ K})\Delta_{rxn}C_p^\circ$$

$$= -802.3 \frac{\text{kJ}}{\text{mol}} + 10.44 \frac{\text{J}}{\text{K mol}} (25 \text{ K})\frac{\text{kJ}}{1000 \text{ J}} = -802.0 \frac{\text{kJ}}{\text{mol}}$$

A common error is to omit the final transformation from joules to kilojoules.

Entropies can be transformed to temperatures other than 298.15 K using Eq. (17) of Chapter 3:

$$\Delta_{rxn}S^\circ(T) = \Delta_{rxn}S^\circ(298.15) + \int_{298.15}^{T} \frac{\Delta_{rxn}C_p^\circ}{T} dT \tag{19}$$

or

$$\Delta_{rxn}S^\circ(T) = \Delta_{rxn}S^\circ(298.15) + \Delta_{rxn}C_p^\circ \ln\left(\frac{T}{298.15}\right) \tag{20}$$

if a temperature-independent $\Delta_{rxn}C_p^\circ$ can be used.

Gibbs free energies can be calculated as

$$\Delta_{rxn}G^\circ(T) = \Delta_{rxn}H^\circ(T) - T\Delta_{rxn}S^\circ(T) \tag{21}$$

A common error is to use $\Delta_{rxn}H^\circ(298.15)$ and $\Delta_{rxn}S^\circ(298.15)$ in Eq. (21) for calculating $\Delta_{rxn}G^\circ(T)$ at a temperature other than 298.15 K.

7.3 Calorimetry

Calorimetry is the determination of thermodynamic properties (enthalpies, free energies, heat capacities, etc.) by measuring the energy released or absorbed in a process. Usually, this is achieved by carrying out the process *adiabatically* (i.e., in a system to which negligible heat is transferred during the measurement). For systems of constant volume, no work is done, and the temperature change is directly related to the change in U of the system.

A diagram of a constant-volume adiabatic calorimeter, called a bomb calorimeter, used for measuring the ΔU of combustion reactions is shown in Fig. 3.

Adiabaticity is achieved by insulating the calorimeter and studying reactions that occur over a short period of time (because time is required for heat transfer). It is also useful to be able to control the starting time of the reaction and study reactions that go to completion, so that the extent of the reaction does not

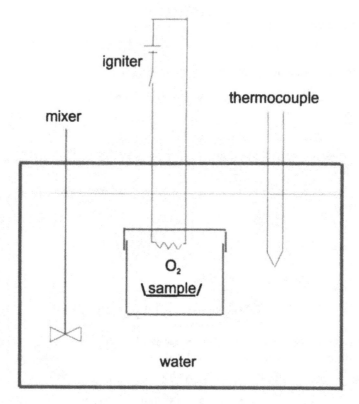

Figure 3   Adiabatic bomb calorimeter.

have to be experimentally determined. Combustion reactions are ideal for study in this manner. They proceed at a negligible rate, until *ignited* by an externally controlled ignition source. When initially pressurized with excess oxygen, the reaction proceeds almost instantaneously, converting hydrocarbons to CO_2 and H_2O (which is almost entirely in the liquid state at the final temperature of the calorimeter). Sulfur is converted to SO_2 and nitrogen to N_2 (and a little nitric acid, which may have to be titrated and corrected for). The calorimetry experiment involves igniting the mixture and measuring the increase in temperature, ΔT, of the calorimeter. Because the calorimeter is adiabatic and does no work, we can write, for an infinitesmal amount of reaction,

$$dU = 0 = U_{\text{prod}+\text{cal}}(T_i + dT) - U_{\text{react}+\text{cal}}(T_i)$$

$$= d_{rxn}U(T_i) + C\,dT \tag{22}$$

$$d_{rxn}U(T_i) = -C\,dT \tag{23}$$

where C is the heat capacity of the calorimeter and the products of the reaction. Measurements are usually performed with temperature changes small enough that heat capacities can be considered constant; thus, we can write

$$\Delta_{rxn} U = -C\Delta T \tag{24}$$

where $\Delta_{rxn} U$ is the internal energy change of the reaction. C can be determined by using a resistor to add a known amount of energy (I^2Rt) to the calorimeter and measuring its temperature increase, or by running the calorimeter with a measured amount of a compound of known heat of combustion. Benzoic acid is convenient for calibration. C is almost all due to the heat capacity of the metallic parts and the water of the calorimeter. In practice, the energy added for ignition and for mixing must be taken into account.

The difference between the heat (enthalpy change) of a reaction and $\Delta_{rxn} U$ results from the work that must be done in moving the atmosphere to make room for gas that is formed in the reaction. We can write

$$\Delta_{rxn} H = \Delta_{rxn} U + \Delta_{rxn}(PV) = \Delta_{rxn} U + RT\Delta_{rxn} n_{gas} \tag{25}$$

where $\Delta_{rxn} n_{gas}$ is the change in the number of moles of gas in the reaction.

An alternative way of determining some thermodynamic functions is by measuring the position and temperature shift of equilibrium. The relation between thermodynamics and the equilibrium of chemical reactions will be explored later in this chapter.

7.4 Estimating the Thermodynamics of Reactions

An alternative reference state from which to tabulate thermodynamic functions is that of isolated atoms. The use of this reference state is shown in Fig. 4, where $\Delta_{bd} H_i$ is a bond dissociation enthalpy in a reactant or product molecule. (Remember that the v_i's are negative for reactants.) The heat of reaction is then given by

$$\Delta_{rxn} H = -\sum_i v_i \Delta_{bd} H_i \tag{26}$$

This equation is for reactions involving only gaseous reactants and products. If some species are in condensed phases, appropriate heats of vaporization of sublimation must be included.

Compared to Eq. (14), the negative sign in Eq. (26) results from bond dissociation energies being enthalpies required to pull molecules apart. Bond dissociation energies are positive. Larger bond dissociation energies of product molecules than reactant molecules are characteristic of exothermic reactions (negative $\Delta_{rxn} H$). The calculation indicated by Fig. 4 is as exact as that indicated

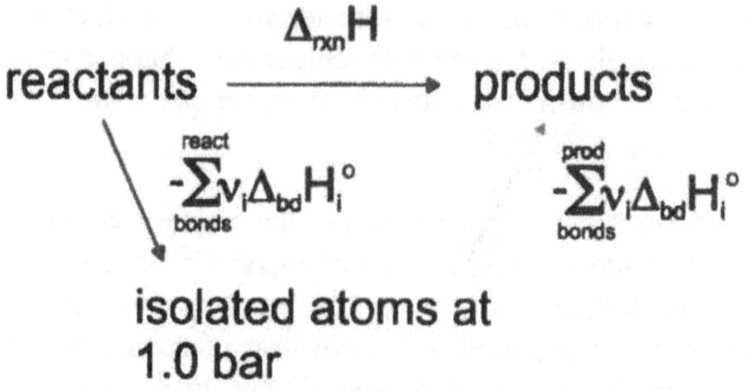

$$\text{reactants} \xrightarrow{\Delta_{rxn}H} \text{products}$$

$$-\sum_{\substack{bonds}}^{react} \nu_i \Delta_{bd}H_i^{\circ} \qquad -\sum_{\substack{bonds}}^{prod} \nu_i \Delta_{bd}H_i^{\circ}$$

isolated atoms at 1.0 bar

Figure 4   Using bond dissociation energies.

by Fig. 2. However, in order to achieve comparable accuracy in calculation, bond dissociation enthalpies specific for the bonds in each particular reactant and product molecule are required. These quantities are not generally available and would require values for every bond in each molecule, a much more difficult task to measure and tabulate than just the ΔH_f of molecules.

Equation (26) is often used with *generic bond dissociation enthalpies*, assuming that it takes the same energy to break the same type (single, double, or triple) bond between the same two atoms. With this assumption, a further simplification in tabulating thermodynamic data is achieved: The heats of a huge number of reactions can be calculated with the few quantities listed in Table 1. Unfortunately, the strength of similar bonds do vary in different molecules, and large errors can be made in using generic bond dissociation enthalpies, as illustrated in Example 3.

Example 3.   Calculate $\Delta_{rxn}H$ of the gaseous reaction

$$2CH_4 \rightarrow C_2H_4 + 2H_2$$

from heats of formation and from generic bond dissociation enthalpies.

TABLE 1   Generic Bond Dissociation Enthalpies (kJ/mol)

H–H	436	O=O	498	C=C	615	N≡N	946
C–H	415	C–O	350	C≡N	890	N–N	159
C–C	344	C=O	725	O–H	463	N=N	418
C=C	615	C–N	292	N–H	391	C–F	441
C≡C	812	C=N	615	O–O	143	C–Cl	328

Solution: With heats of formation taken from Appendix C,

$$\Delta_{rxn}H = 52.26 - 2(-74.81) = 201.88 \text{ kJ/mol}$$

with generic bond dissociation enthalpies,

$$\Delta_{rxn}H = 8(415) - [2(436) + 615 + 4(415)] = 173 \text{ kJ/mol}$$

Obviously, in this example, there is a very significant difference.

Perhaps recognizing that generic bond dissociation enthalpies are so approximate that corrections to energies are unwarranted, the quantities in Table 1 are often called *generic bond dissociation energies* or just *bond dissociation energies*. Reactions that form strong bonds, such as those between C and O or between H and O, release considerable energy in forming product molecules and are likely to be strongly exothermic. Combustion processes are the prototypes of such reactions.

A considerable improvement in the accuracy of estimating thermodynamic quantities is achieved by the use of the method of *group additivities*,[4] which include the effect of nearest-neighbor interactions. In this method, thermodynamic quantities are decomposed into contributions from multibonded atoms (called groups). Each of these atoms is identified by its bonding type (e.g., single-, double-, or triple-bonded carbon atoms) and the identity of the atoms to which it is bonded, which is written in parentheses. For example, $C_d-(C)(H)$ refers to a doubly-bonded carbon, attached to a non-multiply-bonded carbon (sp^3) and a hydrogen. The method will be illustrated for calculation of $\Delta_f H°$ of some hydrocarbons, using the group properties in Table 2. However, group properties for the calculation of all thermodynamic properties of a wide variety of

TABLE 2 Group Values for $\Delta_f H°$ (kJ/mol) of Nonaromatic Hydrocarbons

Group	Value	Group	Value
$C-(H)_3(C)$	−42.68	$C-(C_d)(C)(H)_2$	−19.92
$C-(H)_2(C)_2$	−20.63	$C-(C_d)_2(H)_2$	−7.95
$C-(H)(C)_3$	−7.95	$C-(C_t)(C)(H)_2$	−19.79
$C-(C)_4$	2.09	$C-(C_d)(C)_2(H)$	−6.19
$C_d-(H)_2$	26.19	$C-(C_t)(C)_2(H)$	−7.20
$C_d-(H)(C)$	35.94	$C-(C_d)(C)_2$	7.03
$C_d-(C)_2$	43.26	$C_t-(H)$	112.68
$C_d-(C_d)(H)$	28.37	$C_t-(C)$	115.27
$C_d-(C_d)(C)$	37.15	$C_t-(C_d)$	122.17
$C_d-(C_t)(H)$	28.37	C_a	143.1
$C_d-(C_d)_2$	19.25		

Note: C_d = double-bonded carbon; C_t = triple-bonded carbon; C_a = allenic carbon. Corrections are also provided for non-nearest-neighbor interactions, such as for cis isomers, gauche conformers, and ring systems. (See Note 4.)

compounds are available. The reader is referred to Benson's book for more detailed use of the method. The accuracy of the method is demonstrated by the following example.

Example 4.   Use the group additivity method to estimate the heats of formation of isobutane and 1-butene. Compare your result with the literature values of 134.7 kJ/mol and −0.5 kJ/mol, respectively.

Solution: Isobutane: The contributing groups are $3[C-(C)(H)_3]$ + $C-(C)_3(H)$:

$$\Delta_f H^\circ = 3(-42.68) - 7.95 = 136.0 \text{ kJ/mol}$$

1-Butene: The contributing groups are $C_d-(H)_2 + C_d H(C)$ + $C-(H)_2(C)_2 + C-(H)_3(C)$:

$$\Delta_f H^\circ = 26.19 + 35.94 - 20.63 - 42.68 = -1.18 \text{ kJ/mol}$$

The agreement with the literature values is quite impressive.

7.5   Chemical Equilibrium

We have considered the change of properties of a system in which a chemical reaction occurs with an extent of reaction, ξ. We also need ways to determine just what extent of reaction will occur in a particular system. In many cases, this question will be answered by stoichiometry, namely by assuming that the reaction will proceed until *essentially* all the limiting reagent is converted into product, a situation that is called a reaction *going to completion*.

There are two reasons for reactions not going to completion. The first is that they may be kinetically inhibited, as illustrated by a mixture of H_2 and O_2 gas, which is stable at ambient conditions. We know that such a mixture has a *tendency* to form H_2O, as can be observed by igniting the system with a spark. Without the spark, however, the system reacts at negligible rate. Rates of chemical reactions are the subject of chemical kinetics, which will not be dealt with in this book.

The second reason for a reaction not going to completion is that it proceeds to a state of material equilibrium in which both reactants and products exist. This is the case for all gas-phase reactions, because free energy minimizes in gas mixtures when some of each component is present. (See Problem 16 in Chapter 4.) In fact, even in the case of the spark-initiated H_2-O_2 reaction discussed earlier, some H_2 and O_2 will remain after the reaction. Equilibrium is more evident with the gaseous molecule NO_2, a fraction of which exists as the dimer, N_2O_4, near ambient conditions. This fraction depends on the temperature and pressure and rapidly adjusts to changes in these variables, indicating that there is no kinetic inhibition in this system. The system is described by an *equilibrium constant, K(T)*. Because, once equilibrium is established, there is no further change in the concentrations in the system, the rates of the forward and reverse reaction must be equal at equilibrium, and $K(T)$ can be obtained from this condition. However, equilibrium constants can also be calculated from thermodynamic data, by defining the equilibrium condition as that in which there is no tendency for change. This latter approach will be the subject of this section.

At equilibrium, the change in any thermodynamic property resulting from an infinitesimal change in the extent of a reaction, $d\xi$, can be calculated by the two different paths shown in Fig. 5.

Using the internal energy as an example, for the direct reaction path, no material enters or leaves the system and it may be considered closed. By the first and second laws, for a closed system at (mechanical, thermal, and material) equilibrium, performing no non-PV work, we have

$$dU = \delta q + \delta w \stackrel{\text{m.e.}}{=} T\,dS - P\,dV \tag{27}$$

The system can also undergo the same change by allowing $dn_i = v_i\,d\xi$ of each reactant or product to pass into the system. (Because the v_i are negative for the reactants, the reactants actually pass out of the system.) The system must be open for this to occur, and its change in internal energy is given by Eq. (4) of Chapter 6:

$$dU = T\,dS - P\,dV + \sum_i \mu_i\,dn_i \tag{28}$$

These two expressions describe the same change of state of the system and must be equal, requiring that

$$\sum_i \mu_i\,dn_i = \sum_i \mu_i v_i\,d\xi \stackrel{\text{m.e.}}{=} 0 \tag{29}$$

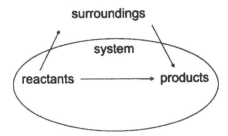

Figure 5   Reaction at equilibrium.

Because $d\xi$ may be brought outside the summation sign and is not zero, this gives as the general condition for chemical equilibrium:

$$\sum_i v_i \mu_i = \Delta_{rxn} G \overset{\text{m.e.}}{=} 0 \tag{30}$$

At equilibrium, a condition such as Eq. (30) must hold for each *independent* reaction that can occur in a system (i.e., for every reaction that cannot be written as a linear combination of other reactions).

7.6 Direction of Chemical Reactions

From Eq. (7) of Chapter 6, for a chemical reaction occurring in a system at constant T and P,

$$dG = \sum_i \mu_i \, dn_i = \left(\sum_i \mu_i v_i \right) d\xi = \Delta_{rxn} G \, d\xi \tag{31}$$

Because processes will be spontaneous in closed systems at constant T and P if they lower the Gibbs free energy ($dG/d\xi < 0$), we see that the criterion for a spontaneous reaction in a system at constant T and P is

$$\sum_i v_i \mu_i = \Delta_{rxn} G < 0 \tag{32}$$

The quantity $A \equiv -\sum_i v_i \mu_i = -\Delta_{rxn} G$ is often called the *affinity* of the chemical reaction. It is the thermodynamic driving force of the reaction.

In a similar manner, a process occurring at constant T and V will be spontaneous if it lowers the Helmholtz free energy, which from Eq. (6) of Chapter 6 also gives $\sum_i \mu_i dn_i < 0$ and Eq. (32). In chapter 10, it is shown that the criterion for spontaneity, Eq. (32), is perfectly general and must hold, regardless of the conditions under which the reaction is conducted. The Gibbs free energy of reaction always decreases in a spontaneous chemical reaction.

7.7 Concentration Dependence of Free Energy

Combining Eq. (11) with Eq. (47) of Chapter 6, we have

$$\Delta_{rxn} G = \sum_i v_i (\mu_i^\circ + RT \ln a_i) = \sum_i v_i \mu_i^\circ + RT \sum_i \ln a_i^{v_i}$$
$$= \Delta_{rxn} G^\circ + RT \ln \prod_i a_i^{v_i} \tag{33}$$

where the ln of a product has been substituted for the sum of ln's. $\Delta_{rxn}G°$ is the Gibbs free-energy change of the reaction under standard condition (the *standard free-energy change of the reaction*), and we define the product as

$$Q_a \equiv \prod_i a_i^{\nu_i} \tag{34}$$

where Q_a is the *proper quotient of activity coefficients for the reaction*. For example, for the reaction: $2CO + O_2 \leftrightarrow 2CO_2$, $Q_a = a_{CO_2}^2/a_{CO}^2 a_{O_2}$. The concentration dependence of activities thus determines $\Delta_{rxn}G$ by the equation

$$\Delta_{rxn}G = \Delta_{rxn}G° + RT \ln Q_a \tag{35}$$

Example 5. For the reaction: $9H_2(g) + C_6H_6(liq) \leftrightarrow 6CH_4(g)$:

(a) Calculate the standard enthalpy change and the standard Gibbs free energy change at 25°C.

(b) Calculate ΔG for the reaction at 25°C when the partial pressure of hydrogen is 0.1 bar and the partial pressure of CH_4 is 50 bar.

Solution:

(a)
$$\Delta H°(298) = 6(-74.81) - (49.1) = -498.0 \text{ kJ/mol}$$
$$\Delta G°(298) = 6(-50.72) - (124.5) = -428.8 \text{ kJ/mol}$$

(b) The ideal gas approximation is very good at 298 K and the pressures of this problem. Activities of CH_4 and H_2 can therefore be taken as their pressures in bar. The activity of liquid benzene can be taken as 1.0. This will be discussed in detail in Section 7.8.1.

$$Q_a = \frac{p_{CH_4}^6}{p_{H_2}^9} = \frac{(50)^6}{(0.1)^9} = 1.56 \times 10^{19}$$

$$\Delta G = \Delta G° + RT \ln Q_a = -428.8 \text{ kJ/mol}$$
$$+ \frac{8.314 \text{ J}}{\text{mol K}}(298 \text{ K})\frac{\text{kJ}}{1000 \text{ J}} \ln(1.56 \times 10^{19})$$
$$= -428.8 + 109.5 = -319.3 \text{ kJ/mol}$$

7.8 Equilibrium Constants

From Eq. (30), at equilibrium, $\Delta_{rxn}G = 0$. (There is no driving force for the reaction in either direction.) We have from Eq. (35), at equilibrium,

$$\Delta_{rxn}G° = -RT \ln K_a \tag{36}$$

where K_a, the *activity equilibrium constant*, is the value of Q_a at equilibrium. It is the product of the equilibrium activities of reactants and products, each raised to their appropriate stoichiometric coefficient. Because the stoichiometric coefficients of the reactants are negative, the reactant terms can be written in the denominator with positive coefficients, as is more commonly done for equilibrium constants. Equation (36) can also be written as

$$K_a = \exp\left(-\frac{\Delta_{rxn}G^\circ}{RT}\right) = \exp\left(\frac{\Delta_{rxn}S^\circ}{R}\right) \exp\left(-\frac{\Delta_{rxn}H^\circ}{RT}\right) \tag{37}$$

which shows that reactions that increase entropy and decrease enthalpy have large equilibrium constants and tend to go to completion.

For equilibria involving ideal gases, whose activities are P/P°, or just the numerical value of the pressure (in bars), we can write

$$\Delta_{rxn}G^\circ = -RT \ln \prod_i \left(\frac{P_i}{P^\circ}\right)^{v_i} = -RT \ln K_P \tag{38}$$

where K_P is an equilibrium constant involving the numerical values of pressures (in bars) (or to a very good approximation, in atmospheres). In Eq. (38), different pressures are used on the left-hand and right-hand sides of the equation. $\Delta_{rxn}G^\circ$ is calculated with all reactants and products at 1.0 bar, the standard pressure at which we choose to tabulate thermodynamic functions. The equilibrium constant, of course, generally involves partial pressures different from 1.0 bar. Although this often creates some confusion, it results from converting free energies from the standard pressure to the equilibrium partial pressures. The free-energy change at the equilibrium pressures does not enter into Eq. (38) because it is zero.

Also, note that because the left-hand side of Eq. (38) is independent of the total pressure of the system, the equilibrium constant must also be pressure independent in the ideal gas regime. This does not mean, however, that the partial pressures of individual components are pressure independent, as shall be seen when we do some calculations with equilibrium constants.

For real gases, we can use Eq. (51) of Chapter 6 to write

$$\Delta_{rxn}G^\circ = -RT \ln \prod_i \left(\frac{f_i}{P^\circ}\right)^{v_i} = -RT \ln K_f \tag{39}$$

where K_f is an equilibrium constant involving the numerical values of fugacity (in bars). Alternatively, using Eq. (55) of Chapter 6, we can write

$$\Delta_{rxn}G^\circ = -RT \ln \prod_i \left(\frac{\phi_i P_i}{P^\circ}\right)^{v_i} = -RT \ln K_P K_\phi \tag{40}$$

where K_ϕ is an equilibrium constant formed from fugacity coefficients. If conditions can be found where measurable amounts of reactants and products coexist at equilibrium, Eqs. (38)–(40) provide a means to determine $\Delta_{rxn}G°$ without making any calorimetric measurements.

In order to avoid confusion, it is very important to clearly state the balanced chemical reaction to which a given equilibrium constant applies. This is illustrated in Example 6.

Example 6. Write down the form of the equilibrium constant that applies to the Haber process for the production of ammonia.

Solution: The Haber process is one of the most important industrial chemical reactions. The reaction describing the process can be written as

$$\tfrac{1}{2}N_2 + \tfrac{3}{2}H_2 \leftrightarrow NH_3 \qquad\qquad\qquad (I)$$

or alternatively as

$$N_2 + 3H_2 \leftrightarrow 2NH_3 \qquad\qquad\qquad (II)$$

The form of K_a for reaction I is $K_a(I) = a_{NH_3}/a_{N_2}^{1/2}a_{H_2}^{3/2}$, whereas that for reaction II is $K_a(II) = a_{NH_3}^2/a_{N_2}a_{H_2}^3 = [K(I)]^2$. (When you double the reaction, you square the equilibrium constant.) In terms of fugacities (which would be used at the high pressures at which the Haber process is carried out), the rate constant is

$$K_f(I) = \frac{f_{NH_3}P°}{f_{N_2}^{1/2}f_{H_2}^{3/2}} \quad \text{and} \quad K_f(II) = \frac{f_{NH_3}^2 P°^2}{f_{N_2}f_{H_2}^3}$$

Both of these forms are in agreement with Eq. (38), because $\Delta_{rxn}G°(II)$ is twice $\Delta_{rxn}G°(I)$.

7.8.1 Equilibria Involving Condensed Phases

Because, at constant temperature, $dG_m = V_m\,dP$ and the molar volumes of condensed phases are very small, it is usually sufficiently accurate to take their molar free energy as pressure independent and the same as that at the 1.0-bar standard state. This is equivalent to setting the activity of pure, condensed phases equal to unity. (See Problem 9.) The activity of a condensed phase is also independent of just how much of the phase is present. As a result of these considerations, no variable describing the condensed phase appears in the equilibrium constant and the equilibrium is independent of just how much condensed phase is present.

Because the activity of condensed phases is not appreciably affected by the amount of the condensed phase or the pressure, equilibria involving condensed phases are achieved by variation of gas-phase concentrations. For example, in the equilibrium $NH_4Cl_{(solid)} \leftrightarrow NH_{3(gas)} + HCl_{(gas)}$, if the amounts of NH_3 and HCl are too low for $a_{NH_3} a_{HCl} = K_a = \exp(-\Delta G^\circ / RT)$, the equilibrium will not hold and there will be no solid NH_4Cl in the system.

Example 7. Show the form of Eq. (36) for the reaction

$$CaCO_3(\text{solid}) \leftrightarrow CaO(\text{solid}) + CO_2(\text{gas})$$

Solution: Because $v_{CaCO_3} = -1$, $v_{CaO} = 1$, and $v_{CO_2} = 1$, $K_a = a_{CO_2} a_{CaO}/a_{CaCO_3}$, CaO and $CaCO_3$ are condensed phases; their activities are unity and they do not enter into the equilibrium expression (as long as some of these materials are present). The activity of CO_2 is $a_{CO_2} = f_{CO_2}/P^\circ$, which, at pressures where CO_2 behaves ideally, is just the partial pressure of CO_2 in units of bar. Equation (36) then becomes

$$\Delta_{rxn}G^\circ = -RT \, \ln\left(\frac{P_{CO_2}}{P^\circ}\right) \quad \text{or} \quad \frac{P_{CO_2}}{P^\circ} = \exp\left(-\frac{\Delta_{rxn}G^\circ}{RT}\right)$$

7.9 General Considerations Involving Multicomponent and Multiphase Equilibrium: The Gibbs Phase Rule

A *component* is a particular chemical substance. A chemical reaction at equilibrium is therefore a *multicomponent system*. In discussing the equilibria of multicomponent systems, it is important to be aware of what constitutes a complete and sufficient specification of the system. If too few variables are specified, the system is not completely defined; if too many are specified, the values of these variables may be inconsistent with equilibrium. It is possible to waste much time considering improperly specified systems.

The *Gibbs phase rule* allows f, the number of *degrees of freedom* of a system, to be determined. f is the number of intensive variables that can and must be specified to define the *intensive state* of a system at equilibrium. By intensive state is meant the properties of all phases in the system, but not the amounts of these phases. Phase equilibria are determined by chemical potentials, and chemical potentials are intensive properties, which are independent of the amount of the phase that is present. The overall concentration of a system consisting of several phases, however, is not a degree of freedom, because it depends on the amounts of the phases, as well as their concentration. In addition to the intensive variables, we are, in general, allowed to specify one extensive variable for each phase in the system, corresponding to the amount of that phase present.

We will consider the number of degrees of freedom, f, of a system consisting of c components and p phases. Two intensive variables, T and P, are needed to specify the state of thermal and mechanical equilibrium of the system. In order to determine the state of material equilibrium of the system, it is necessary to know the chemical potential of each component of every phase. At a given temperature and pressure, chemical potentials are functions of the concentrations in the phase, so we can consider specifying the $c \times p$ concentrations of each component in every phase. However, this value for the degrees of freedom, $c \times p + 2$, is too large. First, the concentrations of the c components of each phase cannot be independently specified, because there is a relation between them, most easily written as the sum of the mole fractions in each phase being unity [Eq. (44) of Chapter 1]. This reduces the number of degrees of freedom by the number of phases, p. Second, we have seen that the general condition for phase equilibrium is that the chemical potential of each component be the same in every phase. For example, if there are three phases, we have

$$\mu_{i\alpha} = \mu_{i\beta}$$

$$\mu_{i\beta} = \mu_{i\gamma} \tag{41}$$

Note that these two restrictions mathematically require that $\mu_{i\alpha} = \mu_{i\gamma}$, which is, therefore, not an independent restriction.

In general, the number of independent restricting equations for phase equilibrium will be one less than the number of phases for each component. The degrees of freedom of the system then become

$$f = c \times p + 2 - p - c \times (p - 1) = c - p + 2 \tag{42}$$

This relation is called the *Gibbs phase rule*. However, as indicated next, c should be limited to the number of *independent components*.

Chemical reactions reduce the number of degrees of freedom of a system. For each independent chemical reaction that can occur in a system, $\Delta_{rxn}G = 0$ is required. If r is the number of independent reactions at equilibrium in the system, this reduces the degrees of freedom by r.

Other restrictions on the system may result from stoichiometric considerations. A very common stoichiometric restriction occurs in ionic solutions where *charge neutrality* must hold. Other restrictions can result if two species are produced only by the decomposition of the same molecule. If the number of stoichiometric restrictions is a, the degrees of freedom of the system become

$$f = c - p + 2 - r - a \tag{43}$$

Usually, the combination $c - r - a$ is called the *number of independent components*, c_{ind}, and the phase rule is written as

$$f = c_{ind} - p + 2 \tag{44}$$

For one-component systems, such as have been considered in Chapter 6, Eq. (44) becomes

$$f = 3 - p \tag{45}$$

Thus, temperature and pressure can both be varied while remaining in a region of a single phase, whereas only temperature or pressure can be varied (e.g., on a vapor–pressure curve) while retaining equilibrium between two phases. The triple point is completely invariant. More interesting applications of the phase rule are obtained with multicomponent systems, as indicated in the following examples.

Example 8.   Ethanol–water.

Discussion: Ethanol does not react with water, so the two components are independent: $c_{ind} = 2$. They are completely miscible, so a single liquid phase exists, which, when in equilibrium with vapor, gives $p = 2$. From Eq. (44), there are two degrees of freedom, usually taken as the temperature and composition, which determine the vapor pressure of the solution. If the solution is exposed to the atmosphere, air is present and is a third component. There are now three degrees of freedom and the pressure is that of the atmosphere.

Example 9.   Sodium chloride–water.

Discussion: There are two ways of treating this system and they should give identical results. If we consider NaCl as a single un-ionized substance, the system has two independent components, which form a solution (assuming that we have not exceeded the solubility limit). There is, thus, two degrees of freedom, just as in the ethanol–water system. Alternatively, we can treat NaCl as an electrolyte, which is completely ionized in solution. Then, there are actually three components, Na^+, Cl^-, and H_2O in the system. However, charge neutrality requires that $[Na^+] = [Cl^-]$, which reduced c_{ind} to 2, giving the same result in this example. If more NaCl is added to the system than will dissolve, a third phase, solid NaCl, appears. The system then has a single degree of freedom, and fixing the temperature determines the concentration of the solution and the vapor pressure of water above the solution.

Example 10.   Acetic acid–water.

Discussion: Acetic acid is soluble in water and partially ionized, resulting in four components: HAc, Ac^-, H^+, and H_2O. However, now there is a reaction, $HAc \leftrightarrow H^+ + Ac^-$, which gives an equilibrium relationship, $K_a = a_{H^+} a_{Ac^-} / a_{HAc}$, as well as the charge neutrality condition. $c_{ind} = 2$, yielding $f = 2$ for the solution in equilibrium with its vapor, the same results as for ethanol–water.

Example 11. Silver chloride–water.

Discussion: The components are H_2O, AgCl, Ag^+ and Cl^-, but the solubility equilibrium of AgCl and charge neutrality reduces c_{ind} to 2. Because AgCl is only slightly soluble, in equilibrium with water vapor this system has three phases and, therefore, only a single degree of freedom. Specification of the temperature, for example, determines all intensive variables in the system (but not the amounts of the phases). Note that considering H^+ and OH^- from the ionization of water introduces two additional components, but also two addition constraints: the ionization equilibrium of water and the requirement that $[H^+] = [OH^-]$, so that c_{ind} and f are unchanged.

Example 12. $CaCO_3 - CaO - CO_2$

Discussion: At high temperature, equilibrium is established among these three components, so this system has $c_{ind} = 3 - 1 = 2$. Usually, crystalline solids exist as pure components, so there will be two solid and one gaseous phase present. With $p = 3, f = 1$. Specification of the temperature will therefore determine the pressure of CO_2, but not its volume, nor the amounts of the two solid phases.

It might occur that CaO and $CaCO_3$ form a solid solution or a mixture in which the particles are so small that it is preferred to consider them as constituting a single solid phase. The system then has two phases and two degrees of freedom. The additional degree of freedom is now the concentration of the solid phase, which is analogous to the relative amounts of the two solid phases when they are considered as separate phases. We are free to set either of these quantities.

In general, the calculated degrees of freedom will be independent of decisions that we make regarding just what substances are dissociated, or whether solids are single or mixed phases.

7.10 Concentrations at Equilibrium

Although free energies are related to equilibrium constants, it is concentrations that are either ultimately desired or experimentally determined. There are many ways to perform equilibrium calculations. We will approach such calculations through the extent of reaction, ξ, defined in Eq. (5), because all concentrations can be expressed in terms of this single variable. In addition, use of the extent of the reaction allows us to perform equilibrium calculations in a very systematic manner. At equilibrium, the extent of reaction becomes ξ_e, the extent of reaction at equilibrium. An intermediate step in the calculation is expressing the mole fractions, x_i, in the gas in terms of ξ_e. The activities of condensed phase species can almost always be taken as unity; therefore, these species do not enter into the equilibrium constant. In terms of mole fractions, the equilibrium constant may be written as

$$K_P(T) = \prod_{\text{gas}} \left(\frac{P_j}{P^o}\right)^{v_j} = \prod_{\text{gas}} \left(\frac{x_j P}{P^o}\right)^{v_j} = \prod_{\text{gas}} \left(\frac{P}{P^o}\right)^{v_j} \prod_{\text{gas}} x_j^{v_j} = \left(\frac{P}{P^o}\right)^{\Delta v_{gas}} K_x(T, P)$$

(46)

Δv_{gas} is the change in the number of moles of gas in the reaction and K_x is the equilibrium constant in terms of mole fractions of gaseous reactants and products, which can be seen to be a function of both temperature and pressure:

$$K_x(T, P) = K_P(T) \left(\frac{P}{P^o}\right)^{-\Delta v_{gas}}$$

(47)

For a reaction with positive gas mole change, Eq. (47) indicates that K_x decreases with pressure. Because ξ_e is a monotonically increasing function of K_x, the equilibrium extent of a reaction with positive Δv_{gas} always decreases as pressure is increased. This is an example of Le Chatelier's principle, which states that a reaction at equilibrium shifts in response to a change in external conditions in a way that moderates the change. In this case, because the reaction increases the number of moles of gas and thus the pressure, the reaction shifts back to reactants. The isothermal compressibility of a reactive system can, therefore, be much greater than that of a nonreactive system. This effect can be dramatic in systems with condensed phases. For example, in the calcium carbonate dissociation discussed in Example 12, if the external pressure is raised above the dissociation pressure of CO_2, the system will compress down to the volume of the solid. Of course, a similar effect is observed in simple vaporization or sublimation equilibrium. As the pressure on water at 100°C is increased above 1.0 atm, all vapor is removed from the system.

The use of ξ_e in equilibrium calculations will be illustrated by a number of examples.

Example 13. Find the fraction of N_2O_4 that is dissociated into NO_2, as a function of the total pressure of the mixture of these two gases at 298 K.

Solution: N_2O_4 dissociates into NO_2 (the compound that sometimes gives the air over Los Angeles its brown color) by the equation

$$N_2O_4 \leftrightarrow 2NO_2$$

At 298 K, we have, from Appendix B,

$$\Delta G° = 2(51.31) - 97.89 = 4.73 \text{ kJ/mol}$$

$$K_P = \exp\left(-\frac{4.73 \text{ kJ}}{\text{mol}}\left|\frac{1000 \text{ J}}{\text{kJ}}\right|\frac{\text{mol K}}{8.314 \text{ J}}\left|\frac{1}{298 \text{ K}}\right.\right)$$

$$= \exp(-1.91) = 0.148$$

The system has two components that are related by one chemical reaction, and, therefore, $c_{\text{ind}} = 1$. There is a single phase and therefore two degrees of freedom. At a fixed temperature, 298 K, the single remaining degree of freedom can be taken as the pressure. We can also specify one extensive variable, which we choose to be n_{init}, the initial number of moles of N_2O_4. We choose $n_{\text{init}} = 1$, so that ξ_e is directly the fraction of N_2O_4 dissociated.

To find ξ_e, we construct a table giving the initial, reacting, and equilibrium number of moles, mole fractions, and partial pressures.

	N_2O_4	NO_2	Total
Initial	1	0	1
Reacting	$-\xi_e$	$+2\xi_e$	ξ_e
Equilibrium	$1 - \xi_e$	$2\xi_e$	$1 + \xi_e$
Mole fraction	$\dfrac{1 - \xi_e}{1 + \xi_e}$	$\dfrac{2\xi_e}{1 + \xi_e}$	1
Partial pressure	$\left(\dfrac{1 - \xi_e}{1 + \xi_e}\right)P$	$\left(\dfrac{2\xi_e}{1 + \xi_e}\right)P$	P

Assuming the pressure is low enough to use unit fugacity coefficients, the equilibrium constant is (with P in bars)

$$K_P = \left(\frac{2\xi_e}{1 + \xi_e}\right)^2 P^2 \left[\left(\frac{1 - \xi_e}{1 + \xi_e}\right)P\right]^{-1} = \frac{4\xi_e^2 P}{1 - \xi_e^2}$$

$$\xi_e = \sqrt{\frac{K_P}{4P + K_P}} = \sqrt{\frac{0.148}{4P + 0.148}}$$

P in this expression includes only the pressures of the "active" species N_2O_4 and NO_2 (see Problem 13). Even in the polluted air of Los Angeles, these are quite low, and N_2O_4 is essentially completely dissociated.

Example 14. A mixture is prepared at 298 K by opening the stopcock between two 500-mL flasks; one containing 600 torr of NO, the other 300 torr of Br_2. Due to the presence of photodissociating light, equilibrium is rapidly established between Br_2, NO, and NOBr. What are the equilibrium partial pressures of each of the compounds?

Solution: There are three components in a single phase. One chemical reaction, $2NO + Br_2 \leftrightarrow 2NOBr$, reduces the number of independent components to two. The number of degrees of freedom is $2 - 1 + 2 = 3$, one of which is the fixed temperature of 298 K. The other two can be taken as the ratio of the initial pressure of NO to that of Br_2 and the total final pressure. First, we calculate the equilibrium constant:

$$\Delta G° = 2(82.4) - 2(86.6) = -8.4 \text{ kJ/mol}$$

$$K_P = \exp\left(\frac{8.4 \text{ kJ}}{\text{mol}} \left| \frac{1000 \text{ J}}{\text{kJ}} \right| \frac{\text{mol K}}{8.314 \text{ J}} \left| \frac{1}{298 \text{ K}} \right. \right) = \exp(3.4) = 30$$

Because we have decided that the result depends only on the ratio of NO to Br_2 and the pressure, we can choose the initial moles of $Br_2 = 1$ and of NO $= 2$. Preparing a table similar to that of Example 13:

	NO	Br_2	NOBr	Total
Initial	2	1	0	3
Change	$-2\xi_e$	$-\xi_e$	$+2\xi_e$	$-\xi_e$
Equilibrium	$2 - 2\xi_e$	$1 - \xi_e$	$2\xi_e$	$3 - \xi_e$
Mole fraction	$\dfrac{2 - 2\xi_e}{3 - \xi_e}$	$\dfrac{1 - \xi_e}{3 - \xi_e}$	$\dfrac{2\xi_e}{3 - \xi_e}$	1
Partial pressure	$\left(\dfrac{2 - 2\xi_e}{3 - \xi_e}\right)P$	$\left(\dfrac{1 - \xi_e}{3 - \xi_e}\right)P$	$\left(\dfrac{2\xi_e}{3 - \xi_e}\right)P$	P

The initial pressure (after the stopcock is opened, but before reaction) is 450 torr. The final pressure is lower due to the reduction in the number of moles by the reaction $\{P = 450 \text{ torr}[(3 - \xi_e)/3]\}$. Substituting into the equilibrium constant and converting pressure to bars gives

$$30 = \left(\frac{2\xi_e}{3 - \xi_e}\right)^2 P^2 \left[\left(\frac{2 - 2\xi_e}{3 - \xi_e}\right)^2 P^2 \left(\frac{1 - \xi_e}{3 - \xi_e}\right)P\right]^{-1}$$

$$= \frac{4\xi_e^2(3 - \xi_e)}{(2 - 2\xi_e)^2(1 - \xi_e)} \frac{760(3)}{450(3 - \xi_e)1.01}$$

or

$$\xi_{se}^2 - 5.98(1 - \xi_{se})^3 = 0$$

This cubic equation can be solved by a number of methods, including trial and error, to give $\xi_{se} = 0.605$. The final total pressure is then 359 torr and the partial pressures are $P_{NO} = 119$, $P_{Br_2} = 59$, and $P_{NOBr} = 181$ torr, values which, when substituted into the equilibrium constant, give the required value of 30.

Example 15. Water gas, a combustible mixture of CH_4 and CO, is made by the reaction of water with coke at elevated temperature. This reaction can be written as $3C_{gr} + 2H_2O_{(g)} \leftrightarrow CH_{4(g)} + 2CO_{(g)}$ and has an equilibrium constant of 118 at 1000°C (see Problem 13). What is the ratio of the partial pressure of CO to that of H_2O in the gas mixture formed at 1000°C and 10 atm pressure by passing steam over coke?

Solution: There are four components in the system related by one chemical equation. There is also a stoichiometric equation resulting from the fact that all of the hydrogen and oxygen come from water and, therefore, $P_{CO}/P_{CH_4} = 2$. The number of independent components is therefore $4 - 2 = 2$. With two phases, there are two degrees of freedom. These can be taken as the temperature and pressure of the system. A table is formed as in the two previous examples. (Graphite, a solid, is assumed to have unit activity and does not enter into the equilibrium constant.) The initial amount of water is arbitrarily chosen as 1.0 mol. (An extensive variable characterizing the amount of gas phase can be specified.)

	H_2O	CH_4	CO	Total
Initial	1	0	0	1
Change	$-2\xi_e$	ξ_e	$2\xi_e$	ξ_e
Equilibrium	$1 - 2\xi_e$	ξ_e	$2\xi_e$	$1 + \xi_e$
Mole fraction	$\dfrac{1 - 2\xi_e}{1 + \xi_e}$	$\dfrac{\xi_e}{1 + \xi_e}$	$\dfrac{2\xi_e}{1 + \xi_e}$	1
Partial pressure	$\left(\dfrac{1 - 2\xi_e}{1 + \xi_e}\right)P$	$\left(\dfrac{\xi_e}{1 + \xi_e}\right)P$	$\left(\dfrac{2\xi_e}{1 + \xi_e}\right)P$	P

Substituting into the equilibrium constant gives

$$K_a = \left(\frac{\xi_e}{1+\xi_e}\right)P\left(\frac{2\xi_e}{1+\xi_e}\right)^2 P^2 \left[\left(\frac{1-2\xi_e}{1+\xi_e}\right)^2 P^2\right]^{-1}$$

$$= \frac{4\xi_e^3 P}{(1+\xi_e)(1-2\xi_e)^2}$$

With $K_a = 118$ and $P = 10$,

$$11.8(1+\xi_e)(1-2\xi_e)^2 = 4\xi_e^3$$

This cubic equation has three real roots: $\xi_e = -1.04$, 0.431, and 0.609. Only the central root gives positive values for all the concentrations. Thus, $\xi_e = 0.431$ and

$$\frac{P_{CO}}{P_{H_2O}} = \frac{2\xi_e}{1-2\xi_e} = 6.2$$

7.11 Temperature Dependence of the Equilibrium Constant

Multiplying Eq. (36) by T^{-1} and taking the derivative with respect to T^{-1},

$$\frac{d(T^{-1}\Delta_{rxn}G^\circ)}{dT^{-1}} = -R\frac{d\ln K_a}{dT^{-1}} \tag{48}$$

Because the derivatives are of pressure-independent quantities, we can use Eq. (29) of Chapter 4 to give

$$\frac{d\ln K_a}{dT^{-1}} = -\frac{\Delta_{rxn}H^\circ}{R} \tag{49}$$

Equation (49) provides a means of determining the heat of a reaction without performing any calorimetric measurements. If an equilibrium can be established for the reaction with measurable amounts of reactants and products over a range of temperature, then $\Delta_{rxn}H^\circ$ can be obtained from the slope of a graph of the logarithm of the equilibrium constant versus T^{-1}. Equation (49) can also be written as

$$\frac{d\ln K_a}{dT} = \frac{\Delta_{rxn}H^\circ}{RT^2} \tag{50}$$

showing that the equilibrium constant of an exothermic reaction decreases with increasing temperature. As heat is added to such a system, the equilibrium shifts back toward reactants (the extent of reaction decreases). The shift of the equilibrium absorbs some of the added heat, and the system's temperature rise is not as large as it would be otherwise. Reaction systems at equilibrium therefore can have very large heat capacities and are useful as heat storage media. This is another example of Le Chatelier's principle, that if a stress (temperature increase, in this case) is put on a system at equilibrium, reactions in the system will shift to moderate the stress (reduce the temperature increase).

The above equations are useful for converting equilibrium constants from one temperature to another. Separating variables and integrating gives

$$\ln K_a(T_2) - \ln K_a(T_1) = -\frac{1}{R} \int_{T_1}^{T_2} \Delta_{rxn} H^\circ \, d(T^{-1}) = \frac{1}{R} \int_{T_1}^{T_2} \frac{\Delta_{rxn} H^\circ}{T^2} \, dT \tag{51}$$

If $\Delta_{rxn} H^\circ$ can be considered temperature independent (either because the heat capacity of the products is close to that of the reactants or because T_1 does not differ much from T_2), it can be taken out of the integral, giving

$$\ln K_a(T_2) = \ln K_a(T_1) - \frac{\Delta_{rxn} H}{R} \left(\frac{1}{T_2} - \frac{1}{T_1} \right) \tag{52}$$

A common occurrence is that the equilibrium constant is known at 298 K, because free energies of formation are tabulated at that temperature. T_1 is then taken as 298 K:

$$\ln K_a(T) = \ln K_a(298) - \frac{\Delta_{rxn} H}{R} \left(\frac{1}{T} - \frac{1}{298} \right) \tag{53}$$

If $|T_2 - T_1|$ is large enough that Eq. (53) is not an adequate approximation, the next level of approximation, assuming temperature-independent heat capacities, is used:

$$\Delta_{rxn} H^\circ(T) = \Delta_{rxn} H^\circ(298) + \Delta_{rxn} C_P(T - 298) \tag{54}$$

Substituting into Eq. (51) and integrating gives

$$\ln K_a(T) = \ln K_a(298) - \frac{\Delta H_{rxn}(298) - 298\Delta_{rxn} C_P}{R} \left(\frac{1}{T} - \frac{1}{298} \right) \tag{55}$$
$$+ \frac{\Delta_{rxn} C_P}{R} \ln \left(\frac{T}{298} \right)$$

In higher-order approximations, the temperature variation of heat capacities (See Table 1 of Chapter 2) is considered.

Example 16. Calculate K_a for the reaction $2SO_2 + O_2 \leftrightarrow 2SO_3$ at 500 K. You may use temperature-independent heat capacities.

Solution: At 298 K, we find, from Appendix B,

$$\Delta_{rxn}H^\circ = 2(-395.72) - [2(-296.83)] = -191.78 \text{ kJ/mol}$$
$$\Delta_{rxn}G^\circ = 2(-371.06) - [2(-300.19)] = -141.73 \text{ kJ/mol}$$
$$\Delta_{rxn}C_P^\circ = 2(50.67) - [2(39.87) + 29.36] = -7.76 \text{ J/mol K}$$

From these, we calculate, from Eq. (36),

$$\ln K_a(298) = -\frac{\Delta_{rxn}G^\circ(298)}{R(298 \text{ K})} = \frac{141.73 \text{ kJ}}{\text{mol}} \frac{1000 \text{ J}}{\text{kJ}} \frac{\text{mol K}}{8.314 \text{ J}} \frac{1}{298 \text{ K}}$$
$$= 57.2$$
$$K_a = 6.9 \times 10^{24}$$

Substituting into Eq. (55),

$$\ln K_a(500) = \ln K_a(298) + \frac{[-191,780 - 298(-7.765)] \text{ J}}{\text{mol}} \frac{\text{mol K}}{8.314 \text{ J}}$$
$$\times \left(\frac{1}{298} - \frac{1}{500}\right)\frac{1}{\text{K}} + \frac{-7.765 \text{ J}}{\text{mol K}} \frac{\text{mol K}}{8.314 \text{ J}} \ln\left(\frac{500}{298}\right)$$
$$\ln K_a(500) = 57.2 - 30.89 - 0.48 = 25.8$$
$$K_a = 1.6 \times 10^{11}$$

For this highly exothermic reaction, the equilibrium shifts strongly toward reactants as the temperature is increased.

7.12 Equilibrium Constant in Terms of Partition Function

For reactions of ideal gases, we can write

$$\Delta G^\circ = \sum_i v_i \mu_i^\circ = \sum_i v_i G_{m,i}^\circ \tag{56}$$

Using Eq. (38) of Chapter 5,

$$\Delta G^\circ = \sum_i v_i \left[U_{0,i}^\circ - RT \ln\left(\frac{q_i^\circ}{N_A}\right)\right] = \Delta U_0^\circ - RT \ln \prod_i \left(\frac{q_i^\circ}{N_A}\right)^{v_i} \tag{57}$$

With Eq. (36), we obtain

$$-RT \ln K_P = \Delta U_0^\circ - RT \ln \prod_i \left(\frac{q_i^\circ}{N_A}\right)^{v_i} \tag{58}$$

from which

$$K_P = \prod_i \left(\frac{q_i^\circ}{N_A}\right)^{\nu_i} \exp\left(-\frac{\Delta U_0^\circ}{RT}\right) \tag{59}$$

Note similarities and differences to Eq. (37).

Questions

1. Write the equation for the Haber process synthesis of ammonia from hydrogen and nitrogen in the form of Eq. (4). For an initial 1 mol of nitrogen and 3 mol of hydrogen and an extent of reaction, $\xi = 0.3$, how much ammonia is formed? How much hydrogen is consumed?

2. How is the equilibrium constant of a reverse reaction (products going to reactants) related to the equilibrium constant of the forward reaction?

3. Can equations analogous to Eq. (12) be written for the following?

    (a) The entropy change of a reaction
    (b) The volume change of a reaction
    (c) The heat released in a reaction

4. What would be an acceptable reference state from which to calculate atomic energies in nuclear physics, where elements can be transmuted?

5. State whether each of the following is true or false in regard to the measured value of the equilibrium constant, K_a, of a gaseous chemical reaction:

    (a) The same value should be obtained whether starting from reactants or products.
    (b) The value obtained should be independent of time.
    (c) The value obtained should be independent of total pressure.
    (d) The value obtained should be independent of temperature.

6. For each of the following reactions, indicate whether ΔH° is greater, less, or just about equal to ΔU°:

    (a) $2H_{2(g)} + O_{2(g)} \rightarrow 2H_2O_{(g)}$
    (b) $H_{2(g)} + Cl_{2(g)} \rightarrow 2HCl_{(g)}$
    (c) The electrolysis of liquid water
    (d) The rusting of iron

7. In a table of standard thermodynamic properties of compounds at 298 K, indicate whether each of the following is always zero, always positive, always negative, or none of the above (do this without looking at such a table and be able to explain your answer):

    (a) ΔG_f° of an element
    (b) S° of an element
    (c) ΔG_f° of an non-elemental compound
    (d) C_P° of an element

8. Decide whether each of the following quantities is usually zero, positive, or negative:

    (a) $\Delta_{rxn}H°$ of a dissociation reaction

    (b) $\Delta_{rxn}H°$ of a combustion reaction

    (c) $\Delta_{rxn}S°$ of a dissociation reaction

    (d) $\Delta_{rxn}C_P°$ of a dissociation reaction

9. As mentioned, the reference-state for elemental bromine changes from liquid to gas at 59°C. Which of the following is discontinuous at 59°C? Why?

    (a) The heat of formation of atomic bromine

    (b) The heat of the reaction $Br_{(g)} + Cl_{2(g)} \rightarrow BrCl_{(g)} + Cl_{(g)}$

10. K_P for the reaction $3O_2 \leftrightarrow 2O_3$ is 10^{-57} at 298 K.

    (a) What is the K_P for the reaction $\frac{3}{2}O_2 \leftrightarrow O_3$?

    (b) What is the K_P for the reaction $O_3 \leftrightarrow \frac{3}{2}O_2$?

    (c) Given the low value of the rate constant in part (a), how could you explain the presence of parts per million concentration of ozone in the stratosphere? This stratospheric ozone layer is largely responsible for shielding life on Earth from the deleterious effects of solar ultraviolet radiation in the range 200–300 nm.

11. Consider a mixture of ice in salt water. How many degrees of freedom has the system under each of the following conditions:

    (a) It is open to the atmosphere,

    (b) It is allowed to come to equilibrium with its vapor in the absence of air.

    (c) In part (b), enough NaCl is added, so that solid NaCl is present.

12. How many degrees of freedom has each of the following systems:

    (a) Solid NH_4Cl in the presence of gaseous NH_3 and HCl

    (b) Solid NH_4Cl heated in a sealed evacuated container

13. How many degrees of freedom has a system consisting of an aqueous solution of acetic acid and sodium acetate, in equilibrium with its vapor?

14. Consider an aqueous titration of NaOH by HCl in a vessel exposed to air (which can be taken as a single component). What are the components of this system? How many independent components and phases are there? How many degrees of freedom does the system have?

15. Show that both exothermic and endothermic reactions follow Le Chatelier's principle and moderate the effect of heat added to a reactive system at equilibrium.

16. Compare chemical reactions with phase transitions for use as thermal storage media for capturing sunlight during the day and releasing heat during the night.

17. Plot $\Delta_{rxn}G$ and A (the affinity) for a reversible reaction as a function of ξ, in a region around ξ_e.

18. For which of the following reactions does K_a increase with temperature?

   (a) The dissociation of ethane into methyl radicals
   (b) The reaction of hydrogen and oxygen to form water
   (c) $3O_2 \rightarrow 2O_3$

19. An aerogel is a solid with a density comparable to a gas. In a reaction involving an aerogel as reactant, should a term involving the aerogel be included in the equilibrium constant?

20. Discuss similarities and differences between the two forms for the equilibrium constant of an ideal gas reaction, Eqs. (37) and (59).

21. Show that Eq. (36) holds for the Haber process, regardless which of the equations of Example 6 are used to describe the process.

Problems

1. In the limit in which gases may be considered ideal and condensed-phase volumes neglected, show that $\Delta_{rxn}V^\circ(T) = \Delta v_{gas}RT/P^\circ$.

2. Write the formation reaction for each of the following compounds at 298 K (indicate the phase of all species): NH_3 (gas), CH_4 (gas), NH_4Br (solid).

3. In a constant-volume bomb calorimeter with a heat capacity of 13.418 kJ/K, 1.17 g of naphthalene, $C_{10}H_8$, is burned. Fifty-two joules of energy are required to ignite the sample. If the temperature rise of the calorimeter is 3.318 K, what is the ΔU and ΔH of combustion of naphthalene? What is the ΔH_f of naphthalene? (You may assume that all water is formed in the liquid phase.)

4. Using the data in Appendix C, calculate $\Delta_{rxn}H^\circ$ from heats of formation, $\Delta_{rxn}S^\circ$ from absolute entropies and $\Delta_{rxn}G^\circ$ from free energies of formation for the reaction $2CO + O_2 \rightarrow 2CO_2$ at 298 K. Verify that your results satisfy $\Delta_{rxn}G^\circ = \Delta_{rxn}H^\circ - T\Delta_{rxn}S^\circ$.

5. Calculate $\Delta_{rxn}H^\circ$ for the reaction $2CO + O_2 \rightarrow 2CO_2$ from generic bond dissociation energies and compare with the value obtained from heats of formation in Problem 4.

6. Compare the values calculated for $\Delta_{rxn}H^\circ$ of the reaction $2C_2H_4 \rightarrow$ 1-butene using heats of formation, generic bond dissociation energies, and group additivities. (See Example 4.)

7. At 298 K, the heat of formation and Gibbs free energy of formation of $CaCl_2$(solid) are -795.8 kJ/mol and -748.1 kJ/mol, respectively.

   (a) Using information in Appendix C, calculate $\Delta_{rxn}H$ and $\Delta_{rxn}G$ for the reaction $CaCl_2$(solid) $+ H_2$(gas) $\leftrightarrow$ 2HCl(gas) $+$ Ca(solid).
   (b) Using your result from part (a), calculate $S^\circ(CaCl_2$(solid)) at 298 K. Compare your result with that given in Appendix C.

8. What is the affinity (driving force) of the reaction $CO + O_2 \rightleftharpoons 2CO_2$ in the atmosphere, where concentrations (by volume) are approximately 21% O_2, 300 ppm (parts per million) CO_2, and 100 pb (parts per billion) CO?

9. Show that the activity of a condensed phase at high pressure is $a = \exp((P - P^\circ)V_m/RT)$. Calculate the activity of water at 100 atm (≈ 100 bar pressure).

10.* Using the results worked out in Example 13 for the equilibrium $N_2O_4 \leftrightarrow 2NO_2$ at 298 K, write an expression for the volume as a function of pressure. You can assume that the gas behaves ideally. Find an expression for the isothermal compressibility, $-(1/V)(\partial V/\partial P)_T$. Show that this quantity is always greater than the compressibility of a constant number of moles of ideal gas.

11. Calculate K_a of the reaction $PCl_3 + Cl_2 \leftrightarrow PCl_5$ at 25°C. Assuming temperature-independent heat capacities, calculate K_a at 300°C.

12. Using your result from problem 11, calculate the fraction of PCl_5 that is dissociated at 25°C and 300°C.

13. Coke and steam react at a high temperature to produce a combustible mixture of CO and CH_4 called water gas. Calculate K_a at 1000°C for this reaction, using the following:

    (a) The approximation of a temperature-independent heat of reaction
    (b) The approximation of temperature-independent heat capacities.

14. One of the most important industrial processes is the Haber process, based on the reaction $N_2 + 3H_2 \leftrightarrow 2NH_3$. Calculate the equilibrium constant of this reaction at 298 K and, using temperature-independent heat capacities, at 1000 K. In light of your result, can you guess why the reaction is carried out at a high temperature?

15. Using the value of K_a (1000 K) calculated in Problem 14 for the reaction $N_2 + 3H_2 \leftrightarrow 2NH_3$, calculate the mole fraction of NH_3 produced at equilibrium in a reactor at 1000 K into which a 3 to 1 mixture of H_2 to N_2 is introduced.

16.* Consider the dissociation of N_2O_4 in the presence of an inert gas pressure P_{inert}. Considering N_2O_4 as an ideal gas, it is obvious that the inert gas cannot influence the extent of its dissociation reaction. By reconstructing the table for Example 13, with an additional column for the inert gas, show that the inert gas changes K_x but not K_P or ξ_{oe}.

17. Fe_2O_3 (hematite) can be reduced to iron by reaction with the reducing agent CO. Calculate the ratio of the partial pressures of CO to CO_2 in a mixture equilibrated by the reaction $Fe_2O_3(solid) + 3CO(gas) \leftrightarrow 2Fe(solid) + 3CO_2(gas)$ for the following:

    (a) at 25°C
    (b) at 1000°C
    (c) Do these results depend on pressure? Explain your answer.

18.* In the blast furnace, the iron oxides Fe_2O_3, Fe_3O_4, and FeO, are successively reduced to iron by CO produced by the reaction of coke with air. Calculate ΔG° at 1500 K for the reaction $FeO(solid) + CO(gas) \leftrightarrow Fe(solid) + CO_2(gas)$, the final stage of reduction of iron oxide in the blast furnace, using the constant heat capacity approximation.

What does this equilibrium constant imply is necessary to reduce FeO(solid) to Fe(solid) at this temperature in the blast furnace? How is this brought about?

19. Consider the following reaction as a route to synthesize methanol:

$$CO(gas) + 2H_2(gas) \leftrightarrow CH_3OH(gas)$$

 (a) Calculate the heat of this reaction and its equilibrium constant at 298 K. What partial pressure of CH_3OH would exist in equilibrium with partial pressures of 5.0 atm each of CO and H_2 at this temperature?

 (b) Using the constant heat capacity approximation, calculate the equilibrium constant of the reaction at 500 K. What partial pressure of CH_3OH would exist in equilibrium with partial pressures of 5.0 atm each of CO and H_2 at this temperature?

20. Calculate $\Delta C_P^o(298)$ for each of the following gas-phase reactions:

 (a) $CO + 2H_2 \leftrightarrow CH_3OH$
 (b) $H_2 + Cl_2 \leftrightarrow 2HCl$
 (c) $N_2O_4 \leftrightarrow 2NO_2$
 (d) $C_2H_4 + H_2 \leftrightarrow 2CH_4$
 (e) From the results of your calculation, can you guess a generalization for when ΔC_P^o might be small enough for a gas-phase reaction, so that the constant heat of reaction approximation could be used in making temperature corrections to the equilibrium constant?
 (f) From what you have learned about heat capacities in Chapter 5, can you offer any explanation for your result?

Notes

1. In earlier compilations, a pressure of 1.0 atm was often chosen as the standard pressure. Because 1.0 atm = 1.01325 bar, the difference between the two compilations is usually negligible for all applications except those requiring the highest accuracy.
2. Some gaseous water will always form, due to its vapor pressure. If the volume is not too large, this may be negligible due to the low density of the vapor.
3. In the *JANAF Thermochemical Tables* [U.S. National Bureau Standards, NSRDS-NBS Pub. 37 (1971 and later)], thermodynamic data are tabulated at many different temperatures.
4. SW Benson. Thermochemical Kinetics. New York: Wiley, 1976.

8

Ideal Solutions

Solution properties can be described by changes in thermodynamic quantities on mixing or by partial molar thermodynamic quantities. The Gibbs–Duhem equation relates changes in the latter quantities for different components of a solution. Different methods for determining partial molar quantities are discussed. The ideal solution applies to solutions of very similar components and gives Raoult's law for the vapor pressure of all components. The ideally dilute solution applies to all solutions in the limit of infinite dilution. In the ideally dilute solution, Raoult's law holds for the solvent and Henry's law for solutes. The boiling points, freezing points, and osmotic pressure of ideally dilute solutions are colligative properties; they depend only on properties of the solvent and the total concentration of solutes. Distribution coefficients for a solute between two solvents are the ratios of the Henry's law constants for these solvents. Vapor-pressure and boiling-point phase diagrams are shown for ideal solutions and may be used to determine the concentrations of coexisting phases. Solid–liquid equilibria, including eutectic mixtures, can also be described by such diagrams.

A solution is a homogeneous multicomponent phase. Of course, the idea of homogeneity depends on the scale of observation. When the nonuniformities are of the order of molecular dimensions, we have the true solutions that will be discussed in this chapter. When they are somewhat larger than the wavelength of light, they can be observed, and then we say that the system is composed of more than a single phase. In between these limits are systems with inhomogeneities in the range from a few nanometers to approximately a micron. Such systems are dominated by surface interactions and are called colloids. Colloidal inhomogeneities generally will not be visible to the naked eye; such systems will appear homogeneous. However, they have special properties and therefore they will be briefly discussed in Chapter 11.

8.1 Measures of Concentration

Most of chemistry and engineering and all of biology deals with multicomponent systems. For systems with more than one component, the relative amounts of the various components are important thermodynamic variables. Measures of concentration may apply to the entire system or to a homogeneous region of the system (i.e., a phase). A multicomponent phase is called a *solution*. Because gaseous mixtures are homogeneous,[1] they are solutions. The component at highest concentration in a solution is usually called the *solvent* (designated by subscript A) and the other components the *solutes* (designated by subscript i or, collectively, by the subscript B). One of the most fundamental measures of concentration is the *mole fraction*, introduced for gaseous solutions in Chapter 1, and defined by

$$x_i \equiv \frac{n_i}{n} \tag{1}$$

Molarity is defined as

$$c_i = \frac{n_i}{V} \tag{2}$$

where V is the volume of the solution in liters. It is a very convenient concentration unit, due to the ease of transferring precise volumes of liquid solutions. For thermodynamics, molarity is less useful because the volume of a given solution undergoes small changes with temperature, changing the molarity of the solution. A closely related measure of concentration, the *molality*, is defined as the moles of solute per kilogram of the solvent:

$$m_i \equiv \frac{n_i}{n_A M_A} \tag{3}$$

where M_A is the molecular weight of the solvent (in kg/mol). m_i does not vary with temperature. Because 1 L of dilute aqueous solution has a mass close to 1 kg and is almost all water, molarities and molalities are numerically similar for such solutions. When concentrations are given as a percent, such as in Sherlock Holmes' seven-percent solution, the mass of solute per mass of solution ($\times 100$) is implied.

It is important to be able to convert among the different concentration units. Conversions between molality and mole fraction are performed by considering a solution containing 1 kg of solvent:

$$x_i = \frac{m_i}{\sum\limits_i m_i + 1/M_A}, \qquad m_i = \frac{n_i}{n_A M_A} = \frac{x_i}{x_A M_A} \tag{4}$$

The conversion between molarity and mole fraction involves the density, ρ, of the solution and is (by calculating the mole fraction of 1 L of solution)

$$x_i = \frac{c_i}{[(\rho - \sum\limits_{i \neq A} c_i M_i)/M_A] + \sum\limits_{i \neq A} c_i} \tag{5}$$

8.2 Partial Molar Quantities

Extensive properties of multicomponent phases (solutions) are related to the amount of material in the phase, but may not be just the sum of the properties of the constituent components. Probably the best known example of this difference is the observation that mixing 1.0 L of ethanol with 1.0 L of water at standard temperature and pressure (STP) produces 1.93 L of water–ethanol solution. We define the difference between an extensive property of the solution and the sum of the properties of its pure components as the property change of mixing for the solution:

$$\Delta_{\text{mix}} X \equiv X - \sum\limits_i n_i X_{i,m}^* \tag{6}$$

where $X_{i,m}^*$ is a molar property of a pure component. There is a volume change of mixing, $\Delta_{\text{mix}} V$ of -0.07 L for the above water–ethanol solution. The word "change" is often omitted in discussing these quantities. Thus, we use the terminology "volume of mixing" and the "Gibbs free energy of mixing." The "heat of mixing" is taken to mean the enthalpy change on mixing. In ideal gas mixtures, where interactions are negligible, we have seen that $\Delta_{\text{mix}} V$ is zero. However, in Chapter 4, we saw that, even for ideal gases, $\Delta_{\text{mix}} S$ is not zero, and this also holds for $\Delta_{\text{mix}} A$ and $\Delta_{\text{mix}} G$, to which $\Delta_{\text{mix}} S$ contributes. In order to use Eq. (6), we must develop methods of calculating extensive properties of solutions.

To calculate a general extensive property, X, of a solution, we start with a differential relation for X:

$$dX = \left(\frac{\partial X}{\partial P}\right)_{T,n_i} dP + \left(\frac{\partial X}{\partial T}\right)_{P,n_i} dT + \sum_i \left(\frac{\partial X}{\partial n_i}\right)_{P,T,n_{j \neq i}} dn_i \tag{7}$$

For c components in a phase, this equation has $c + 2$ terms, one more than given by the phase rule, because its integration requires the size of the phase as well as its intensive variables. The quantity $(\partial X / \partial n_i)_{P,T,n_{j \neq i}}$ is called the *partial molar X with respect to i* and given the symbol $\bar{X}_i$:

$$\bar{X}_i \equiv \left(\frac{\partial X}{\partial n_i}\right)_{P,T,n_{j \neq i}} \tag{8}$$

These partial molar quantities are the key quantities used to describe the properties of solutions.

We have previously introduced one particular partial molar quantity. In Eq. (9) of Chapter 6, the chemical potential was given as

$$\mu_i = \left(\frac{\partial G}{\partial n_i}\right)_{P,T,n_{j \neq i}} \equiv \bar{G}_i \tag{9}$$

This shows that the chemical potential of a component is just its partial molar Gibbs free energy. Note that the definitions of the chemical potential in terms of other thermodynamic variables, given in Chapter 6, Eq. (8), are not partial molar quantities because pressure and temperature are not the variables held constant in these derivatives.

Partial molar quantities are ratios of two extensive variables and are, therefore, intensive variables. As intensive variables they are independent of the size of the system and dependent only on other intensive variables. We can take our partial derivatives to be dependent on P, T, and $c - 1$ concentrations and write Eq. (7) as

$$dX = \left(\frac{\partial X}{\partial P}\right)_{T,n_i} dP + \left(\frac{\partial X}{\partial T}\right)_{P,n_i} dT + \sum_i \bar{X}_i(P,T,c_j) dn_i \tag{10}$$

Consider a process beginning with a very small system at given P, T, and c_j's and, holding P and T constant, adding the components of the system in the proper ratio to keep the system concentrations constant. Integrating Eq. (10) for this process, the first two terms drop out, giving

$$X_f - X_0 = \sum_i \int_0^f \bar{X}_i \, dn_i \tag{11}$$

The partial molar quantities are constant during the process and may be taken out of the integrals, and the limit may be taken in which the initial size of the system (and the extensive variables X_0 and the $n_{i,0}$'s) go to zero:

$$X = \sum_i \bar{X}_i n_i \tag{12}$$

where the subscript f has been dropped for simplicity. Equation (11) of Chapter 6 is just Eq. (11) written for the Gibbs free energy.

A useful relation can be derived by considering the change in X upon arbitrary additions of material to the system at constant T and P. Because concentrations change upon such addition, we must, in using Eq. (12) consider changes in the partial molar quantities as well as changes in the number of moles:

$$dX = \sum_i \bar{X}_i \, dn_i + \sum_i n_i \, d\bar{X}_i \tag{13}$$

However, Eq. (10) tells us that a perfectly general expression for dX at constant T and P is

$$dX = \sum_i \bar{X}_i \, dn_i \tag{14}$$

The only way that these two relations can be reconciled is if

$$\sum_i n_i \, d\bar{X}_i = 0 \tag{15}$$

Equation (15) is called the general *Gibbs–Duhem relation*. Because it tells us that there is a relation between the partial molar quantities of a solution, we will learn how to use it to determine a $\bar{X}_i$ when all other $\bar{X}_{j \neq i}$ have been determined. (In a two-component system, knowing $\bar{X}_{\text{solvent}}$ determines $\bar{X}_{\text{solute}}$.) This type of relationship is required by the phase rule because, at constant T, P, and c components, a single-phase system has only $c - 1$ degrees of freedom.

Many relations that hold between extensive thermodynamic variables also hold between the corresponding partial molar quantities. In particular, we have

$$\bar{G}_i = \mu_i = \bar{H}_i - T\bar{S}_i \tag{16}$$

$$\left(\frac{\partial \mu_i}{\partial P} \right)_{T,c_j} = \bar{V}_i \tag{17}$$

$$\left(\frac{\partial \mu_i}{\partial T} \right)_{P,c_j} = -\bar{S}_i \tag{18}$$

The first relation follows directly by taking the partial derivative of $G = H - TS$ with respect to n_i (holding P, T, and the other n_j's constant). The latter two relations can be proved by reversing the order of differentiating with respect to n_i and with T and P respectively.

Use of Eq. (12) in Eq. (6) allows us to calculate changes of thermodynamic function on mixing:

$$\Delta_{\text{mix}}X = \sum_i n_i \bar{X}_i - \sum_i n_i X_{m,i}^* = \sum_i n_i(\bar{X}_i - X_{m,i}^*) \tag{19}$$

This shows that nonzero mixing properties of solutions result from partial molar quantities not being equal to the molar properties of the pure components. From a microscopic point of view, nonzero volumes of mixing could be due to the components of a solution fitting each other better or more poorly than they fit themselves. Nonzero energy and enthalpy of mixing result if the forces between molecules of different components are different than those between molecules of the same component. For S, A, and G, an additional term results from the "randomness" of the solution compared to the pure components, which, in Chapter 4, was shown to give an entropy of mixing of $-nR \sum_j X_j \ln X_j$.

It can easily be shown that simple thermodynamic relationships also hold for changes of thermodynamic functions on mixing. For example,

$$\Delta_{\text{mix}}G = \Delta_{\text{mix}}H - T\Delta_{\text{mix}}S \tag{20}$$

$$\left(\frac{\partial \Delta_{\text{mix}}G}{\partial P}\right)_{T,c_j} = \Delta_{\text{mix}}V \quad \text{and} \quad \left(\frac{\partial \Delta_{\text{mix}}G}{\partial T}\right)_{P,c_j} = -\Delta_{\text{mix}}S \tag{21}$$

8.3 Measurement of Partial Molar Quantities

The measurement of partial molar quantities will be illustrated with reference to partial molar volumes. We can measure absolute volumes of solution and, thus, can determine partial molar volumes directly from its definition:

$$\bar{V}_i(P, T, c_j) \equiv \left(\frac{\partial V}{\partial n_i}\right)_{P,T,n_{j \neq i}} \tag{22}$$

For binary solutions, this is just the slope of the plot at constant T and P of the volume of the solution versus its molality. This is illustrated in Fig. 1 for an aqueous $MgSO_4$ solution. Note that at low concentration, the partial molar volume of $MgSO_4$ in aqueous solution is *negative*.

An alternative definition of $\bar{V}_i$ is the change of volume of a solution upon adding 1 mol of component i in the limit in which the solution volume is so large that its concentration is unchanged by the addition. Using such large volumes is usually impractical, but smaller additions and volumes can be used. This method avoids the difficulty of measuring slopes of curves.

In a binary solution (two components), the partial molar volumes of the components can be determined from a plot of the molar volume of the solution versus mole fraction, as shown in Fig. 2.

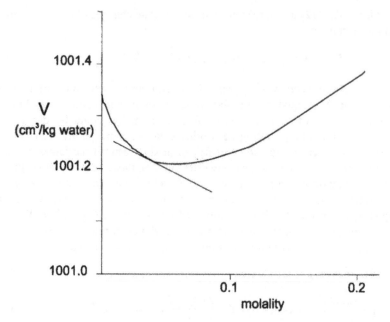

Figure 1   Volume of the solution versus its molality.

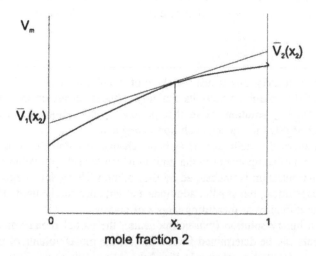

Figure 2   Molar volume of the solution versus mole fraction.

Using Eq. (12), we have

$$V_m = \frac{V}{n_1 + n_2} = \frac{n_1 \bar{V}_1 + n_2 \bar{V}_2}{n_1 + n_2} \tag{23}$$

Writing this in terms of mole fractions, we have

$$V_m(x_2) = x_1 \bar{V}_1(x_2) + x_2 \bar{V}_2(x_2) = \bar{V}_1(x_2) + x_2[\bar{V}_2(x_2) - \bar{V}_1(x_2)] \tag{24}$$

where, in this expression, we have explicitly indicated that the partial molar volumes are functions of the concentration of the binary solution. The tangent line to the curve of Eq. (24) at x_2 has slope $\bar{V}_2(x_2) - \bar{V}_1(x_2)$ and its intercepts with $x_2 = 0$ and $x_2 = 1$ are $\bar{V}_1(x_2)$ and $\bar{V}_2(x_2)$, respectively, as shown in Fig. 2.

8.3.1 Partial Molar Gibbs Free Energies

Partial molar Gibbs free energies (chemical potentials) can most easily be measured by taking advantage of the equality of chemical potentials in different phases at equilibrium. The chemical potential of a component of a solution is its chemical potential in the vapor that is in equilibrium with the solution. Thus, if the component has a measurable vapor pressure, it can be used to determine its chemical potential in the solution. Assuming that the vapor can be considered an ideal gas, its chemical potential is determined by its vapor pressure:

$$\mu_{i,l} = \mu_{i,g} = \mu_{i,g}^0 + RT \ln\left(\frac{P_i}{P^0}\right) \tag{25}$$

where $\mu_{i,g}^0$ is the chemical potential of the 1.0-bar pure component in the gas phase. Because zeros of energy are arbitrary, this quantity is usually fixed by setting the chemical potentials of the *elements* in their reference states[2] equal to zero. A similar relation holds for the pure component:

$$\mu_{i,l}^* = \mu_{i,g}^* = \mu_{i,g}^0 + RT \ln\left(\frac{P_i^*}{P^0}\right) \tag{26}$$

where P_i^* is the vapor pressure of pure i. Subtracting Eq. (24) from Eq. (23) gives

$$\mu_{i,l} - \mu_{i,l}^* = \bar{G}_i - G_{m,i}^* = RT \ln\left(\frac{P_i}{P_i^*}\right) \tag{27}$$

For nonideal vapors, fugacities must be used. From Eq. (51) of Chapter 6,

$$\mu_{i,l} - \mu_{i,l}^* = \bar{G}_i - G_{m,i}^* = RT \ln\left(\frac{f_i}{f_i^*}\right) \tag{28}$$

Strictly speaking, this holds only for both the solution and pure component under the vapor pressure of component i. However, because the molar volumes of solutions are small, the result is little changed at other pressures.

In a binary solution, the Gibbs-Duhem relation [Eq. (15)] determines the variation of a partial molar property of one component in terms of the variation of the partial molar quantity of the other component. This relation is useful for obtaining chemical potentials in binary solutions when only one of the components has a measurable vapor pressure. Applying Eq. (15) to chemical potentials in a binary solution,

$$n_B \, d\mu_B = -n_A \, d\mu_A \tag{29}$$

Dividing by $n = n_A + n_B$ and using $x_B = 1 - x_A$ gives

$$d\mu_B = -\frac{x_A}{1 - x_A} \, d\mu_A \tag{30}$$

Integration of this equation will be discussed in Chapter 9.

8.4 The Ideal Solution

In our study of gases, the ideal gas model was both simple and useful. It neglects all interactions between molecules. Obviously, a model neglecting all intermolecular interactions can have little validity for liquid or solid solutions, which are held together by cohesive forces. The simplest model for such solutions, the ideal solution, allows interactions between molecules, but it assumes that they are the same between any two species in the solution. Thus, for a binary solution, $A–A$, $B–B$ and $A–B$ interactions are equal. A and B molecules are also required to have similar size and shape. This model rarely provides a good fit to data. (Very similar species, such as benzene–toluene, and mixtures of isomers and isotopic variations of molecules are a few examples where it does fit.) However, it does provide a good reference point against which the behavior of real solutions can be measured. Deviations from ideal solution behavior can tell us something about the interactions in systems.

Because thermodynamics describes macroscopic behaviors, we need a macroscopic definition of the ideal solution in addition to the microscopic description given above. We define an ideal solution as one that, for each of its components, at all T and P and over the entire range of concentrations,

$$\mu_i \overset{\text{i.s.}}{=} \mu_i^* + RT \ln x_i \tag{31}$$

Equation (31) implies that the properties of the solution change smoothly into those of pure component i as $x_i \to 1$. The pure components of an ideal solution must, therefore, have the same phase as the solution. Standard states of an ideal liquid solution are thus just the pure liquid components at the given T and P. Comparing Eq. (31) with the normal form of the chemical potential given in Eq. (47) of Chapter 6, $\mu = \mu^0 + RT \ln a$, we see that, for the ideal solution, the activity of every component is just its mole fraction. The other partial molar quantities of the components of an ideal solution are as follows:

$$\bar{V}_i = \left(\frac{\partial \mu_i}{\partial P}\right)_{T,n_i} \overset{\text{i.s.}}{=} \left(\frac{\partial \mu_i^*}{\partial P}\right)_{T,n_i} = V_{m,i}^* \tag{32}$$

$$\bar{S}_i = -\left(\frac{\partial \mu_i}{\partial T}\right)_{P,n_i} \overset{\text{i.s.}}{=} -\left(\frac{\partial \mu_i^*}{\partial T}\right)_{P,n_i} - R \ln x_i = S_{m,i}^* - R \ln x_i \tag{33}$$

$$\bar{H}_i = \mu_i + T\bar{S}_i = \mu_i^* + RT \ln x_i + T(S_{m,i}^* - R \ln x_i) = H_{m,i}^* \tag{34}$$

From Eq. (19), the Gibbs free energy of mixing of an ideal solution is

$$\Delta_{\text{mix}}G = \sum_i n_i(\mu_i - \mu_i^*) \overset{\text{i.s.}}{=} RT \sum_i n_i \ln x_i \tag{35}$$

and from Eqs. (32)–(34), we obtain

$$\Delta_{\text{mix}}V \overset{\text{i.s.}}{=} \Delta_{\text{mix}}H \overset{\text{i.s.}}{=} 0 \tag{36}$$

$$\Delta_{\text{mix}}S \overset{\text{i.s.}}{=} -R \sum_i n_i \ln x_i = -nR \sum_i x_i \ln x_i \tag{37}$$

Expression (37) is the same result that holds for the ideal gas, where the entropy of mixing results entirely from the increase in randomness of the mixture. It is necessarily positive because all $x_i < 1$ and $\ln x_i < 0$. All of the thermodynamic properties of mixing of an ideal solution result from this randomness; there are no energy effects.

In order to find the vapor pressure of a component of an ideal solution, we equate its chemical potential in the solution with that in the vapor in equilibrium with the solution. Assuming that the vapor is ideal, this gives

$$\mu_{i,l} \overset{\text{i.s.}}{=} \mu_i^* + RT \ln x_i = \mu_{i,g,} = \mu_i^0 + RT \ln (P_i/P^0) \tag{38}$$

where μ_i^0 is the chemical potential of 1.0 bar vapor. We do the same for the pure liquid component:

$$\mu_i^* = \mu_{i,g}^* = \mu_i^0 + RT \ln(P_i^*/P^0) \tag{39}$$

Substituting Eq. (39) into Eq. (38) gives

$$P_i \overset{\text{i.s.}}{=} x_i P_i^* \tag{40}$$

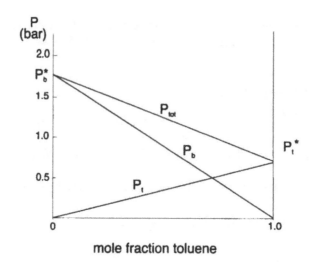

Figure 3   Vapor-pressure diagram of benzene–toluene solutions.

which is known as *Raoult's law*. The total vapor pressure of the components of a solution obeying Raoult's law is $\sum_i P_i = \sum_i x_i P_i^*$. For a binary solution, this becomes

$$P = x_A P_A^* + x_B P_B^* = P_A^* + x_B(P_B^* - P_A^*) \tag{41}$$

The vapor pressure of such a solution varies linearly with x_B from P_A^* to P_B^*. This behavior is illustrated in the *vapor-pressure diagram* of benzene–toluene solutions shown in Fig. 3.

8.5   The Ideally Dilute Solution

Whereas the ideal solution model applies over the entire range of concentrations, but only for very similar components, the ideally dilute solution model applies to any solution, but only over a very limited range of concentrations. From a microscopic point of view, the ideally dilute solution holds as long as solute molecules are almost always completely surrounded by solvent molecules and rarely interact with other solute molecules.

The applicable concentration range of the ideally dilute approximation depends on the size of the solute molecules, because, obviously, large polymer molecules will interact at much lower concentrations than will smaller species. For ionic solutes, the range of applicable concentrations is so small that it is practically useless, and even in this range, allowance must be made for the ionization of the solute. Ionic solutions will be considered in detail in Chapter 10. For small, nonionized solutes, the ideally dilute solution approximation can usually be used up to total solute mole fractions of 0.01–0.1, depending on the accuracy required.

Calling the solvent component A and a solute component i, we define the ideally dilute solution as one that has, in a range around infinite dilution,

$$\mu_A \stackrel{\text{id.s.}}{=} \mu_A^0 + RT \ln x_A \tag{42}$$

$$\mu_i \stackrel{\text{id.s.}}{=} \mu_i^0 + RT \ln x_i \tag{43}$$

Comparison with the standard form for the chemical potential, $\mu = \mu^0 + RT \ln a$ [Eq. 47 of Chapter 6], shows that in the ideally dilute solution activities are equal to mole fractions for both solvent and solute. In order to find the standard state of the solvent in the ideally dilute solution, we note that at $x_A = 1$ (infinite dilution, within the range of applicability of the model), we have $\mu_A^0 = \mu_A^*$. The standard state of the solvent in the ideally dilute solution is pure solvent, just like the standard states of all components in an ideal solution. The solvent in the ideally dilute solution behaves just like a component of the ideal solution. Although it is also true that μ_i^0 becomes μ_i^* at $x_i = 1$, this is clearly outside the realm of applicability of Eq. (43). In order to avoid this difficulty, in determining μ_i^0 we make measurements at very low values of x_i and extrapolate to $x_i = 1$ using $\mu_i^0 = \mu_i - RT \ln x_i$, as if the high dilution behavior held to $x_i = 1$. In other words, our standard state for a solute in the ideally dilute solution is the *hypothetical* state of pure solute with the behavior of the solute in the infinitely dilute solution.

It is more common to use molality or molarity as the concentration unit for the solute in the ideally dilute solution. The relevant equations are then

$$\mu_i \stackrel{\text{id.s.}}{=} \mu_i^0 + RT \ln m_i \tag{44}$$

$$\mu_i \stackrel{\text{id.s.}}{=} \mu_i^0 + RT \ln c_i \tag{45}$$

where m_i and c_i are the numerical values of the solute molality and molarity, respectively.[3] The standard state for the solute in these cases is the *hypothetical* state of a $1.0\,m$ or a $1.0\,M$ solution, with the solute behaving as it does in the infinitely dilute solution.

As shown in Example 1, the chemical potential of a solute can be of a more complicated form than given by Eqs. (43)–(45), even though the solution shows ideally dilute behavior. This can result from a transformation of the substance when it dissolves in the solution.

Example 1. N_2O_4 almost completely dissociates into NO_2 in dilute aqueous solution. What is the chemical potential and activity of N_2O_4 in such solutions?

Solution: The reaction is $N_2O_4 \leftrightarrow 2NO_2$. If the nominal molality of the solution is $m_{N_2O_4}$, the solution actually contains $2m_{N_2O_4}$ of NO_2 and practically no N_2O_4. We then have

$$\mu_{N_2O_4} = 2\mu_{NO_2} = 2(\mu_{NO_2}^0 + RT \ln m_{NO_2}) = 2(\mu_{NO_2}^0 + RT \ln 2m_{N_2O_4})$$
$$= 2\mu_{NO_2}^0 + RT \ln(2m_{N_2O_4})^2 = \mu_{N_2O_4}^0 + RT \ln(2m_{N_2O_4})^2$$

This shows that the activity of N_2O_4 in this solution is $(2m_{N_2O_4})^2$.

We will see in Chapter 10, when we deal with the dissociation of electrolytes into ions, that, in general, the chemical potential of a dissociating solute is the chemical potential of its component parts, whereas the activity of the solute is the product of the component parts.

In order to find the vapor pressures above the ideally dilute solution, we equate the chemical potentials of the components in the solution with those in the vapor. Because the solvent in the ideally dilute solution behaves just like a component of an ideal solution, its vapor pressure follows Raoult's law. For the solute in an ideally dilute solution, we obtain

$$\mu_{i,l} \stackrel{id.s.}{=} \mu_{i,l}^0 + RT \ln x_i = \mu_{i,g} = \mu_{i,g}^0 + RT \ln(P_i/P^0) \tag{46}$$

which can be rearranged to

$$RT \ln\left(\frac{P_i}{x_i P^0}\right) = \mu_{i,l}^0 - \mu_{i,g}^0 \tag{47}$$

or

$$\frac{P_i}{x_i P^0} = \exp\left(\frac{\mu_{i,l}^0 - \mu_{i,g}^0}{RT}\right) \tag{48}$$

The standard chemical potential $\mu_{i,l}^0$ depends on T and P, and $\mu_{i,g}^0$ depends on T, but both are independent of concentration. Equation (51) can thus be written as

$$P_i = x_i K_i \quad \text{with } K_i = P^0 \exp\left(\frac{\mu_{i,l}^0 - \mu_{i,g}^0}{RT}\right) \tag{49}$$

This equation is known as *Henry's law* and K_i is known as the Henry's law constant (more exactly as the *mole fraction-based Henry's law constant*). K_i depends on the particular solute–solvent combination being considered, as well as the temperature and the total pressure. However, it does not depend on the solution concentration. Using Eq. (4), the partial pressure can be expressed in terms of the molality as

$$P_i = K_i \frac{m_i}{m_i + 1/M_A} \tag{50}$$

Because the ideally dilute solution approximation only holds in the limit of very dilute solutions (i.e., where $m_i \ll 1/M_A$), this becomes

$$P_i = K_i M_A m_i = K_{m,i} m_i \tag{51}$$

where $K_{m,i}$ is the *molality-based Henry's law constant* of component i in the given solution. In a similar manner, a Henry's law constant can also be defined in terms of concentration expressed as molarity.

Henry's law is also used to find the amount of a gas that dissolves at equilibrium in a solvent at a known partial pressure of the gas. The solubility of the gas is proportional to the inverse of the Henry's law constants, as we have defined them in Eqs. (49) and (51). The reader is cautioned, however, that sometimes the Henry's law constant is defined so that solubilities are directly proportional to the Henry's law constants. The units listed for the constant indicate the definition used in any tabulation. Some Henry's law constants are given in Table 1.

Solubilities of gases usually decrease as the temperature increases, indicating that solubility is exothermic. Solutes with higher boiling points generally have greater solubility. (See Question 14.) Unusually large solubility may result in cases in which there are specific (e.g., hydrogen-bonding) interactions between solute and solvent. Helium shows much greater solubility than expected from its boiling point due to its ability to fit into cavities in solvents.

TABLE 1 Mole-fraction-Based Henry's Law Constants K_i (10^9 Pa) at 298 K

Solute	Solvent	
	Water	Benzene
Argon	3.73	
Carbon dioxide	0.167	0.0114
Helium	14.93	
Hydrogen	7.12	0.367
Hydrogen sulfide	0.057	
Nitrogen	8.68	0.239
Oxygen	4.40	
Methane	3.74	
Benzene	0.0300	
Naphthalene	0.0024	

Source: Data from A James, M Lord. VNR Index of Chemical and Physical Data. New York: Van Nostrand Reinhold 1992, p 495.

8.6 Freezing-Point Depression, Boiling-Point Elevation and Osmotic Pressure

We often have a situation in which the solvent in a solution can equilibrate with another phase, whereas the solutes are confined to the solution. This will occur with vaporization if the solutes are non-volatile, it will occur with freezing if the solute is excluded from the crystal lattice of the solid solvent, and it will occur with osmosis if only solvent can pass through a semipermeable membrane. For vaporization and freezing, when the solvent is pure ($x_A = 1$), these equilibrations occur at the normal boiling point (T_b^*) and freezing point (T_f^*) of the solvent, whereas for osmosis, equilibration across a semipermeable membrane occurs when pure solvent is at the same pressure on both sides of the membrane.

If solute is added to a solvent, the mole fraction of solvent and its escaping tendency from the liquid, as measured by its chemical potential or activity, is reduced. We will see that in order to reestablish equilibrium with the other phase, there must be a reduction in freezing point, an elevation of boiling point or an increase in osmotic pressure. Because the addition of solute reduces the escaping tendency of the solvent from a solution, it extends the realm of stability of the solution. Thus, the addition of antifreeze (ethylene glycol) to water in a car's radiator both lowers the freezing point and raises the boiling point of the liquid in the radiator, both of which are beneficial for the cooling system.

Considering freezing, equilibrium requires that the chemical potential of the solvent in solution equal that of the (pure) solid solvent:

$$\mu_{A,s}^* = \mu_{A,l} = \mu_{A,l}^* + RT \ln a_A \tag{52}$$

or

$$\ln a_A = \frac{\mu_{A,s}^* - \mu_{A,l}^*}{RT} = -\frac{\Delta_{fus}G^*}{RT} \tag{53}$$

Taking the derivative with respect to T^{-1} and using Eq. (29) of Chapter 4,

$$\left(\frac{\partial \ln a_A}{\partial T^{-1}}\right)_P = -\frac{1}{R}\left(\frac{\partial(T^{-1}\Delta_{fus}G^*)}{\partial T^{-1}}\right)_P = -\frac{\Delta_{fus}H^*}{R} \tag{54}$$

Separating variables and integrating from the pure solvent (with $a_A = 1$ and $T_f = T_f^*$) gives

$$\ln a_A = -\frac{\Delta_{fus}H^*}{R}\left(\frac{1}{T_f} - \frac{1}{T_f^*}\right) \approx -\frac{\Delta_{fus}H^*}{R}\left(\frac{\theta_f}{(T_f^*)^2}\right) \tag{55}$$

where we have substituted T_f^* for T_f in the denominator in the final step because $T_f^* - T_f \equiv \theta_f$, the *freezing-point depression*, is quite small. Equation (55) also gives the activity of solvent in solution in equilibrium with solid A at various temperatures below the freezing point. A similar analysis of the boiling point gives, in terms of the *boiling-point elevation*, $\theta_b = T_b - T_b^*$:

$$\ln a_A = \frac{\Delta_{vap}H^*}{R}\left(\frac{1}{T_b} - \frac{1}{T_b^*}\right) \approx -\frac{\Delta_{vap}H^*}{R}\left(\frac{\theta_b}{(T_b^*)^2}\right) \tag{56}$$

Osmosis is the selective passage of particular components of a solution through a *semipermeable membrane*. Usually, it is the solvent that passes through the membrane, because the solute is blocked. However, some membranes also allow small solute molecules to pass through as well and only block the passage of macromolecular solute molecules. The *osmotic pressure* of a solution is the pressure difference produced at equilibrium across the membrane, with the solution on one side of the membrane and pure solvent on the other side. As shown in Fig. 4, the reduced activity of the solvent in solution is compensated for by an increase in the pressure of the solution:

$$\mu_A^*(P) = \mu_A^*(P') + RT \ln a_A \tag{57}$$

where P is the pressure on the pure solvent side of the membrane and P' is the pressure on the solution side of the membrane. This can be written as

$$\ln a_A = \frac{\mu_A^*(P) - \mu_A^*(P')}{RT} = \frac{1}{RT}\int_{P'}^{P}\left(\frac{\partial \mu_A^*}{\partial P}\right)_T dP = -\frac{\Pi V_m^*}{RT}. \tag{58}$$

In the last step, the excellent approximation of constant molar volume of the solvent has been made. Π, the osmotic pressure, is $P' - P$. We see that the activity of the solvent in solution can be determined by measuring the freezing-

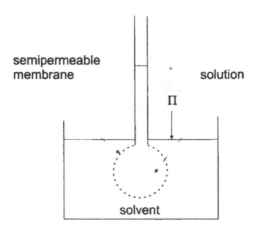

Figure 4 Osmotic pressure.

point depression, boiling-point elevation, or osmotic pressure of the solution, as well as by measuring the vapor pressure of solvent above the solution.

Particularly simple forms of the equations for the freezing-point depression, boiling-point elevation, and osmotic pressure are obtained when the solution is ideal or when it is sufficiently dilute, so that the ideally dilute solution approximation is appropriate. In both of these cases, the activity of the solvent is equal to its mole fraction, so that

$$\ln a_A = \ln x_A = \ln(1 - \sum x_i) \equiv \ln(1 - x_B) \tag{59}$$

The $\sum x_i$ extends over all the solute species and is designated as x_B, which we have previously used to designate the mole fraction of the single solute in a binary solution. At a concentration small enough to make the ideally dilute solution approximation, it is usually sufficient to use only the first term in the Taylor series approximation of $\ln(1 - x_B)$,[4]

$$\ln a_A = -x_B \tag{60}$$

Equation (55) for the freezing-point depression then becomes

$$\theta_f = \frac{R(T_f^*)^2 x_B}{\Delta_{fus} H^*} \tag{61}$$

whereas Eq. (56) for the boiling-point elevation becomes

$$\theta_b = \frac{R(T_b^*)^2 x_B}{\Delta_{vap} H^*} \tag{62}$$

Equation (58) for the osmotic pressure becomes

$$\Pi = \frac{RT x_B}{V_m^*} \tag{63}$$

Freezing-point depression, boiling-point elevation and osmotic pressure are known as *colligative properties*, because they are dependent on the properties of the *solvent* and the total mole fraction of all solutes, but are independent of any particular property of the solutes. Equations (61)–(63) are usually written in terms of m_B, the sum of the molalities of all the solutes, which for ideally dilute solutions is related to x_B by

$$x_B = \frac{n_B}{n_A + n_B} \approx \frac{n_B}{n_A} = m_B M_A$$

where M_A is the molecular weight of the solvent (in units of kg/mol). This gives

$$\theta_f = K_f m_B, \qquad K_f = \frac{R(T_f^*)^2 M_A}{\Delta_{fus} H^*} \tag{64}$$

$$\theta_b = K_b m_B, \qquad K_b = \frac{R(T_b^*)^2 M_A}{\Delta_{vap} H^*} \tag{65}$$

$$\Pi = \left(\frac{RTM_A}{V_m^*}\right) m_B \tag{66}$$

m_B is the total molality of the solution. K_f and K_b, the molal freezing-point depression and boiling-point elevation constants, respectively, depend only on the properties of the *solvent* and are usually evaluated by calibration, as in the following example.

Example 2. The freezing point of pure benzene is 5.45°C. The addition of 0.52 g of cyclohexane to 100.0 g of benzene forms a solution with freezing point 5.13°C. If 0.81 g of an unknown compound is added to 100.0 g of benzene, the resulting freezing point is 5.21°C. What is the molecular weight of the unknown compound?

Solution: First, we calculate the molality of the cyclohexane solution:

$$\frac{0.52\ g_{chex}}{100\ g_{benz}} \frac{mol_{chez}}{84\ g_{chez}} \frac{1000\ g_{benz}}{kg_{benz}} = 0.062\ m$$

The freezing depression constant of benzene is then

$$K_f = \frac{(5.45 - 5.13)°C}{0.062\ m} = 5.2°C/m$$

The molality of the solution of unknown is then calculated:

$$\frac{(5.45 - 5.21)°C}{5.2°C/m} = 0.046\ m = 0.046\ mol_x/kg_{benz}$$

From this, the molecular weight of the unknown compound is calculated:

$$\frac{0.81\ g_x}{100\ g_{benz}} \frac{kg_{benz}}{0.046\ mol_x} \frac{1000\ g_{benz}}{kg_{benz}} = \frac{180\ g_x}{mol_x}$$

When doing calculations with colligative properties, it is usually advantageous to make measurements at a number of concentrations and extrapolate the results to zero concentration, where the ideally dilute solution theory is clearly applicable.

A few values of K_f and K_b are given in Table 2. For a macromolecular solution, the ideally dilute approximation holds only up to such low molality that freezing-point depression and boiling-point elevation are useless for determining

TABLE 2. Freezing-point depression and boiling-point elevation constants

Solvent	K_f (K kg/mol)	K_b (K kg/mol)
Water	1.86	0.51
Benzene	5.12	2.53
Naphthalene	5.94	5.8
Carbon tetrachloride	30.0	4.95

polymer molecular weights by the method of Example 2. Osmotic pressure, however, is measurable even for very dilute solutions.

Equation (63) for osmotic pressure is usually rewritten by substituting $x_B \approx n_B/n_A$ and then taking the volume of the solvent, $n_A V_m^*$, as approximately equal to the volume of the solution:

$$\Pi = \frac{RTn_B}{V} \tag{67}$$

This equation is easy to remember because it has the same form as the ideal gas law. Note that it depends only on the concentration of the solution and is independent of any specific properties of both the solute and the solvent. However, if the osmotic pressure is established by a rise of the level of the solution, as in Fig. 4, this rise will be dependent on the density of the solution (approximately the density of the solvent). Example 3 shows how even small concentrations of macromolecular solutes in solution can produce readily measurable osmotic pressures.

Example 3. What is the osmotic pressure at 25°C of a solution having a density of 0.8 g/mL and containing 1.5 g of a polymer of molecular weight 12,000 per liter of solution? How high will the solution rise to establish this pressure?

Solution:

$$\pi = \frac{0.082l \text{ atm}}{\text{mol K}} 298 \text{ k} \frac{1.5 \text{ g}}{L} \frac{\text{mol}}{12,000 \text{ g}} = 3.0 \times 10^{-3} \text{ atm}$$

The rise of the solution is

$$h = (3.0 \times 10^{-3} \text{ atm}) \frac{760 \text{ mm Hg}}{\text{atm}} \frac{13.6 \text{ mm solution}}{0.8 \text{ mm Hg}}$$

$$= 31 \text{ mm solution}$$

which is easily measurable.

Synthetic polymers are often produced with a range of degree of polymerization and molecular weight. As shown next, a molecular weight determined for such a polymer by measurement of osmotic pressure is a *number-average* molecular weight.

In the ideally dilute limit, colligative properties depend on the *total* solute mole fraction or molality. If there are a number of different solutes present, this can be expressed as (for the case of freezing-point depression)

$$\theta_f = K_f m_B = K_f \sum_i m_i \tag{68}$$

Expressing m_B in terms of w_B, the total mass of all solutes added to a 1 kg of solvent,

$$m_B = \frac{w_B}{M_B} = \sum_i m_i \tag{69}$$

where M_B is the appropriate "average" molecular weight of the solutes. From this, we obtain

$$M_B = \frac{w_B}{\sum_i m_i} = \frac{\sum_i m_i M_i}{m_B} = \sum_i f_i M_i \tag{70}$$

where $f_i = m_i/m_B$ is the fraction of the moles (or molecules) of the solute that have molecular weight M_i. The type of average given in Eq. (70) is a number average. Thus, when a molecular weight is determined from a colligative property, as illustrated in Examples 2 and 3, a *number-average molecular weight* is obtained (i.e., all molecules contribute equally). This average is to be distinguished from molecular weights determined by physical methods, such as light scattering, in which the contributions of different molecules are proportional to their size, giving a *mass-average molecular weight*. By measuring different averages of the molecular weight, information about the distribution of the masses of molecules can be obtained.

8.7 Distribution of a Solute Between Two Solvents

Liquid-phase extraction is a procedure by which some fraction of a solute is taken out of solution by shaking the solution with a different solvent (in which the solute usually has greater solubility). The analysis of this process assumes that the shaking is sufficient so that equilibrium is established for the solute, i, between the two solutions. At equilibrium, the chemical potentials of the solute in the two solutions are equal. Assuming ideally dilute solutions, we can write

$$\mu_{i,1} = \mu_{i,1}^0 + RT \ln x_1 = \mu_{i,2} = \mu_{i,2}^0 + RT \ln x_2 \tag{71}$$

Defining K_{12}, the *distribution coefficient* of solute between solute 1 and solute 2, as

$$K_{12} \equiv \frac{x_1}{x_2} \tag{72}$$

we obtain

$$\ln K_{12} = -\frac{(\mu_{i,1}^0 - \mu_{i,2}^0)}{RT} = -\frac{\Delta_{trans} G^\circ}{RT} \tag{73}$$

where $\Delta_{trans} G^\circ$ is the standard free-energy change for transfer of one mole of i from solvent 2 to solvent 1. Equation (73) can be written as

$$K_{12} = \exp\left(\frac{\mu_{i,2}^0 - \mu_{i,1}^0}{RT}\right) = \exp\left(\frac{\mu_{i,2}^0 - \mu_{i,g}^0}{RT}\right) \exp\left(\frac{\mu_{i,1}^0 - \mu_{i,g}^0}{RT}\right)^{-1} \tag{74}$$

where $\mu_{i,g}^0$ is the standard (1.0 bar) chemical potential of i in the gas phase. Referring to Eq. (49), this becomes

$$K_{12} = \frac{K_2}{K_1} \tag{75}$$

(i.e., the distribution coefficient is the ratio of Henry's law coefficients for the solute in the two solvents). Note that the solute has a greater tendency to be distributed into the solvent in which it has the lower Henry's law constant (because it has a lower escaping tendency from that solvent).

Equation (75) may be derived more simply by using an alternative measure of escaping tendency, the partial pressure of i above the solution. This partial pressure must be equal above the two solutions between which the distribution of i is equilibrated (Problem 18). Extraction is not a colligative property because it is the solute, not the solvent, that is distributing between the two phases. The distribution coefficient, like the Henry's law constants, depends on the identity of the solutes, as well as that of the solvent.

8.8 Phase Diagram of a Binary Ideal Solution

In Chapter 6, we found that phase diagrams were an effective way of communicating information concerning phase transformations of single-component systems. These diagrams are even more essential when dealing with the more complicated phase transformations of multicomponent systems. A binary system has two components. For a binary solution in equilibrium with its vapor, there are two degrees of freedom ($f = 2 - 2 + 2 = 2$). If we investigate the phase diagram at a particular temperature, there is only one additional degree of freedom. A thermostat for preparing such a system is shown in Fig. 5. The container

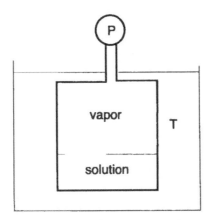

Figure 5 Thermostat for preparing a binary system.

containing the solution in equilibrium with its vapor is equipped with a pressure gauge and it is assumed that all of the air has been removed from the container.[5] Because there is only a single degree of freedom, reading the pressure on the gauge determines the concentration, x_B, of the liquid phase. For an ideal solution, Eq. (41) gives the relation between these two intensive variables. With only a single degree of freedom, the pressure measurement must also determine the concentration of the vapor phase. In fact, the vapor-phase mole fraction of B, $y_B = P_B/P$, can be easily calculated (see Problem 6). Both the liquid and vapor mole fractions as a function of P are plotted in Fig. 6, known as a *pressure versus concentration liquid–vapor-phase diagram*, or more simply as a *vapor-pressure diagram*. Following convention, pressure is plotted as the y-axis of this graph.

The concentrations of both the liquid phase, x_B, and the vapor phase, y_B, are determined by the pressure. The line relating vapor pressure to liquid-phase concentration is called the *bubble-point line*, because when the pressure on the liquid is reduced to this value, bubbles appear. The lower line, which relates vapor pressure to the vapor-phase concentration, is called the *dew-point line* because when vapor is compressed to give this pressure, liquid droplets appear on surfaces.

The overall concentration of the system, z_B, is not determined by the pressure because the overall concentration depends on the relative amounts of the two phases and is not an intensive degree of freedom of the system. Nevertheless, the overall concentration, being the weighted average of the liquid and vapor-phase concentrations, must lie between the bubble-point line and the dew-point line on a horizontal line at P. This line is known as the *tie line* (because it "ties together" the coexisting phases) and it lies in the region between the liquid and vapor curves, which is called the *two-phase region*. In this book, two-phase

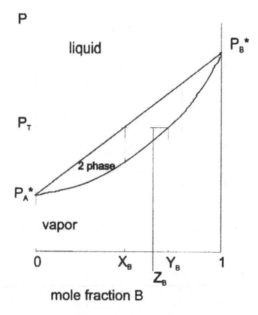

Figure 6 Lever rule.

regions will be shaded, but often in the literature they are not distinguished from single-phase regions. If the system consists mainly of the liquid phase, the point representing the overall concentration lies very close to the liquid curve on the tie line; whereas if it is mainly vapor phase, the point lies very close to the vapor line. In fact, the mole-weighted deviation of the mole fractions of these phases from that of the overall mole fraction balances around the system mole fraction.[6] This gives rise to the *lever rule*, which, referring to Fig. 6, can be written as

$$n_g(y_B - z_B) = n_l(z_B - x_B) \tag{76}$$

A more formal proof of this simple method of determining the relative amounts of the phases from the concentrations of the phases and the system is treated in Problem 7. Tie lines are not provided on phase diagrams, but are drawn by the user of the diagram at the pressure of interest.

It is important to note that the system consists of two phases only in the two-phase region. If the point representing the overall concentration and the pressure lies above the liquid curve in Fig. 6, the system consists only of liquid with $x_B = z_B$. With only a single phase, the system has three degrees of freedom, which are most conveniently taken as T, P, and $x_B = z_B$. Thus, at a given temperature, pressure and composition can both be varied, as long as the system remains in this one-phase region. In a similar manner, at pressures that put the point representing the system below the vapor curve in Fig. 6, only a single vapor phase, with $y_B = z_B$, exists.

An alternative method of studying two-component systems is by holding the pressure constant and allowing the temperature to vary. An apparatus for doing this is shown in Fig. 7. The constant pressure is completely established by the mass, M, on the weightless piston (and by atmospheric pressure). The system temperature is measured by the thermocouple. The *temperature–concentration liquid-vapor phase diagram* which describes this binary system at a given pressure is called a *boiling-point diagram* and is shown in Fig. 8. Note that the liquid region is at the bottom of the boiling-point diagram, whereas it is at the top of the vapor-pressure diagram. The two-phase region is enclosed by the bubble-point curve on the bottom and the dew-point curve on the top. Neither of these curves is a straight line, as there is no Raoult's law for boiling points. Note that in this ideal solution, the addition of small amounts of the more volatile component (B) to the less volatile component (A) *lowers* the boiling point of the solution. This is in contrast to the boiling-point elevation discussed regarding colligative properties, where it was assumed that only one component of the solution (the solvent) was volatile. Because the temperature is uniform throughout the system, when liquid and vapor coexist at equilibrium, these two phases are connected by a horizontal tie line. Tie lines and the lever rule may be used to determine the relative amounts of coexisting phases on boiling-point diagrams, just as they are used on vapor-pressure diagrams.

The tie line between liquid and vapor concentration shows that the vapor is enriched in the more volatile component. This is the basis for the process of

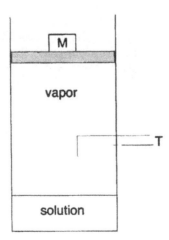

Figure 7 Apparatus for the two-component system allowing the temperature to vary.

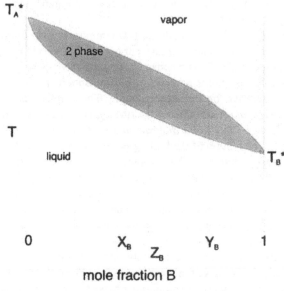

Figure 8   Boiling-point diagram.

distillation, where a solution is separated into two solutions: one enriched in the more volatile component (obtained by condensing the vapor) and the other enriched in the less volatile component. If these two solutions are then each subjected to additional distillation steps, further separation of the original solution into its two components is possible. In Fig. 9, this process is illustrated for the case in which, at each stage of distillation, half of the volume of solution is converted into vapor. $x_1, x_2, x_3, \ldots$ are the successive liquid-phase concentrations and $y_1, y_2, y_3, \ldots$ are the successive vapor phase concentrations. As can be seen, after a number of steps, separation into nearly pure components is achieved. Successive distillation is extremely tedious and practically never performed. *Fractional distillation* is a continuous method for carrying out successive distillations by continuously condensing and evaporating solutions in a distillation column. Industrial distillation columns may achieve more than 100 successive liquid–vapor equilibrations (each liquid–vapor equilibrium is known as a *theoretical plate*; see Problem 21), with resulting excellent separation of binary solutions into their components.

8.8.1 Solid–Liquid Equilibrium

Phase diagrams can present information on liquid–solid equilibrium, as well as on liquid–vapor equilibrium. At fixed pressure, a pure liquid freezes at a definite

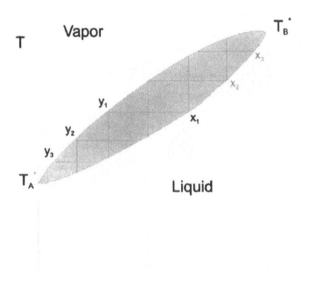

mole fraction B

Figure 9 Distillation process.

temperature. When *B* is added to liquid *A*, forming a very dilute solution, the freezing point of the solution is lowered. This behavior is shown in Fig. 10, where on the left-hand side of the diagram, it is assumed that the solid that freezes out is pure *A*. (A solid solution is not formed). The two-phase region represents equilibrium between pure *A* and the liquid solution.

This phenomenon can also be looked at in terms of the dissolution of solid *A* in the solution (even though *A*, because it is present at a greater concentration, would usually be considered the solvent of this solution). As the temperature is lowered, the solubility of *A* in the solution decreases. If the solution is ideal, we can use Eq. (55) with $a_A = x_A$ to calculate the solubility of *A* in the solution at temperatures below the freezing point of *A*:

$$\ln x_A = \frac{\Delta_{\text{fus}} H_A^*}{R} \left(\frac{1}{T_{f,A}^*} - \frac{1}{T} \right) \tag{77}$$

Equation (77) provides solubilities at temperatures much lower than $T_{f,A}^*$ with reasonable accuracy only in rare cases in which solutions follow ideal behavior.

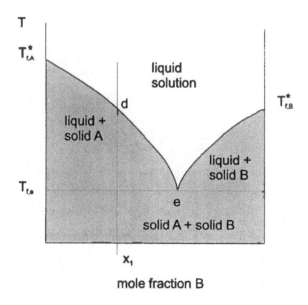

Figure 10   Solid–liquid phase diagram.

However, the general form of the curve on the left-hand side of Fig. 10 is followed even in nonideal systems. Looking at the right-hand side of Fig. 10, we see that as A is added to pure B, the freezing point of the solution is also lowered. (We have also assumed that B does not form solid solutions.) Alternatively, this can be described as reducing the solubility of B in the solution as the temperature of the solution is lowered below the freezing point of pure B. At an intermediate concentration, the two solubility curves in Fig. 10 meet at point e and a solution with the lowest freezing point is obtained. This solution is known as the *eutectic* and the concentration and temperature at point e, is known as the eutectic concentration and temperature.

It is interesting to consider what happens as solutions of different concentrations are cooled. The pure components, of course, freeze at a single temperature, and temperature remains constant until all of the liquid is frozen. As the solution of concentration x_1 is cooled, pure A begins to freeze out at point d, and the solution becomes more enriched in B. Microscopic investigation shows that fairly large crystals of A are formed at this point. With further cooling, the concentration of the remaining solution moves along the curve d–e. The relative amounts of pure A and the solution can be obtained by drawing the tie line at the appropriate temperature and using the lever rule. When the eutectic temperature at point e is reached, pure solid B begins to appear. There are now three phases in equilibrium: pure solid *A*, pure solid *B*, and the solution of the eutectic composition. Because pressure is fixed, the phase rule tells us that there are $f = 2 - 3 + 2 - 1$ (pressure) $= 0$ degrees of freedom. On a two-component phase diagram, the condition of three phases in equilibrium is represented by a horizontal line on the diagram.[7] No further reduction of temperature can occur until all the remaining eutectic solution is converted into *A* and *B*.

Microscopic investigation of the solid formed when a eutectic freezes indicates that it is a mixture of very fine particles of pure *A* and pure *B*, rather than a solid solution of *A* and *B*. If a solution that has the eutectic composition is cooled, freezing occurs at a single concentration, just like a pure component. A eutectic can be distinguished from a pure component either by microscopic inspection of the solid or by adding one of the pure components. In the eutectic case, the freezing point of the solution will be increased by this addition, rather than decreased, as occurs for a pure substance.

The structure of phase diagrams can be investigated by *thermal analysis*. This is performed by measuring *cooling curves*, which are temperature versus time diagrams for the cooling of the substance from the molten state. Assuming a constant rate of heat loss from the system, the slope of the cooling curve is inversely proportional to the heat capacity of the system. The cooling curve for pure component *A* for the solution of composition x_1, and for a solution of the eutectic composition are shown in Fig. 11. The cooling curves of the pure component and the eutectic both show horizontal segments as the liquid is

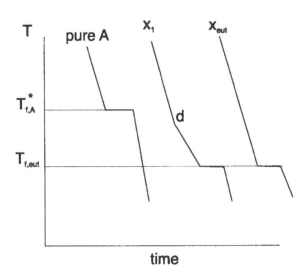

Figure 11 Cooling curve for component A, solution of composition x_1, and solution of the eutectic composition.

converted into solid at the melting point and the *eutectic halt*, respectively. The system can be considered to have infinite heat capacity at these temperatures. The solution of concentration x_1 first shows a *break* (a sudden change in slope) in its cooling curve at point d, as additional heat must be removed for solidification, and then a halt, as the remaining eutectic is converting into a solid. By determining the cooling curves at different concentrations, the phase boundaries can be determined.

Questions

1. Which of the following properties of a given solution will change as the temperature of the solution is increased?

    (a) Density
    (b) Molarity
    (c) Molality
    (d) Mole fraction
    (e) Partial pressure (of a gaseous solution)

2. Dissolving a gas in a liquid can be thought of as a two-step process: First, cavities are made in the liquid and then the gas molecules are placed in the cavities. Wolfenden and Radzicka (Science 265:936, 1994) have noted that the solubility of water in cyclohexane at 26°C is equal to the concentration of water vapor in equilibrium with liquid water at this temperature. Show that this observation implies that the Gibbs free energy to make water-sized cavities in cyclohexane is equal to the Gibbs free energy of attraction of water molecules to cyclohexane in these cavities.

3. For which of the following solutions at 298 K would you expect Raoult's law to be a good approximation for all components:

    (a) $C_2H_5{}^{35}Cl–C_2H_5{}^{37}Cl$
    (b) Benzene–toluene
    (c) Benzene–naphthalene
    (d) Benzene–CCl_4

4. For a solid, there is often a "solubility" (i.e., a maximum amount that will dissolve in a solvent at a given temperature). Is there a "solubility" for a gas in a solvent? What is the difference?

5. The maximum pressure in an auto radiator is determined by the release pressure of the radiator cap. Do a phase-rule analysis and determine whether each of the following statements is true or false:

    (a) With just water in the radiator, the maximum temperature is completely determined by the release pressure.
    (b) With a water–ethylene glycol (antifreeze) solution in the radiator, the maximum temperature can be varied by changing the concentration of the solution.

6. The heat of solution for NaCl in water is 3.9 kJ/mol. Because this process in endothermic, how can you explain the large solubility of NaCl in water? Will this solubility increase or decrease as the temperature is increased?

7. For an ideal solution, is each of the following greater than, less than, or equal to zero or is there no way to tell?

    (a) The partial pressure of the less volatile component minus the vapor pressure of pure less volatile component

    (b) The partial pressure of the more volatile component minus the vapor pressure of pure less volatile component

    (c) The total pressure above the solution minus the vapor pressure of the less volatile component

    (d) The total pressure above the solution minus the vapor pressure of the more volatile component

8. Show that in the ideal solution limit, for a group of solvents obeying Trouton's law, the fractional change of boiling point produced by a given mole fraction of solute is the same for all the solvents.

9. The Henry's law constant of N_2 in water reaches a maximum value at $\sim 60°C$. What does this imply will happen as water in equilibrium with the N_2 of air is heated from room temperature to $60°C$?

10. Referring to Fig. 5, if at fixed temperature there are two nonreacting, nonvolatile solutes in an ideally dilute solution,

    (a) does measuring the pressure of the solution determine x_B?

    (b) does measuring the pressure completely determine the system?

11. Referring to Fig. 6, if the pressure over a liquid of mole fraction of B of 0.4 is slowly reduced, what is the mole fraction of the first drop of vapor that is formed? If the reduction in pressure continues so that the liquid is continuously converted into gas, what is the mole fraction of B in the last drop of liquid to evaporate?

12. Before ethylene glycol [melting point (mp) $= -13°C$; boiling point (bp) $= 199°C$] became widely adopted as automobile radiator antifreeze, methanol (bp $= 65°C$) was used for this purpose. Methanol is less expensive. However, unlike ethylene glycol, it is not an "all-season" coolant and must be removed from the radiation in hot weather. Explain this difference.

13. Assuming that you have a very accurate and sensitive thermocouple, how would you use it to determine the freezing point of a solution?

14. Although Raoult's law is not accurate for the solubility of gases in solvents, its use can explain some rough trends in solubility. Using Raoult's law show that the solubility of a gas in a liquid is

    (a) proportional to the partial pressure of the gas

    (b) reduced as temperature is increased

    (c) greater for gases with higher boiling points.

Problems

1. An amount of NaCl is dissolved in water at 25°C to make a 0.0200 M solution. What is the molarity of this solution at 80°C? You can take the thermal expansion coefficient as that of water, 2.1×10^{-4} K^{-1}, over the temperature range of this problem. What is the molality of the solution at 25°C and 80°C?

2. An aqueous solution containing 45.02 g of NaCl per liter of solution has a density of 1.029 kg/L. Calculate the molarity, molality, and mole fraction of the solution.

3. Derive Eq. (5) and find an expression for c_i in terms of the mole fractions in the solution.

4. In 1.0 M methanol in water solution, how far apart, on average, are the methanol molecules? Treating the water molecules as spheres of 0.3 nm diameter, how many water molecules on the average are between two methanol molecules?

5. Derive Eq. (21).

6. Show that the mole fraction of B in the vapor phase above an ideal binary solution consisting of components A and B with vapor pressures P_A^* and P_B^* is $y_B = P_B^*(P - P_A^*)/P(P_B^* - P_A^*)$, when the solution vapor pressure is P.

7. Prove Eq. (76). (Suggestion: Write the number of moles of B in the system, n_B, in two different ways: as z_B times the total moles of the system and as the sum of the moles of B in each phase. Equate these two expressions for n_B.)

8. Derive Eq. (54).

9. Assume that benzene and naphthalene form an ideal solution and that the solids are pure components (no solid solutions are formed) with melting points at 1.0 atm pressure of 5.5°C and 80.5°C. Estimate the composition and melting point of the benzene–naphthalene eutectic using Eq. (77).

10. What is the entropy change when 20 g of benzene and 20 g of toluene are mixed to form an ideal solution?

11. The vapor pressures of benzene and toluene at 90°C are 1034 and 413 torr, respectively. What is the mole fraction of benzene in the benzene–toluene solution that boils at 90°C?

12. A trimeric protein, P_3, dissociates completely into three monomer units in solution. What is the chemical potential and activity of P_3 in terms of those of the monomer?

13. Reverse osmosis is a process whereby seawater can be purified by passing the water molecules through a semipermeable membrane. What is the minimum amount of work that must be done to obtain 1.0 L of pure water by reverse osmosis, starting with seawater that is 0.7 M NaCl? Compare this work with the energy required to vaporize the water in a distillation process.

14. An air-breathing creature might be able to breathe with its lungs filled with a liquid, if concentrations (mol/L) of oxygen in the liquid were comparable to the concentration of oxygen in ambient air. What Henry's law constant for oxygen would be required for a

liquid so that it contains an oxygen concentration equal to the air when it is equilibrated with ambient air?

15.   Show that for a substance obeying Trouton's rule, the boiling-point elevation constant is proportional to the boiling point.

16.   Pure benzene freezes at 5.50°C and has a density of 0.876 g/mL. A solution of 1.7 g of nitrobenzene in 250 mL benzene freezes at 5.18°C. What is the molality-based freezing-point depression constant of benzene and at what temperature does a solution containing 3.2 g of bromobenzene in 250 mL of benzene freeze? (You may make the ideally dilute approximation for both these solutions.)

17.   The Henry's law constants of N_2 and O_2 in water at 0°C are 5.2×10^{-4} atm^{-1} and 2.8×10^{-4} atm^{-1}, respectively. What is the freezing point of water in equilibrium with 1.0 atm air (78% N_2 and 21% O_2)?

18.   Derive Eq. (75) by equating the partial pressure of solute above the two solutions between which it is equilibrated.

19.   Show that

$$\frac{d \ln K_{12}}{dT} = \frac{\Delta_{\text{trans}} H^\circ}{RT^2}$$

where K_{12} is the distribution coefficient of a solute between solvents 1 and 2 and $\Delta_{\text{trans}} H^\circ$ is the standard molar enthalpy change for transfer of the solute from solvent 2 to solvent 1.

20.   Show that the chemical potentials for the ideal solution satisfy the Gibbs–Duhem relation [Eq. (29)].

21.   A *theoretical plate* is defined as the degree of separation attained for an infinitesmal vaporization at equilibrium (i.e., the concentration of liquid in a theoretical plate is that of the first bit of vapor to be formed from the liquid in the previous one). Using this definition, approximately how many theoretical plates would be required to achieve a separation into a vapor with $x_B = 0.1$ and a liquid with $x_B = 0.9$ for the system described by the boiling-point diagram in Fig. 9.

22.*   In the ideal macromolecular solution, the solute molecules consist of a large number, x, of segments, but each segment has size and attractive interactions similar to that of the solvent molecules. Several statistical theories have given as the entropy of mixing of such a solution from n_A mol of solvent and n_B mol of macromolecule

$$\Delta_{\text{mix}} S = -R\left[n_A \ln\left(\frac{n_A}{n_A + xn_B}\right) + n_B \ln\left(\frac{n_B}{n_A + xn_B}\right)\right]$$

(See PJ Flory. Principles of Polymer Chemistry. Ithaca, NY Cornell Press, (1953, pp 495–503.) $\Delta_{\text{mix}} H$ is zero.

(a) Use the Gibbs free energy of this solution to show that

$$\mu_A = \mu_A^0 + RT \ln\left(\frac{n_A}{n_A + x n_B}\right) + RT\left(1 - \frac{n_A + n_B}{n_A + x n_B}\right)$$

$$\mu_B = \mu_B^0 + RT \ln\left(\frac{x n_B}{n_A + x n_B}\right) + RT\left(1 - \frac{x(n_A + n_B)}{n_A + n_B}\right)$$

(b) Show that these chemical potentials satisfy the Gibbs–Duhem relation [Eq. (29)].

(c) Show that for a very dilute solution, $\ln a_A \approx -x_B$, as would be required to give Eqs. (61)–(63) for the colligative properties of such a solution.

23. What are $\Delta_{mix}H$, $\Delta_{mix}S$, and $\Delta_{mix}G$ of an ideally dilute solution?

Notes

1. There are some exceptions to this for fluids above the critical temperature. See RP Gordon. J Chem Educ 49:249, 1972.
2. As discussed in Chapter 7, the reference states of elements are chosen as the state that the element is stable in at 1.0 bar pressure and the temperature of interest.
3. Some authors write this as m_1/m^0 and c_i/c^0, where $m^0 = 1.0$ *molal* and $c^0 = 1.0$ *molar*.
4. $\ln(1 - x_B) = -x_B - \frac{1}{2}x_B^2 - \cdots$.
5. From the result obtained in Chapter 6, we do not expect the presence of atmospheric pressure to appreciably affect the vapor pressure of the solution.
6. In much the same way that the average grade in a three-student class with grades of 60, 60, and 90 is 70 [2(70 − 60) = 90 − 70].
7. This should distinguished from a horizontal line drawn in a two-phase region of the diagram by the reader to connect two phases that are at equilibrium.

9

Nonideal Solutions

If you're not part of the solution, you're part of the precipitate.

Harry J. Tillman

Nonideal solutions are described by activity coefficients referred to either the ideal solution or the ideally dilute solution. Activity coefficients may be determined from partial pressures and distribution coefficients. Often, the activity coefficient of the solvent is determined from a colligative property, and that of the solute by integrating the Gibbs–Duhem equation, most conveniently by using the osmotic coefficient. Liquid–vapor-phase diagrams of nonideal systems can show maximum or minimum boiling-point azeotropes. Liquids may be partially miscible. Solids can form solid solutions and compounds, which may melt congruently or incongruently. Ternary systems can be described by triangular phase diagrams.

9.1 Activity Coefficients

Unlike the model of the ideal gas, which is an accurate representation of the behavior of real gases under usual laboratory conditions, both the ideal solution and ideally dilute solution models have limited practical applicability. The ideal solution model provides sufficient accuracy to describe real solutions only in very special cases. These are solutions in which the components are very similar in size, shape, and interaction with one another. The ideally dilute solution model, on the other hand, although applicable to all solutions, has practical use only at very low concentrations, where solutes only interact with surrounding solvent molecules. In fact, for the case of ions in solution, interactions between charged particles extend to such a large separation that the ideally dilute solution is not useful under almost all conditions. Both the ideal and ideally dilute solution models, however, provide particularly simple forms for the variation of the thermodynamic properties of solutions with concentration. In our discussion of real solutions, we will maintain this simple form for these variations.

Because all thermodynamic functions can be obtained from the chemical potential, we will concentrate on maintaining the form of μ. In Eq. (47) of Chapter 6, the chemical potential was expressed in terms of activity as

$$\mu = \mu^0 + RT \ln a \tag{1}$$

In both the ideal and the ideally dilute solution models, the activity can be set equal to the mole fraction, $a = x$, for all components. In order to maintain the form of Eq. (1) for real solutions, we define an *activity coefficient*, γ, so that

$$a = \gamma x \tag{2}$$

$$\mu = \mu^0 + RT \ln \gamma x \tag{3}$$

holds for all components of a solution. The value of γ depends on that for μ^0, the chemical potential in the standard state. The two references used for comparison with the properties of real solutions are the ideal solution and the ideally dilute solution.

When the ideal solution is used as the reference for real solutions, thermodynamic properties are designated by (RL) for *Raoult's law reference*. This reference is often used in solutions in which all solutes are liquids at the temperature of interest, especially when the compositions of components are varied over a considerable range. In this case, for every component, we write Eq. (3) as

$$\mu_i(RL) = \mu_i^* + RT \ln \gamma_i x_i \tag{4}$$

(i.e., the chemical potential in the standard state is that of the pure component). Because the solution approaches pure i as $x_i \rightarrow 1$,

$$\gamma_i(RL) \rightarrow 1 \quad \text{as } x_i \rightarrow 1 \tag{5}$$

At other concentrations, deviations of $\gamma_i(RL)$ from unity measure deviations of the solutions from ideal behavior.

When the ideally dilute solution is used as the reference for real solutions, thermodynamic properties are designated by (HL) for *Henry's law reference*. This reference is always used when some components are not liquids at the temperatures employed and may also be used if they are all liquids, but only very dilute solutions are being considered. For this reference, we treat the solvent in the same manner as for the (RL) reference:

$$\mu_A(HL) = \mu_A^* + RT \ln \gamma_A x_A \tag{6}$$

with

$$\gamma_A \to 1 \quad \text{as } x_A \to 1 \tag{7}$$

For the solutes in the (HL) reference, we compare behaviors with that in the infinitely dilute solution ($x_i \to 0$, $x_A \to 1$):

$$\mu_i(HL) = \mu_i^0 + RT \ln \gamma_i x_i \tag{8}$$

with

$$\gamma_i \to 1 \quad \text{as } x_i \to 0 \tag{9}$$

The standard state for solutes in the (HL) reference is therefore the hypothetical state of pure solute ($x_i = 1$), but with solute molecules interacting only with solvent molecules ($\gamma_i = 1$). Practically, chemical potentials in the standard state are obtained by making measurements at very low concentrations and extrapolating them to $x_i = 1$, assuming that Henry's law continues to hold to this concentration. At nonzero concentration of solutes, activity coefficients in the (HL) reference measure deviations of the solution from ideally dilute behavior.

Activities in the (HL) reference are more usually expressed in terms of molality:

$$a_i = \gamma_{m,i} m_i \tag{10}$$

$$\mu_i(HL) = \mu_{m,i}^0 + RT \ln \gamma_{m,i} m_i \tag{11}$$

with

$$\gamma_{m,i} \to 1 \quad \text{as } m_i \to 0 \tag{12}$$

Alternatively, molarity can be used:

$$a_i = \gamma_{c,i} c_i \tag{13}$$

$$\mu_i(HL) = \mu_{c,i}^0 + RT \ln \gamma_{c,i} c_i \tag{14}$$

with

$$\gamma_{c,i} \to 1 \quad \text{as } c_i \to 0 \tag{15}$$

The standard states for the solute in these expressions are the hypothetical state of $1.0\,m$ or $1.0\,M$, respectively, but with solute molecules interacting only with solvent molecules at these concentrations. In both Eqs. (11) and (14), it is the numerical value of the molality or molarity of the solution that is used, because it is not permissible to take the logarithm of quantities with units. At higher concentrations, the activity coefficients in Eqs. (11) and (14) contain a contribution from the nonproportionality of molality and molarity to mole fraction [see Eqs. (4) and (5) of Chapter 8, as well as that due to the system not following the ideally dilute law [Eq. (43) of Chapter 8].

9.2 Excess Thermodynamic Functions

An alternative way of expressing the deviation from ideal solution behavior is by means of excess thermodynamic functions. These are defined as the difference between a thermodynamic property of a solution and the thermodynamic property it would have if it were an ideal solution:

$$X^E = X - X^{id} \tag{16}$$

Adding and subtracting the thermodynamic property for the unmixed components,

$$X^E = (X - \sum_i n_i X_{m,i}) - (X^{id} - \sum_i n_i X_{m,i}) = \Delta_{mix} X - \Delta_{mix} X^{id} \tag{17}$$

Thus, an excess thermodynamic property is also the difference between the thermodynamic property for mixing the real and ideal solutions. For the Gibbs free energy, this becomes, using Eq. (3) and Eq. (35) of Chapter 8,

$$G^E = RT \sum_i n_i \ln \gamma_i x_i - RT \sum_i n_i \ln x_i = RT \sum_i n_i \ln \gamma_i = H^E - TS^E \tag{18}$$

where the (RL) reference is used. Because $\Delta_{mix} H^{id} = \Delta_{mix} V^{id} = 0$,

$$H^E = \Delta_{mix} H \tag{19}$$

$$V^E = \Delta_{mix} V \tag{20}$$

In the usual manner, we can show that

$$H^E = \left(\frac{\partial (T^{-1} G^E)}{\partial T^{-1}} \right)_{P,x} \tag{21}$$

$$S^E = - \left(\frac{\partial G^E}{\partial T} \right)_{P,x} \tag{22}$$

$$V^E = \left(\frac{\partial G^E}{\partial P} \right)_{T,x} \tag{23}$$

9.3 Determining Activity Coefficients

9.3.1 Activity Coefficients from Partial Pressures of Vapors

Because activities, activity coefficients, chemical potentials, and partial pressures are all measures of escaping tendency, one of the most straightforward ways of measuring activity coefficients is by measuring partial pressures of vapors in equilibrium with solutions. We will assume that the vapors may be treated as ideal gases, because deviations from ideal behavior are generally much smaller in gases than in solutions. Equating the chemical potential of a component of the vapor in equilibrium with a solution with that of the component in the solution, we have for the (RL) reference state

$$\mu_i(RL) = \mu_{i,l}^* + RT \ln \gamma_i x_i = \mu_{i,g} = \mu_{i,g}^0 + RT \ln(P_i/P^0) \tag{24}$$

Doing the same for the pure component,

$$\mu_{i,l}^* = \mu_{i,g}^0 + RT \ln(P_i^*/P^0) \tag{25}$$

Combining these two equations gives

$$RT \ln \gamma_i x_i = RT[\ln(P_i/P^0) - \ln(P_i^*/P^0)] \tag{26}$$

or

$$\gamma_i = \frac{P_i}{x_i P_i^*} \tag{27}$$

In other words, the activity coefficient is just the ratio of the partial pressure of the component to the partial pressure expected if the solution were ideal.

In Fig. 1, the partial pressures of two different nonideal solutions are contrasted with ideal behavior ($P_i = x_i P_i^*$), shown as the central straight line. In the upper curve, the partial pressure is greater than an ideal solution. This is called *positive deviation* from ideal behavior and gives an activity coefficient greater than unity, and, from Eq. (1), an activity greater than the mole fraction. This behavior gives a positive contribution to the excess Gibbs free energy. Positive deviation for a component results when it feels less attraction to molecules of other components of the solution than it does to molecules of itself. In the lower curve, representing *negative deviation* from ideal behavior, the escaping tendency of the component is lower than in the ideal solution, and this is indicated by $P_i < x_i P_i^*$, $\gamma_i < 1$, $a_i < x_i$, and a negative contribution to the excess Gibbs free energy. It is noteworthy that as $x_i \rightarrow 1$, both the lower and upper curves become tangent to the line representing ideal solution behavior, because in

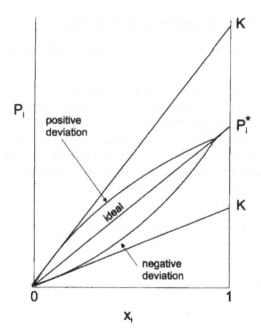

Figure 1   Positive and negative deviations for ideal behavior.

a practically pure liquid, molecules interact predominantly with similar molecules.

Example 1.   Partial pressures of chloroform and acetone solutions (in torr at 35.2°C) above a solution of these components are given below. Calculate the Raoult's law reference activity coefficients for each component at each composition.

x_{chl}	0	0.060	0.184	0.263	0.361	0.424	0.508	0.581
P_{chl}	0	9	32	50	73	89	115	140
P_{Ac}	345	323	276	241	200	174	138	109

x_{chl}	0.662	0.802	0.918	1.00
P_{chl}	170	224	266	293
P_{Ac}	79	38	13	0

(Data from JH Hildebrand, RL Scott. Solubilities of Nonelectrolytes. New York: Reinhold, (1950, p 181.) These original data are plotted in Fig. 2.

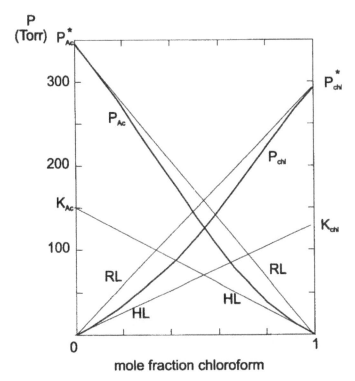

Figure 2 Acetone–chloroform solution partial pressures.

Solution: We use Eq. (27) for both components. A sample calculation is

$$x_{chl} = 0.424, \qquad x_{Ac} = 1 - 0.42 = 0.576$$

$$\gamma_{chl}(RL) = \frac{89}{0.424(293)} = 0.72$$

$$\gamma_{Ac}(RL) = \frac{174}{0.576(345)} = 0.87$$

The results are plotted in Fig. 3. As can be seen, with the Raoult's law reference, the acetone–chloroform system shows negative deviation from ideal behavior. This is unusual and is due to there being some tendency to form hydrogen bonds between acetone and chloroform. Note that as the system approaches either of the pure components, the vapor-pressure curve of that component becomes tangent to its Raoult's law line.

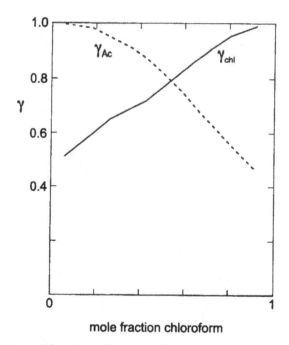

mole fraction chloroform

Figure 3 Acetone–chloroform solution Raoult's law activity coefficients: solid curve: γ_{chl}; dashed curve: γ_{Ac}.

For the (HL) reference state, the chemical potential of the solvent is the same as in the (RL) reference state, so that Eq. (27) holds for the solvent. For solutes, with the (HL) reference state, we have

$$\mu_i(HL) = \mu_{i,l}^0 + RT \ln \gamma_i x_i = \mu_{i,g} = \mu_{i,g}^0 + RT \ln(P_i/P^0) \quad (28)$$

or

$$\exp\left(\frac{\mu_{i,l}^0 - \mu_{i,g}^0}{RT}\right) = \frac{P}{P^0 \gamma_i x_i} \quad (29)$$

Using Eq. (49) of Chapter 8,

$$\gamma_i = \frac{P}{K_i x_i} \quad (30)$$

In other words, the activity coefficient is just the ratio of the partial pressure of the component to the partial pressure expected if Henry's law held to the concentration of interest. In order to use Eq. (30), the Henry's law constant for the particular solute–solvent combination is needed. This can be obtained from vapor pressure measurements on solutions dilute enough so that γ_i can be taken as unity.

Example 2. Using Fig. 2, what is the Henry's law constant treating acetone as the solute and what is the Henry's law constant treating chloroform as the solute? Calculate Henry's law reference activity coefficients for both of these cases.

Solution: The Henry's law constants are found by drawing the tangent lines to the vapor-pressure curves and measuring their slopes. Because these lines begin at the origins ($x_i = 0$ and $P_i = 0$), the slopes are just the intercept with the $x_i = 1$ axis, or from Fig. 2, $K_{Ac} = 150$ torr and $K_{chl} = 140$ torr. The accuracy of this procedure is limited by how well the initial slopes can be determined from limited data. (See Problem 6.)

A sample calculation of the Henry's law reference activity coefficients is

$$x_{chl} = 0.424, \qquad x_{Ac} = 1 - 0.42 = 0.576$$

$$\gamma_{chl}(RL) = \frac{89}{0.424(140)} = 1.50$$

$$\gamma_{Ac}(RL) = \frac{174}{0.576(150)} = 2.01$$

The Henry's law reference activity coefficients are plotted in Fig. 4. Note that this system shows positive deviation with respect to the ideally dilute behavior of Henry's law.

9.3.2 Activity Coefficients from Distribution Coefficients

When a solute is distributed at equilibrium between two solvents, its vapor pressure, which is a measure of its escaping tendency, must be equal above the two solvents:

$$P_i = \gamma_1 K_1 x_1 = \gamma_2 K_2 x_2 \tag{31}$$

The distribution coefficient is then

$$K_{12} \equiv \frac{x_1}{x_2} = \frac{\gamma_2 K_2}{\gamma_1 K_1} \tag{32}$$

This result holds even if the partial pressure of the solute above the solvents is immeasurably small.

Example 3. K_{12} of I_2 between glycerine and CCl_4 (x_{gly}/x_{CCl_4}) is 0.50 at $x_{gly} = 0.0001$ and is 0.40 at $x_{gly} = 0.002$. Assuming that both solutions are ideally dilute at the lower concentration and that the

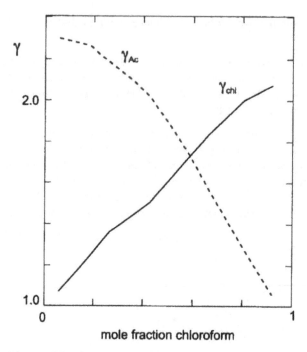

Figure 4 Acetone–chloroform solution Henry's law activity coefficients: solid curve for chloroform with acetone as the solvent, dashed curve for acetone with chloroform as the solvent.

CCl_4 solution is also ideally dilute at the higher concentration, find γ_{gly} at $x_{gly} = 0.002$.

Solution: At $x_{gly} = 0.0001$, both solutions are ideal so $K_{CCl_4}/K_{gly} = 0.50$. At $x_{gly} = 0.002$, $0.40 = (1/\gamma_{gly})0.50$, where $\gamma_{gly} = 0.50/0.40 = 1.25$.

Note that if there is some mutual solubility of the two solvents, activity coefficients measured by this method are actually for solvents which are the dilute solutions of one of the solvents in the other.

9.3.3 Activity Coefficients from Colligative Properties

The phenomena of freezing-point depression, boiling-point elevation, and osmotic pressure all result from the reduction in the escaping tendency of solvent in a solution due to the addition of solute. They can all be used to determine the activity of the solvent. For example, as long as pure solvent freezes out of the solution, Eq. (55) of Chapter 8 for the freezing point of a solution holds and can be rearranged to

$$\gamma_A = \frac{1}{x_A} \exp\left[\frac{\Delta_{fus}H^*}{R}\left(\frac{1}{T_f^*} - \frac{1}{T_f}\right)\right] = \frac{1}{x_A} \exp\left(-\frac{\Delta_{fus}H^*\theta_f}{R(T_f^*)^2}\right) \tag{33}$$

Likewise, if only the solvent in the solution is volatile, Eq. (56) of Chapter 8 for the boiling point of a solution holds and can be rearranged to

$$\gamma_A = \frac{1}{x_A} \exp\left[\frac{\Delta H^*_{vap}}{R}\left(\frac{1}{T_b} - \frac{1}{T_b^*}\right)\right] = \frac{1}{x_A} \exp\left(-\frac{\Delta H^*_{vap}\theta_b}{R(T_b^*)^2}\right) \tag{34}$$

Finally, if only the solvent passes through a semipermeable membrane, Eq. (58) of Chapter 8 holds and can be rearranged to

$$\gamma_A = \frac{1}{x_A} \exp\left(-\frac{\Pi V_m^*}{RT}\right) \tag{35}$$

Example 4. The freezing point of a 20% by weight aqueous solution of ethanol is 10.92°C. What is the activity coefficient of water in this solution?

Solution:

$$x_A = 80\,g\frac{mol}{18\,g}\left(80\,g\frac{mol}{18\,g} + 20\,g\frac{mol}{46\,g}\right)^{-1} = 0.911$$

From Eq. (64) of Chapter 8 and Table 1 of Chapter 8, we have, for water,

$$\frac{R(T_f^*)^2}{\Delta_{fus}H^*} = \frac{1.86\,K\,kg}{mol}\frac{mol}{0.018\,kg} = 103\,K$$

From Eq. (33),

$$\gamma_A = \frac{1}{0.911}\exp\left(-\frac{10.92\,K}{103\,K}\right) = 0.987$$

9.3.4 Activity Coefficient of the Solute from That of the Solvent

Quite often, the solute in a solution is not volatile and, thus, its escaping tendency cannot be directly measured. However, in a *binary* solution, changes in the escaping tendency of the solute and solvent are related by the Gibbs–Duhem equation [Eq. (15) of Chapter 8], which, when written for partial molar Gibbs free energies (chemical potentials), becomes (at constant temperature)

$$n_A \, d\mu_A + n_B \, d\mu_B = n_A RT \, d \, \ln a_A + n_B RT \, d \, \ln \, a_B = 0 \tag{36}$$

or

$$n_A \left(d \, \ln \, \gamma_A + \frac{dx_A}{x_A} \right) + n_B \left(d \, \ln \, \gamma_b + \frac{dx_B}{x_B} \right) = 0 \tag{37}$$

Because $x_A + x_B = 1$ and $dx_A + dx_B = 0$ in a binary solution, we have

$$d \, \ln \, \gamma_B = -\frac{x_A}{x_B} \, d \, \ln \, \gamma_A \tag{38}$$

A relation such as Eq. (38) between the activity coefficient of the solvent and solute of a two-component solution is expected, because such a solution can only have a single degree of freedom at fixed T and P. The negative sign in this equation indicates that the activity coefficients of the solvent and solute change in the opposite direction and that the deviation from ideal behavior must both be positive or both be negative for the two components of the solution. The magnitude of the change is larger for the component present in a lower concentration. This is illustrated in Fig. 3 for the acetone–chloroform system.

Substituting $x_B = 1 - x_A$ and integrating for a Raoult's law reference where $x_A = 0$, $x_B = 1$, and $\gamma_B = 1$,

$$\ln \, \gamma_B = - \int_{x_A=0}^{x_A} \frac{x_A}{1-x_A} \, d \, \ln \, \gamma_A \tag{39}$$

Example 5. Use the Raoult's law reference activity coefficients calculated for chloroform in Example 1 to calculate the activity coefficient of acetone at $x_{chl} = 0.508$ in a chloroform–acetone solution.

Solution: The calculation is shown in the following table. The integral in Eq. (36) of Chapter 6 is evaluated stepwise, using the trapezoidal rule. A is chloroform.

x_A	0	0.060	0.184	0.263	0.361	0.424	0.508
$\frac{x_A}{1-x_A}$	0	0.064	0.225	0.357	0.565	0.736	1.032
γ_A	0.50	0.512	0.594	0.649	0.690	0.716	0.772
Area		0.00077	0.0214	0.0259	0.0281	0.0241	0.0663

Total area $= 0.167$ $\ln \, \gamma_B = -0.167$ $\gamma_B = 0.846$

This compares with $\gamma_{Ac} = 0.813$ calculated directly from the partial pressure of acetone at this mole fraction. The difference is probably primarily due to the large intervals used in evaluating the integral.

If the solute is a solid, we will usually want to use a molality-based Henry's law reference, with $a_B = \gamma_B m$. Equation (15) of Chapter 8 then becomes

$$d \ln a_B = d \ln \gamma_B + d \ln m = -\frac{n_A}{n_B} d \ln a_A \tag{40}$$

To avoid singularities in the integrand when integrating Eq. (40), it is advantageous to define a quantity ϕ, called the *osmotic coefficient*, as

$$\phi \equiv -\frac{n_A}{n_B} \ln a_A \tag{41}$$

ϕ can be determined by any of the methods that we have discussed for measuring activity coefficients of the solvent. From Eq. (41), we have

$$d \ln a_A = -\frac{n_B}{n_A} d\phi - \phi d\left(\frac{n_B}{n_A}\right) \tag{42}$$

Substituting into Eq. (40) gives

$$d \ln \gamma_B = d\phi + \phi d\left(\frac{n_B}{n_A}\right)\left(\frac{n_B}{n_A}\right)^{-1} - \frac{dm}{m} \tag{43}$$

Because n_B/n_A is proportional to m, Eq. (43) can be written

$$d \ln \gamma_B = d\phi + \frac{(\phi - 1)}{m} dm \tag{44}$$

Integrating with a lower limit of the infinitely dilute solution (where $\gamma_B = 1$, $\ln \gamma_B = 0$, $a_A = x_A = 1$, $\phi = 1$),[1]

$$\ln \gamma_B = (\phi - 1) + \int_0^m \frac{(\phi - 1)}{m'} dm' \tag{45}$$

(A prime has been used to distinguish the integration variable from the molality of the solution.)

9.4 Equilibrium Constants

Equation (2) can be combined with Eq. (36) of Chapter 7 to give

$$\Delta_{rxn} G^\circ = -RT \ln K_a = -RT \ln \prod_i a_i^{\nu_i} = -RT \ln \prod_i \gamma_i^{\nu_i} x_i^{\nu_i} \tag{46}$$

Performing the products over activity coefficients and mole fractions separately,

$$\Delta_{rxn}G° = -RT \ \ln\left(\prod_i \gamma_i^{\nu_i} \ \prod_i x_i^{\nu_i}\right) = -RT \ \ln K_\gamma K_x \tag{47}$$

Because K_γ depends on concentrations and the product $K_\gamma K_x$ is concentration independent, K_x must also depend on concentration. This shows that the simple equilibrium calculations usually carried out in first courses in chemistry are approximations. Actually such calculations are often rather poor approximations when applied to solutions of ionic species, where deviations from ideality are quite large. We shall see that calculations using Eq. (47) can present some computational difficulties. Concentrations are needed in order to obtain activity coefficients, but activity coefficients are needed before an equilibrium constant for calculating concentrations can be obtained. Such problems are usually handled by the method of successive approximations, whereby concentrations are initially calculated assuming ideal behavior and these concentrations are used for a first estimate of activity coefficients, which are then used for a better estimate of concentrations, and so forth. $\Delta_{rxn}G°$ is calculated with the standard state used to define the activity. If molality-based activity coefficients are used, the relevant equation is

$$\Delta_{rxn}G° = -RT \ \ln K_\gamma K_m \tag{48}$$

whereas if molarity-based activity coefficients are used,

$$\Delta_{rxn}G° = -RT \ \ln K_\gamma K_c \tag{49}$$

is employed.

9.5 Phase Diagrams of Binary Nonideal Systems

9.5.1 Liquid–Vapor-Phase Diagrams

Positive deviation from ideal behavior is the usual occurrence for solutions of volatile components, and it results either when solute–solute and solvent–solvent interactions are stronger than solute-solvent interactions or when the addition of the solute breaks up "structure" (usually due to hydrogen-bonding) in the solvent. A case of mildly positive deviation is illustrated by the diethyl ether–ethanol system shown in Fig. 5. Here, the resulting total vapor pressure of the solution increases continuously as the concentration of the more volatile component (diethyl ether) is increased.

A more extreme case of positive deviation is illustrated by the ethanol–benzene system shown in Fig. 6, where the addition of benzene breaks up the hydrogen-bond interactions in ethanol. The total pressure in this system shows a

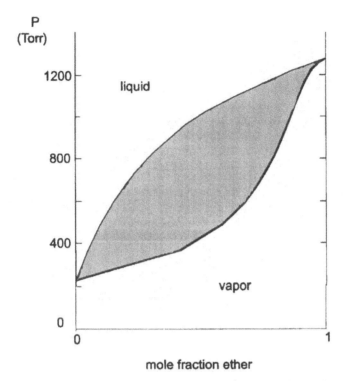

Figure 5 Ethyl ether–ethyl alcohol vapor-pressure diagram at 50°C. (Data from J Timmermans. Physiochemical Constants of Binary Systems in Concentrated Solutions. Volume 2. New York: Interscience, 1960, p 401.)

maximum as a function of concentration. This maximum is called an *azeotrope* and the concentration at the maximum is known as the azeotropic concentration.

The boiling-point diagrams, corresponding to Fig. 6 is shown in Fig. 7. Corresponding to the maximum vapor pressure of the azeotrope is its minimum boiling point, which may not occur at exactly the same concentration, depending on the pressure for which the diagram is drawn. Another example of a minimum-boiling-point azeotrope is the water–ethanol system, which forms an azeotrope with 4% water (by mass).

In systems with negative deviation from ideal behavior, maximum-boiling-point azeotropes can occur. This is illustrated in Fig. 8 for the chloroform–acetone system, treated in Example 1. This system shows negative deviation from ideal behavior due to the possibility of hydrogen bonds between chloroform and acetone, which cannot occur with the pure components.

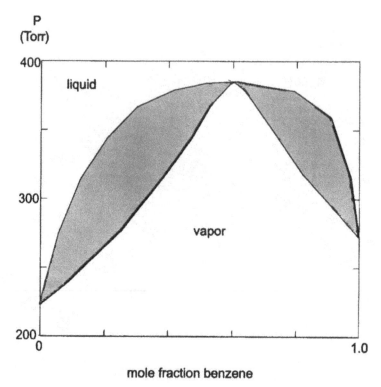

Figure 6 Benzene–ethyl alcohol vapor pressure diagram at 50°C. (Data from J Timmermans. Physiochemical Constants of Binary Systems in Concentrated Solutions. Volume 2. New York: Interscience, 1960, p 60.)

For both minimum- and maximum-boiling-point azeotropes, fractional distillation can only separate a solution into a pure component and the azeotrope, not into the two pure components. It is not possible to distill past an azeotropic concentration. "Absolute alcohol" cannot be prepared by simply distilling aqueous solutions obtained by fermenting grains.

9.5.2 Liquid–Liquid-Phase Diagrams

If positive deviation from ideal behavior becomes great enough, a solution may separate into two liquid phases. We say that the components of this system are *partially miscible*. An example of this behavior is the water–ethyl ether system, which separates into water-rich and ether-rich phases at room temperature. By contrast, water and ethanol are *completely miscible* and do not separate. A phase

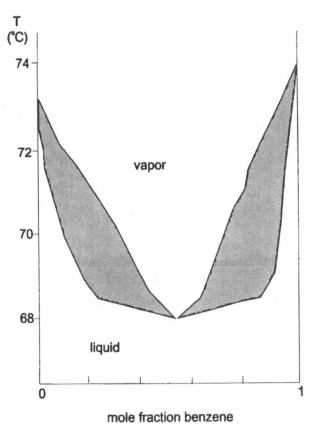

Figure 7 Benzene–ethyl alcohol boiling-point diagram at 1.0 atm. Data from same source as Fig. 6.

diagram for the liquid portion of the water–phenol system, indicating the two-phase region where separation occurs, is shown in Fig. 9.

In Fig. 9, the curve on the left-hand side of the two-phase region gives the solubility of phenol in water as a function of temperature, whereas that on the right-hand side of the region gives the solubility of water in phenol. Because dissolving one component in another increases entropy and $\Delta G = \Delta H - T\Delta S$, we can generally expect solubility to increase with temperature. The two solubility curves move toward each other and finally meet at T_c, the *critical solution temperature*, which is 65.8°C for water–phenol. This is a critical temperature in the same sense as the critical temperature of a single-component gas–liquid system. Above this temperature, no liquid-phase separation is

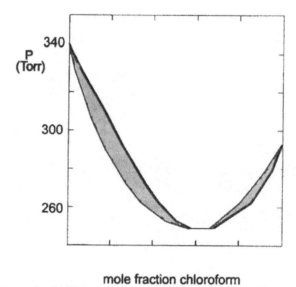

mole fraction chloroform

Figure 8 Acetone–chloroform vapor pressure diagram at 35.2°C. (Data from Example 1.)

observed. Cases are also known in which the solubility of two liquids increases as temperature is lowered. This is due to an enthalpic effect, such as the formation of hydrogen bonds between solute and solvent. An example of this behavior occurs with water and triethylamine. In this case, T_c is at the bottom of the two-phase region on the phase diagram. Some systems are known in which liquid-phase separation occurs, but with solubility increasing as temperature both increases and decreases, giving both upper and lower critical solution temperatures.

9.5.3 Solid–Liquid Phase Diagrams

In Chapter 8, the simple case of totally immiscible solids, exhibiting a minimum melting eutectic, was discussed. There are a variety of other behaviors that can be demonstrated in solid–liquid equilibria. For example, a solid solution may be formed. In a solid solution, the arrangement of atoms shows some degree of randomness on the molecular level. This occurs in a *substitutional solid solution*, where the components are very similar and can substitute for each other in the solid lattice. Although the lattice is regular, which atoms in the lattice are substituted is random. (If the substitution were periodic, the system would be a *compound*.) Copper and nickel illustrate this behavior and form a substitutional solid solution at all concentrations. Another type of solid solution is an *interstitial*

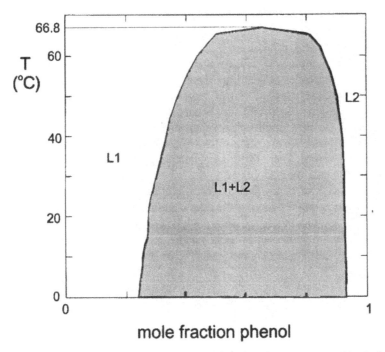

mole fraction phenol

Figure 9 Water–phenol liquid-phase diagram. (Data from J Timmermans. Physiochemical Constants of Binary Systems in Concentrated Solutioins. Volume 4. New York: Interscience, 1960, p 324.)

solid solution, which occurs when the atoms of one component are small enough to fit into the interstitial holes in the lattice of the other component. Although the lattice is regular, just which holes are filled is random. (If all the holes were filled, the system would be a compound.) Many types of steel are at least partially composed of a solid solution of up to 4% carbon in iron.

If in both the solid and liquid phases, the components are completely miscible, the resulting phase diagram will look much like one of those already discussed for the equilibrium of gases with liquid solutions. For example, if both the liquid and solid solutions approach ideal behavior, the phase diagram will look like Fig. 8 of Chapter 8. Positive and negative deviations from ideal behavior can occur in both phases and combine to give a maximum or minimum in the melting point, similar to the gas–liquid diagrams shown in Figs. 7 and 8. Phase diagrams, however, only present properties at equilibrium and give no information about how rapidly equilibrium is attained. Because the atoms in a solid are largely fixed in their positions,[2] convection in a solid is negligible, diffusion is extremely slow, and the equilibrium arrangement may not be approached for many thousands or millions of years. As a result, the properties of solids are as much determined by their thermal history (their rates of heating and cooling) as by their equilibrium properties.

A very common occurrence is that, in the liquid phase, the components are completely miscible, whereas in the solid phase, the components are only partially miscible, usually in small ranges around the pure components. This is illustrated in Fig. 10. Except for the single-phase solid solution regions in the vicinity of the pure solid components, this diagram is similar to Fig. 10 of Chapter 8. It shows a eutectic, which freezes to a mixture of fine crystals of the two solid solutions. These three coexisting phases are represented by a horizontal line on the phase diagram.

A number of binary systems form solid compounds of fixed stoichiometry. The phase diagram for an A–B system, which forms a compound A_2B, is shown in Fig. 11. Although there are three possible components: A, B, and A_2B, in this system, they are connected by a chemical equilibrium relationship, and so there are only two independent components. The phase diagram looks like two separate

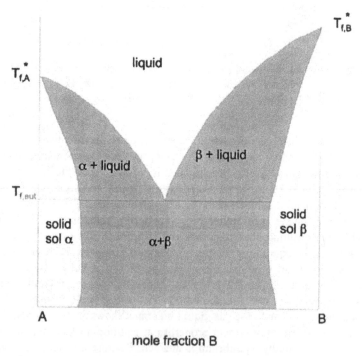

Figure 10   Solid–liquid-phase diagram with solid solution formation.

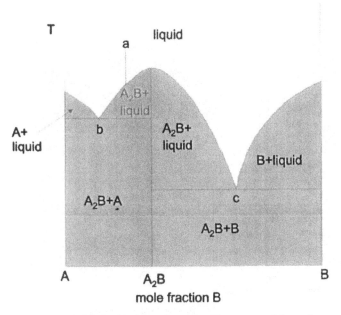

Figure 11 Solid–liquid-phase diagram with compound formation.

diagrams, one for A–A_2B and the other for B–A_2B, side by side. For example, when cooling a solution of composition a, A_2B begins to precipitate out. Further cooling precipitates more A_2B and moves the liquid concentration toward the left-hand eutectic at point b. At the eutectic temperature, there is a halt in the cooling while the remainder of the liquid converts into a very fine mixture of A and A_2B crystals. Note that a second eutectic at point c exists in this system, but it is never accessed starting with solutions containing mole fractions of B less than 0.33.

A_2B melts to a liquid of the same composition as the compound. We say that it *melts congruently*. Some compounds *melt incongruently*, (i.e., to liquids of a different composition). This behavior is illustrated by the compound $NaOH \cdot 2H_2O$ in the NaOH–water phase diagram shown in Fig. 12.[3]

When solid $NaOH \cdot 2H_2O$ is heated to 10°C, it converts to a mixture of $NaOH \cdot H_2O$ and solution of composition a. During melting, this solution exists at equilibrium with the two solid phases, $p = 3$, and the system is invariant (at the pressure of the diagram). This is similar to the behavior of a eutectic solution, but because the solution of composition a is of lower concentration than both of the solids, it is known as a *peritectic solution*, with a corresponding peritectic temperature and peritectic composition.

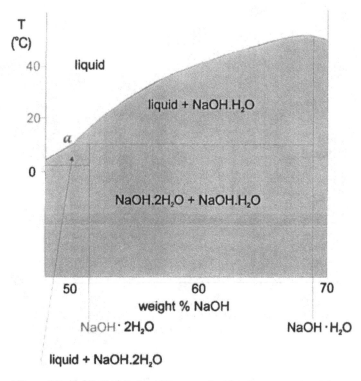

Figure 12 Solid–liquid-phase diagram, showing incongruent melting.

9.6 Phase Diagrams of Ternary Systems

Ternary systems have three components. We will only briefly discuss the use of phase diagrams to describe such systems. If a single phase is present, a three-component system has $f = 3 - 1 + 2 = 4$ degrees of freedom. Because the maximum number of variables that can be represented in a plane is two, two intensive variables must be held constant. These are usually taken as temperature and pressure. The three concentration variables are related and are usually plotted on a triangular diagram. The equilateral triangle is particularly useful for this purpose because of its symmetry. A triangular diagram for the A–B–C system is shown in Fig. 13. The apexes of the triangle represent the pure components, and the perpendicular distance from an apex to the opposite side represents 100 units (mol% or wt%) of that component. The sum of the perpendicular distances to the three sides from any interior point of the triangle equals 100 units. Because there are no phase boundaries in Fig. 13, it represents a completely soluble (or

B

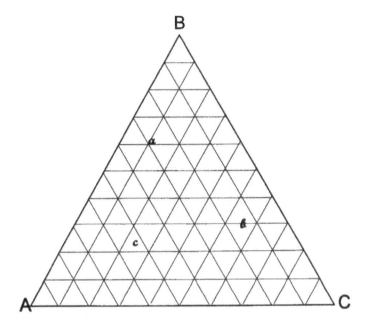

A   C

Figure 13   Equilateral-triangle ternary-phase diagram.

insoluble) system. The coordinates lines shown in Fig. 13 are usually omitted for
more complicated systems in order to avoid cluttering the diagram.

Example 6.   Give the percent concentrations of A, B, and C corresponding to points *a*, *b*, and *c* in Fig. 13 and show that they sum to 100
for each of the points.

Solution:

a: A = 30, B = 60, C = 10; A + B + C = 100
b: A = 15, B = 30, C = 55; A + B + C = 100
c: A = 53, B = 23, C = 24; A + B + C = 100

In Fig. 14, the phase diagram for the three-component $NaNO_3$–KNO_3–
H_2O, system at 36°C is shown. An aqueous solution of two electrolytes has only
three components if the electrolytes have an ion in common. Point C on this
diagram represents the solubility of KNO_3 in water. To the right of point C on the
bottom line of the triangle, the system consists of a saturated solution of KNO_3
and solid KNO_3. As $NaNO_3$ is added to these solutions, the solubility of KNO_3
decreases, but the solution is not yet saturated with $NaNO_3$. When the overall
concentration of the system lies in the region C–B–KNO_3 (e.g., point *a*), the

relative amount of the solution and solid KNO_3 can be determined from the lever rule on the tie line. At point B, the solution is saturated both with $NaNO_3$ and KNO_3. Region B–$NaNO_3$–KNO_3 is a three-phase region, with the saturated solution existing with both solid $NaNO_3$ and solid KNO_3. Figure 14 thus contains one-phase, two-phase, and three-phase regions. In inspecting ternary diagrams, these usually can be easily distinguished because the one-phase and two-phase regions have at least one side that is curved, whereas all of the sides of the three-phase region are straight lines. (Unfortunately, curve A–B in this diagram looks very much like a straight line.)

The phase diagram of the N_2H_4–NaOH–H_2O system shown in Fig. 15 includes a two-phase region, A–P–B, in which the system separates into two liquid solutions. Because there are no isotherms or isobars to connect coexisting phases on a ternary-phase diagram (temperature and pressure are constant for the entire diagram), the tie lines for coexisting solutions are determined experimentally. Several of these tie lines are drawn in the two-phase region. In this diagram,

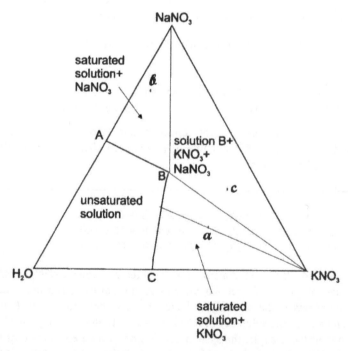

Figure 14 Phase diagram for $NaNO_3$–KNO_3–H_2O, system at 50°C. (Adapted from FF Purdon, VW Slater. Aqueous Solutions and the Phase Diagram. London: Edward Arnold, 1946, p 32.)

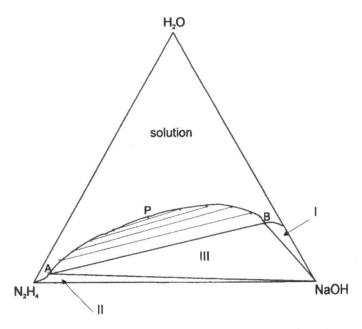

Figure 15 Phase diagram of the N_2H_4–NaOH–H_2O system at 100°C. (Adapted from RA Penneman, and LF Audrieteh. J Am Chem Soc 71: 1644, 1949.)

region III is a three-phase region, where solutions A and B coexist with solid NaOH.

The ternary-phase diagrams presented here illustrate only a small sample of the variety of equilibrium behaviors observed in solid–liquid equilibria in three-component systems. This subject, along with kinetic considerations make up much of the subject matter of a variety of fields, including metallurgy and geology.

Questions

1. Why would a Raoult's law reference be inappropriate for aqueous sugar solutions, no matter how concentrated?

2. Indicate whether each of the following in true or false (A is solvent, i is a solute)

   (a) $\gamma_A(\text{RL}) \to 1$ as $x_A \to 1$
   (b) $\gamma_A(\text{HL}) \to 1$ as $x_A \to 1$
   (c) $\gamma_i(\text{HL}) \to 1$ as $x_i \to 1$
   (d) $\gamma_i(\text{HL}) \to 0$ as $x_i \to 1$
   (e) $\gamma_i(\text{HL}) \to 1$ as $x_i \to 0$

3.   Why do we not choose the state of infinite dilution as the standard state of the solute in a solution?

4.   Benzene and toluene form an ideal solution, which follows Raoult's law, and has the activity coefficient on the mole fraction scale equal to 1 at all concentrations. If we described this solution by a molality-based Henry's law, would the activity coefficient be equal to 1 at all concentrations. If not, why?

5.   Sketch the partial pressures above a solution in which both solute and solvent show positive deviation from ideal solution behavior. Using Raoult's law reference for both solute and solvent, sketch the activity coefficients for this solution.

6.   Indicate whether each of the following is always greater than zero (GT), always equal to zero (EQ), always less than zero (LT), or none of the above (NA):

   (a)   The boiling point of a solution that shows positive deviation from ideal behavior minus the boiling point of the same solution, if it were ideal
   (b)   The excess enthalpy of a solution that shows positive deviation from ideal behavior
   (c)   The mole fraction of the more volatile component of a maximum-boiling-point azeotrope minus the mole fraction of the vapor in equilibrium with the azeotrope
   (d)   The entropy of a solid solution of two elements minus the entropy of a compound of the same two elements with approximately the same composition as the solution

7.   List five ways of determining the activity coefficient of a nonionized solute.

8.   Use Fig. 5 to estimate the purity of ethanol that is obtained from three distillations of an equimolar ethanol–diethyl ether mixture. Assume that, in each stage of the distillation, half the solution is evaporated and condensed.

9.   Why are the concentrations of the azeotropes in Figs. 6 and 7 not exactly the same?

10.   Draw the liquid portion of a phase diagram for a binary system that shows both an upper and a lower critical solution temperature.

11.   Describe the $NaNO_3-KNO_3-H_2O$ system at the concentrations given by points b and c in Fig. 14.

12.   For the $N_2H_4-NaOH-H_2O$ system described by Fig. 15, what phases exist when the system concentration lies in region I? In region II?

Problems

1.   Prove that $G^E = H^E - TS^E$.

2.*   L'Hospital rule states that if two functions, $f(x)$ and $g(x)$, both approach zero as x approaches zero, the ratio $f(x)/g(x)$ can be evaluated as the ratio of the slopes of the curves as they approach zero,

$$\lim_{x \to 0} \frac{f(x)}{g(x)} = \lim_{x \to 0} \left[\left(\frac{df}{dx} \right) \left(\frac{dg}{dx} \right)^{-1} \right]$$

Show that L'Hospital rule gives for the limit of the osmotic coefficient.

$$\lim_{m \to 0} \phi = - \lim_{m \to 0} \left[(\ln a_A) \left(\frac{n_i}{n_A} \right)^{-1} \right] = 1.$$

3. At 30°C the vapor pressure of CS_2 is 430 mm Hg. A solution of CS_2 in acetone with $x_{CS_2} = 0.040$ at this temperature is in equilibrium with a partial pressure of CS_2 of 80 mm Hg. This solution can be considered ideally dilute. The partial pressure of CS_2 over an equimolar solution of CS_2 and acetone is 330 torr. Calculate γ_{CS_2} for the equimolar solution, using both the Raoult's law and Henry's law reference states.

4. At 30°C, the density of carbon tetrachloride is 1.57478 g/mL and the density of cyclohexane is 0.76918 g/mL. The density of a carbon tetrachloride–cyclohexane solution with $x_{CCl_4} = 0.6506$ is 1.27034 g/mL. What is V^E for 1 mol of a solution of this concentration?

5. A benzene–methanol solution with $x_{CH_3OH} = 0.5191$ and $y_{CH_3OH} = 0.5571$ has a total vapor pressure of 292.5 torr. The vapor pressure of methanol is 150 torr and that of benzene is 210 torr at the temperature of the solution. Calculate the Raoult's law reference activity coefficients for benzene and methanol in this solution.

6. It is difficult to draw tangent lines to partial pressure curves for estimating Henry's law constant with accuracy. Show that an alternative way of obtaining the Henry's law constant is as $K_i = \lim_{x_i \to 0} \gamma_i(RL)P_i^*$. From the data for the acetone–chloroform system given in Example 1, evaluate K_{Ac} and K_{chl}, and compare the values obtained with those determined by drawing tangent lines.

7. The vapor pressure at 25°C over a solution of nonvolatile solute B dissolved in solvent A are given as follows:

x_B:	0	0.01	0.02	0.03
P (torr):	420	406.2	392.7	379.3

(a) Calculate γ_A for each of the four solutions.
(b) The density of pure A is 0.90 g/cm^3 and its molecular weight is 60 g/mol. What is the osmotic pressure at 25°C of the solution with $x_B = 0.03$ in contact with pure A through a membrane that only allows passage of A.

8. Calculate G^E for 1 mol of a chloroform–acetone solution for each of the concentrations given in Example 1. Sketch how G^E depends on the concentration of the solution.

9. Calculate $\Delta_{mix}G$ for preparing 1 mol of a chloroform–acetone solution with $x_{chl} = 0.424$ from its components. Compare your result to what would have been obtained if the solution were ideal.

10. The freezing-point depression of a 32% by weight aqueous solution of ethylene glycol is 16.23°C. What is the activity coefficient of water in this solution?

11. Derive a relationship between the vapor pressure of the solvent above a solution and the osmotic pressure of the solution.

12. Show that the Gibbs–Duhem equation requires that in a binary solution, both solvent and solute must show positive deviation from ideal behavior or must both show negative deviation from ideal behavior.

13. Using the results of Example 1, calculate the osmotic coefficient for a chloroform–acetone solution with $x_{chl} = 0.424$, treating acetone as the solvent.

Notes

1. ϕ approaches the indeterminate form $0/0$ as $x_i \rightarrow 0$. It can, however, be shown to be unity by L'Hospital rule. See Problem 2.
2. It is the vibration of the atoms around these fixed positions that allows diffusion to occur.
3. Only part of this phase diagram is shown because the full diagram is quite complicated. See FF Purdo, VW Slater. Aqueous Solution and the Phase Diagram. London: Arnold, 1946, p 18.

10

Ionized Systems

Eminent British Politician: *"What good is electricity?"*

Michael Faraday: *"I don't know yet, but some day you'll tax it."*[1]

In solutions of electrolytes, dissociation into ions and long-range interactions of these ions must be considered. Because solutions containing a single ion cannot be prepared, mean ionic concentrations and activity coefficients are considered. In very dilute ionic solutions, the mean ionic activity coefficient can be estimated by the Debye–Huckel theory or extensions of this theory. Higher ionic concentrations, polyvalent ions, and solvents of low dielectric constants may require the consideration of ion pairs. In electrochemical systems, the electrostatic potential of ions due to charging of phases is added to the chemical potential to give the electrochemical potential. Different types of electrode can be combined to form electrochemical cells. Cells with more than one liquid phase are not reversible. Standard cell potentials can be conveniently obtained from a table of standard electrode potentials. The Nernst equation gives the concentration dependence of the potential of galvanic cells and may be used to extrapolate measured potentials to provide standard free-energy changes of chemical reactions involving ions.

Activity coefficients of ions and the pH's of solutions can also be obtained from electrochemical measurements.

10.1 Ionic Solutions

Solutions of electrolytes, substances that are ionized to some degree in solution, differ greatly from those of nonelectrolytes. The most obvious difference is the greater electrical conductivity of electrolyte solutions. The conductivity of electrolyte solutions span a wide range, depending in part on whether the electrolyte is completely ionized (a strong electrolyte, e.g., NaCl) or only partially ionized (a weak electrolyte, e.g., acetic acid). Although conductivity measurements provide a number of parameters used in treating the thermodynamics of electrolyte solutions, such measurements will not be discussed in this volume.

The ionization of electrolytes is clearly manifest in the thermodynamic properties of their solutions. For example, in the ideally dilute solution limit, a solution of a strong electrolyte behaves as ions, rather than molecules, interacting with solvent molecules. A NaCl solution of molality m behaves, in the limit of infinite dilution, as an ideally dilute solution of concentration $2m$, as 2 mol of ions are produced from each mole of NaCl dissolved in solution. A general strong electrolyte, dissociating by the equation

$$M_{v_+}X_{v_-}(\text{solid}) \rightarrow v_+M^{z+} + v_-X^{z-}, \tag{1}$$

behaves as a solution of molality $(v_+ + v_-)m$, where m is the nominal molality of the solution (calculated from the mass of the solid dissolved per kilogram of solvent).

Another difference between solutions of electrolytes and nonelectrolytes is that the ideally dilute solution model holds for strong electrolyte solutions only up to extremely low concentrations. This maximum concentration is considerably lower when the electrolyte dissociates to polyvalent ions (z_+ or $|z_-|$ greater than 1). Because, in the ideally dilute limit, solute particles interact only with solvent molecules, deviations from ideally dilute behavior must result from interactions between solute particles. These deviations are much larger in solutions of strong electrolytes, because the solute particles are ions, which interact by Coulomb's law. The energy of the electrostatic interaction is inversely proportional to the separation of the ions, as compared with the interaction between neutral particles, in which energy generally falls off as the sixth power of the separation between the particles. The electrostatic interaction is also proportional to the product of the charges on the ions, explaining the larger deviations observed in solutions of polyvalent ions. When considering thermodynamic properties of ionic solutions, it is almost always necessary to deal with activities, rather than concentration.

10.2 Mean Ionic Activity Coefficients

In a solution of a strong electrolyte of molality m, we can write the contribution of the solute to the Gibbs free energy per 1.0 kg of solvent as

$$G_i = m_i\mu_i = m_i\nu_+\mu_+ + m_i\nu_-\mu_- \tag{2}$$

where the last equation expresses the free energy as the contribution of the ions produced by the electrolyte. (For a strong electrolyte, no un-ionized molecules exist in solution.) The chemical potential of the electrolyte can therefore be written as

$$\mu_i- = \nu_+\mu_+ + \nu_-\mu_- \tag{3}$$

The chemical potential of an individual ion in solutions can never be measured because it depends on the other ions in the solution with which it interacts. We can, however, discuss $\mu_\pm \equiv \mu_i/\nu$, the *mean ionic chemical potential*, or the average chemical potential of an ion produced from the electrolyte. Writing μ_+, μ_-, $\mu_\pm$, and μ_i each in the form $\mu = \mu^0 + RT \ln a$ gives

$$\begin{aligned}\mu_i &= \mu_i^0 + RT \ln a_i = \nu(\mu_\pm^0 + RT \ln a_\pm) \\ &= \nu_+(\mu_+^0 RT \ln a_+) + \nu_-(\mu_-^0 + RT \ln a_-)\end{aligned} \tag{4}$$

With

$$\mu_i^0 = \nu\mu_\pm^0 = \nu_+\mu_+^0 + \nu_-\mu_-^0 \tag{5}$$

we get

$$a_i = a_\pm^\nu = a_+^{\nu_+} a_-^{\nu_-} \tag{6}$$

$a_\pm$ is the *mean ionic activity*. Writing each activity in the form $a = \gamma m$,

$$\gamma_i m_i = \gamma_\pm^\nu m_\pm^\nu = \gamma_+^{\nu_+} m_+^{\nu_+} \gamma_-^{\nu_-} m_-^{\nu_-} \tag{7}$$

As a result, $m_\pm$, the *mean ionic molality*, is given by

$$m_\pm = (m_+^{\nu_+} m_-^{\nu_-})^{1/\nu} = [(\nu_+ m_i)^{\nu_+}(\nu_- m_i)^{\nu_-}]^{1/\nu} = \nu_\pm m_i \tag{8}$$

where $\nu_\pm = (\nu_+^{\nu_+}\nu_-^{\nu_-})^{1/\nu}$

Note that for a 1:1 electrolyte (e.g., NaCl, CaSO$_4$, $\nu_\pm = 1$.

$\gamma_\pm$, the *mean ionic activity coefficient*, is from Eq. (7):

$$\gamma_\pm = (\gamma_+^{\nu_+}\gamma_-^{\nu_-})^{1/\nu} \tag{9}$$

The overall expression for μ_i is

$$\mu_- = \mu_i^0 + \nu RT \ln \gamma_\pm m_\pm = \mu_i^0 + \nu RT \ln \gamma_\pm\nu_\pm m_i \tag{10}$$

Note that $a_\pm$, $m_\pm$, $\nu_\pm$, and $\gamma_\pm$ are geometric means, whereas $\mu_\pm$ is an arithmetic mean. Because, at infinite dilution, the solution follows the ideally dilute model,

$\gamma_+ = \gamma_- = \gamma_\pm \to 1$ as $m_\pm \to 0$. The standard state of the electrolyte is thus the hypothetical state of unit mean ionic molality, assuming that the solution remains ideally dilute to this concentration.

Example 1. For $0.02\,m$ $ZnCl_2$, find the following:

(a) $m_\pm$
(b) $\gamma_\pm$ in terms of γ_+ and γ_-

Solution:

(a) $m_\pm = \nu_\pm m_i = (\nu_+^{\nu_+}\nu_-^{\nu_-})^{1/\nu}m_i = (1 \times 2^2)^{1/3}(0.02) = 0.0318\ \text{mol/kg}$
(b) $\gamma_\pm = (\gamma_+\gamma_-^2)^{1/3}$

10.3 Mean Ionic Activity Coefficients from Experimental Data

Just as we discussed in Chapter 9, we can use measured activities of solvents (determined from vapor pressure, freezing-point depression, boiling-point elevation, or osmotic pressure) to determine activity coefficients of electrolytes in solution. For an ionic substance, the Gibbs–Duhem equation is

$$n_A\,d\mu_A + n_i\,d\mu_i = n_A RT\,d\ln a_A + n_i\nu RT\,d\ln \gamma_\pm m_\pm = 0, \tag{11}$$

which can be rearranged to

$$d\ln\gamma_\pm = -d\ln m_\pm - \frac{n_A}{\nu n_i}d\ln a_A = -d\ln m_\pm - \frac{1}{\nu m_i M_A}d\ln a_A. \tag{12}$$

To circumvent the last term blowing up when integrating this equation from infinite dilution, we define the *osmotic coefficient for an electrolyte* as

$$\phi \equiv -\frac{n_A}{\nu n_i}\ln a_A = -\frac{1}{\nu m_i M_A}\ln a_A \tag{13}$$

Using this to substitute for $d\ln a_A$ gives

$$d\ln a_A = -\nu M_A(m_i\,d\phi - \phi\,dm_i) \tag{14}$$

Equation 12 then becomes[2]

$$d\ln\gamma_\pm = d\phi + \frac{(\phi-1)}{m_i}dm_i \tag{15}$$

which can be integrated without difficulty to

$$\ln \gamma_{\pm} = (\phi - 1) + \int_0^{m'} \frac{(\phi - 1)}{m'} \, dm' \tag{16}$$

because ϕ approaches unity in the infinitely dilute limit. (See Problem 3.)

10.4 Calculation of Mean Ionic Activity Coefficient

As we have noted, electrolyte solutions deviate from ideally dilute behavior even at a very low concentration because of the long-range Coulombic interaction between charged particles. However, because these solutions are overall electrically neutral, the escaping tendency of ions from these solutions must result from the arrangement of the ions in the solution. Ions can be distributed in such a way as to lower their energies and escaping tendencies from a system. The most obvious example of such a distribution is the regular arrangement of ions in crystals, in which the nearest neighbors of positive ions are negative ions and vice versa. Of course, in solution, the positions of ions are not fixed, and thermal motion opposes their tendency to attain the most energetically favorable distribution. In addition, solvent molecules lie between charged particles in solution, and the orientation of solvent dipoles reduces the forces between the charged particles. This latter effect is measured by the dielectric constant of the solvent and is quite large in aqueous solution. (The dielectric constant of water is 78.4 at 25°C.) Nevertheless, at short distances around a positive ion in solution, a small excess of negative charge exists, and around a negative ion, the opposite occurs. The effect of the nonuniform distribution of ions in solution on their energy (and thus on their activity coefficients) was first treated by Debye[3] and Huckel in 1923. The result of the theory will be stated and discussed, but not derived. The theory is illustrated in Fig. 1.

In the *Debye–Huckel theory,* an ion in solution is treated as a conducting sphere. The distance of closest approach of two ions is a.[4] The solution beyond a

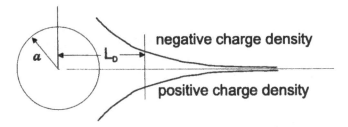

Figure 1 Distribution of charge around a positive ion.

is treated as a medium with relative dielectric constant $\varepsilon_{r,A}$, containing continuous distributions of positive and negative ions. The charge density in the solution around an ion is determined by a Boltzmann distribution in the energy of the electrostatic interaction with both the charge of the ion and the other charged particle in the solution. The distributions (charge densities) of positive and negative charge are shown in Fig. 1. Around a positive charge, there is excess negative charge in the solution and depletion of positive charge in the solution. The ionic character of the solution is characterized by its *concentration-scale ionic strength*, I_c, defined by

$$I_c \equiv \frac{1}{2} \sum_j z_j^2 c_j \tag{17}$$

with the sum extending over all the ions in the solution. The charge imbalance caused by an ion dies out beyond a distance L_D, the *Debye length* of the solution, which is defined by

$$L_D \equiv \sqrt{\frac{\varepsilon_0 \varepsilon_{r,A} RT}{2\mathscr{F}^2 I_c}} \tag{18}$$

ε_0 and $\varepsilon_{r,A}$ are the permittivity of free space and the dielectric constant (relative permittivity) of the solvent, respectively, and $\mathscr{F}$ is the Faraday constant (the charge on 1 mol of hydrogen ions, 94,485 C). Charge imbalances die out in a shorter distance if there is more charge in the solution, the temperature is lower, and dielectric constant is smaller.

For the activity coefficient of an ion, the theory gives

$$\ln \gamma = -\frac{z^2 A \sqrt{I_c}}{1 + Ba\sqrt{I_c}} \tag{19}$$

where z is the charge on the ion. The constants A and B are independent of concentrations in the solution and are given by

$$A = \frac{(2N_A)^{1/2}}{8\pi} \left(\frac{e^2}{\varepsilon_0 \varepsilon_{r,A} kT}\right)^{3/2} \quad \text{and} \quad B = e\left(\frac{2N_A}{\varepsilon_0 \varepsilon_{r,A} kT}\right)^{1/2} \tag{20}$$

N_A, e, and k are universal constants (Avogadro's number, electronic charge, and Boltzmann's constant, respectively).

Solutions with a single ion cannot be prepared, and because ions influence each other, only the mean ionic activity coefficient can be measured. For a binary electrolyte (one producing only two types of ions in solution, i.e., M_{v_+}, X_{v_-}), Eq. (18) leads to

$$\ln \gamma_\pm = -\frac{z_+ |z_-| A \sqrt{I_c}}{1 + Ba\sqrt{I_c}} \tag{21}$$

In the lowest level of the theory, the B term in the denominator of Eq. (18) is dropped, to give the *Debye–Huckel limiting law* (DHLL),

$$\ln \gamma_\pm = -z_+ |z_-| A \sqrt{I_c} \tag{22}$$

a relation with no adjustable parameters. At this level of the theory, the mean ionic activity coefficient depends only on the charge state of the ion and not on the chemical nature of the ion. Thus, at the same concentration, all $1:1$ electrolytes would have the same mean ionic activity coefficient. For aqueous solutions at 25°C ($\varepsilon_{r,A} = 78.4$). A has the value 1.17 $(L/mol)^{1/2}$. It is more usual to use a logarithm to the base 10, which gives with I_c in units of moles/liter:

$$\log_{10} \gamma_\pm \stackrel{aq.25°C}{=} -0.51 z_+ |z_-| \sqrt{I_c} \tag{23}$$

It is noteworthy that in the Debye–Huckel theory, $\ln \gamma_\pm$ is always negative, corresponding to $\gamma_\pm$ less than 1. This is expected because the ions are distributed in a manner that lowers their energy and escaping tendency in solution and produces negative deviation from ideal behavior. The effect is larger for electrolytes that produce multiply-charged ions, and it is the ionic strength of *all* of the ions in solution that is a measure of the attraction for an ion of an electrolyte. Ionic strength enters in a square-root dependence (i.e., less than linear) because the larger I_c, the smaller the region around any ion that is disturbed from electrical neutrality (as indicated by L_D). Inspection of the constant A shows that increased temperature acts to reduce the ordering of ions in solution, as does increased dielectric constant. Large dielectric constants occur in polar solvents, the molecules of which align, so that their dipoles cancel some of the ion-generated electric field.[5]

The Debye–Huckel theory is accurate in solutions in which the interactions between ions are not too great (i.e., at low ionic strengths in solutions of monovalent ions in solvents with large dielectric constants). In Fig. 2, the predictions of Eq. (21) are compared with experimental data for some strong electrolytes with different ionic charges.

It can be seen that Eq. (21) shows the proper limiting behavior as concentrations approach zero. However, its quantitative predictions are reliable only up to $I_c \approx 0.01\ M$. Better agreement is obtained between experiment and the full theoretical result, Eq. (21). However, there does not seem to be a good way to choose a, the ionic size, other than fitting experimental data to Eq. (21). Typical

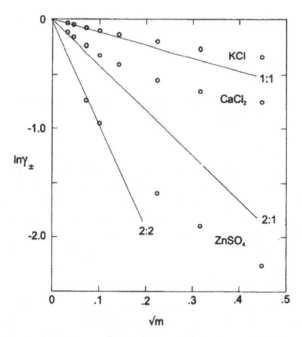

Figure 2 Comparison of measured mean ionic activity coefficients with those predicted by the Debye–Huckel limiting law. (Data from Handbook of Chemistry and Physics, 77th ed. A James, M Lord. VNR Index of Chemical Physical Data. New York: Van Nostrand Reinhold, 1992.

values of a found for small inorganic ions by this procedure range from 0.3 to 0.9 nm, which are physically reasonable values.

Some workers, noting that in aqueous solutions at 25°C, $B = 3.3 \times 10^9$ $(\text{L/mol})^{1/2}\ m^{-1}$, take $a = 0.3$ nm $= 0.3 \times 10^{-9}\ m$, giving $Ba = 1$, and for Eq. (21),

$$\log_{10} \gamma_\pm \overset{\text{aq.25°C}}{=} -0.51 z_+ |z_-| \frac{\sqrt{I_c}}{1 + \sqrt{I_c}} \tag{24}$$

The addition of a term linear in I_c to Eq. (24) has been found to improve its accuracy, and an empirical equation proposed by Davies,

$$\log_{10} \gamma_\pm \overset{\text{aq.25°C}}{=} -0.51 z_+ |z_-| \left(\frac{\sqrt{I_c}}{1 + \sqrt{I_c}} - 0.3 I_c \right) \tag{25}$$

with no parameters, has been shown to give good agreement with experiment up to $I_c = 0.1\ M$, for singly-charged ions. In Fig. 3, mean ionic activity coefficients are shown for a number of compounds to high concentrations.

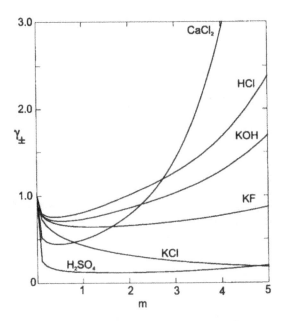

Figure 3 Mean ionic activity coefficients to high concentrations. (Data from Handbook of Chemistry and Physics. 77th ed.)

It can be seen that $\gamma_\pm$ usually increases at high concentrations and, in some cases, becomes larger than 1, in disagreement with the Debye–Huckel theory. Some of the reasons proposed for the failure of the Debye–Huckel theory are ion pairing (loose association of oppositely charged ions in solution) and hydration of ions, with accompanying reduction of the amount of free solvent.

Example 2. Calculate the activity of a 0.05 m HCl solution using the following:

(a) The Debye–Huckel limiting law [Eq. 22)]
(b) The Debye–Huckel theory in the form of Eq. (24)
(c) The Davies equation [Eq. (25)]

Solution:

$$a_i = a_\pm^\nu = m_\pm^\nu \gamma_\pm^\nu = m_+^{\nu_+} m_-^{\nu_-} \gamma_\pm^\nu$$

In this case, $\nu_+ = \nu_- = 1$, $m_+ = m_- = m_\pm = 0.05$, and $I_c = 0.05$. We obtain the following

(a) $\log_{10}\gamma_\pm = -0.51(1)(1)\sqrt{0.05} = -0.114$, $\gamma_\pm = 0.769$, and $a_i = [(0.05)(0.769)]^2 = 1.47 \times 10^{-3}$

(b) $\log_{10}\gamma_\pm = -0.114/1.223 = -0.0932$, $\gamma_\pm = 0.807$, and $a_i = [(0.05)(0.807)]^2 = 1.63 \times 10^{-3}$

(c) $\log_{10}\gamma_\pm = -0.51(1)(1)[\sqrt{0.05}/(1 + \sqrt{0.05}) - 0.3(0.05)] = -0.0856$, $\gamma_\pm = 0.821$, and $a_i = [(0.05)(0.821)]^2 = 1.68 \times 10^{-3}$.

Note that because HCl produces two ions, in each case the activity varies as the square of the HCl molality, modified by the activity coefficient.

The very important chemical quantity pH[6] is defined as[7]

$$pH \equiv -\log a_{H^+} \tag{26}$$

a_{H^+} includes all of the hydrated forms of the hydrogen ion. Because of its logarithmic nature, nonideality has a smaller effect on pH than it does on a_{H^+}; thus, the particular method of correcting for nonideal behavior is less important.

Example 3. Calculate the pH of a 0.5 *m* HCl solution, assuming it is ideally dilute and also using each of the approximations for $\gamma_\pm$ employed in Example 2.

Solution: pH $= -\log a_{H^+} = -\log(\gamma_+ m_{H^+}) = -\log(\gamma_\pm m_{H^+})$, where, in the last step, we take the result of the Debye–Huckel theory for a 1 : 1 electrolyte.
Assuming ideally dilute, pH $= -\log(0.05) = 1.301$.
With DHLL, pH $= -\log[(0.769)(0.05)] = -\log(0.0384) = 1.415$.
With Eq. (24), pH $= -\log[(0.807)(0.05)] = -\log(0.0403) = 1.394$.
With the Davies equation, pH $= -\log[(0.821)(0.05)]$
$= -\log(0.4105) = 1.387$.

10.5 Ionic Equilibrium

In calculations of ionic equilibria, the required activity coefficients depend on ionic strength and, thus, on the equilibrium. As a result, such calculations are usually carried out by the method of successive approximations. Fortunately, convergence to a unique solution is usually achieved very rapidly in these problems.

Example 4. Calculate m_{H^+} and the pH of an aqueous 1.0 *m* acetic acid solution at 25°C. For acetic acid, $K_a = 1.75 \times 10^{-5}$. Use the DHLL approximation for activity coefficients.

Solution: The equilibrium constant is

$$K_a = K_m K_\gamma = \frac{m_{H^+} m_{Ac^-}}{m_{HAc}} \frac{\gamma_\pm^2}{\gamma_{HAc}}$$

For this solution, γ_{HAc} can be taken as unity. For the first approximation, we also take $\gamma_{\pm} = 1$. Choosing as a basis 1.0 kg of solvent, we have for the reaction $HAc \leftrightarrow H^+ + Ac^-$,

$$m_{H^+} = m_{Ac^-} = \xi_e, \quad m_{HAc} = 1.0 - \xi_e, \quad \frac{\xi_e^2}{1.0 - \xi_e} = 1.75 \times 10^{-5}$$

The solution of this quadratic equation is $\xi_e = m_{H^+} = m_{Ac^-} = 0.00417\ m$, which is also the ionic strength. $\log \gamma_{\pm} = -0.51(1)(1)\sqrt{0.00417} = -0.0329$, $\gamma_{\pm} = 0.927$, and $K_{\gamma} = 0.859$. This gives

$$\frac{\xi_e^2}{1 - \xi_e} = \frac{1.75 \times 10^{-5}}{0.859} = 2.04 \times 10^{-5}$$

resulting in $\xi_e = m_{H^+} = 0.00451$.

Recalculating $\gamma_{\pm}$, $\log \gamma_{\pm} = -0.51(1)(1)\sqrt{0.00451} = -0.0342$ for $\gamma_{\pm} = 0.924$, which is a negligible change. Following Example 2,

$$pH = -\log[(0.00451)(0.924)] = -\log(0.00417) = 2.38.$$

Note that if we neglected the nonideality of the solution, we would also have $pH = -\log(0.00417) = 2.38$. This is because pH depends on the *activity* of the hydrogen ions, which, in the limit of small percent dissociation, is almost independent of the activity coefficient.

Example 5. Calculate m_{H^+} and the pH of an aqueous solution at 25°C that is $0.05\ m$ in acetic acid and $0.3\ m$ in NaCl.

Solution: Because NaCl is a strong electrolyte, the Na^+ and Cl^- ionic concentrations will determine the ionic strength.

$I_c = 0.3$: For HAc, $\log \gamma_{\pm} = -0.51(1)(1)\sqrt{0.3} = -0.279$, $\gamma_{\pm} = 0.526$, and $K_{\gamma} = 0.277$.

Following Example 4,

$$K_m = \frac{\xi_e^2}{1 - \xi_e} = \frac{1.75 \times 10^{-5}}{0.277} = 6.3 \times 10^{-5},$$

$$\xi_e = m_{H^+} = 0.0079$$

The considerable increase in the hydrogen ion concentration produced by the addition of NaCl is called *salting in*.

Following Example 2,

$$pH = -\log[(0.0079)(0.526)] = 2.38$$

Once again, the pH is almost independent of the activity coefficient.

10.6 Ion Pairs and Ion Solvation

The Debye–Huckel theory deals with long-range interaction between ions. In more detailed consideration of ionic activity coefficients, short-range interactions between ions, or *ion pairs*, must also be considered. An ion pair is a transient association of opposite-charged ions held together by Coulombic force, rather than by specific chemical or coordination bonding. Ion pairing is strongest when the forces between ions are the largest (i.e., between polyvalent ions at high concentrations in solvents of low dielectric constants). The oppositely charged ions do not have to be in contact at the distance of closest approach, but their energy of attraction has to be greater than or comparable to their thermal energy, so that they remain associated for an appreciable time. The Bjerrum theory of ion pairs requires that the energy of the attractive interaction be at least $2kT$.[8] Simple monovalent ions, such as K^+ and Cl^-, are too large to approach to distances where their interaction reaches this value in aqueous solution. Therefore, these ions are considered to not form ion pairs.

The magnitude of ion-pair formation in other cases is described by means of a dissociation constant, K_{dis}, in much the same manner as for the treatment of weak electrolytes. These constants may be obtained from Bjerrum's theory or from measurements of solution conductivity. A selection of pK_{dis} values is given in Table 1. The larger pK_{dis}, the smaller the dissociation constant and the larger the tendency for ion-pair formation. Interestingly, K^+ and NO_3^- are found to form ion pairs. Because the size of NO_3^- is larger than Cl^-, treatment of polyatomic ions as spheres cannot be correct. Most probably, K^+ interacts with charge localized on one of the oxygen atoms of NO_3^-. Ion-pair formation lowers the escaping tendency of ions from solution.

Example 6. Find the fraction of ion pairs in 0.1 m aqueous KNO_3 at 25°C.

TABLE 1 Dissociation constants (pK_{dis}) for ion pairs at 25°C in aqueous solution

Substance	pK_{dis}
KNO_3	−0.2
NaOH	−0.7
$Ca(OH)_2$	1.30
$CaSO_4$	2.3

Note: More extensive data in CW Davies. In WJ Hamer, ed. The Structure of Electrolytic Solutions. New York: Wiley, 1959.

Solution: From Table 1,

$$K_{dis} = 10^{0.2} = 1.58 = K_m K_\gamma = \frac{m_{K^+} m_{NO_3^-}}{m_{K^+ NO_3^-}} \gamma_\pm^2$$

where we have used the assumption that the activity coefficient of the ion pair is unity, because it is uncharged.

First iteration: Take $\gamma_\pm = 1$, for the reaction $K^+ NO_3^- \leftrightarrow K^+ + NO_3^-$,

$$1.58 = \frac{\xi_{e}^2}{0.1 - \xi_{e}}, \qquad \xi_{e} = 0.0943, \qquad I_c = 0.0943$$

Because this is fairly high, we use Eq. (24) to calculate $\gamma_\pm$:

$$\log \gamma_\pm = -0.51 \frac{0.307}{1 + 0.307} = -0.12, \qquad \gamma_\pm = 0.76$$

Second iteration:

$$\frac{1.58}{(0.76)^2} = 2.73 = \frac{\xi_{e}^2}{0.1 - \xi_{e}}, \qquad \xi_{e} = 0.0966$$

which is close enough to our first iteration that further calculation is not necessary. Note that the fraction of the ions that form ion pairs is only $(0.1 - 0.0966)/0.1 = 0.034$.

The fraction of ions that are associated to ion pairs is much greater than calculated in Example 6, when concentrations are higher, with polyvalent ions and in solvents of low dielectric constants.

For a $1:1$ strong electrolyte with ion pairing, the concentration of both ions is αm_i, where α is *the fraction of ion pairs that are dissociated.* ($\alpha m_i = \xi_e$, the extent of the dissociation reaction). The chemical potential and activity of such electrolytes are easily calculated, because, at equilibrium,

$$\mu_{IP} \overset{1:1}{=} \mu_+ + \mu_- \tag{27}$$

$$\mu_i = \alpha \mu_+ + \alpha \mu_- + (1 - \alpha)\mu_{IP} = \mu_+ + \mu_- \tag{28}$$

Note that Eq. (28) is identical to Eq. (3) for a strong electrolyte with no ion-pair formation. However, for ion pairing, we have

$$\mu_i \overset{1:1}{=} \mu_i^0 + RT \ln a_i$$

$$= \mu_+ + \mu_-$$

$$= (\mu_+^0 + RT \ln \gamma_+ \alpha m_i) + (\mu_-^0 + RT \ln \gamma_- \alpha m_i) \tag{29}$$

$$= (\mu_+^0 + \mu_-^0) + RT \ln(\gamma_\pm \alpha)^2 m_i^2$$

In the limit of infinite dilution, $\gamma_\pm \to 1$, $\alpha \to 1$, and $a_i \to m_i^2$. Therefore, at noninfinite dilution,

$$\gamma_i = \gamma_\pm^2 \alpha^2 \tag{30}$$

Ions also interact at short distances with solvent molecules. (The hydration of the proton to H_3O^+ and higher-hydrated species in aqueous solutions is well known.) Hydration of ions reduces the concentration of free water and can result in an increase in the escaping tendency of ions from solution, because their "effective concentration" is greater than their actual concentration.

10.7 Electrochemical Systems

In using the chemical potential of ions as a measure of their escaping potential from a phase, we are implicitly assuming that the phase is electrically neutral. In fact, due to electrostatic forces, the escaping tendency of a charged particle from a phase can be dramatically affected by the charge state of the phase.[9] This is illustrated by the phenomenon of field emission, where sufficient numbers of electrons are forced onto a sharply pointed metallic surface, so that some of them are ejected into the surrounding vacuum by their mutual electrostatic repulsion. Electrostatic attraction and repulsion are very strong and long range, so it takes only a tiny net charge transfer to a phase to have an appreciable effect.

Charge will spontaneously develop at the interface between two phases when there is a difference in the ease with which particles with charge of opposite sign can be transferred across the phase boundary. One example of this is at the interface between a metal and a solution, where metallic ions, but not electrons, can dissolve in the solution.[10] Another example is at the interface between two metals, where electrons, but not ions, undergo rapid transfer. In the latter case, the electron transfer depends on temperature and forms the basis for measuring temperature differences by means of thermocouples.

The variable used to describe the differences in escaping tendency of charged particles from charged phases is the *electrostatic potential*, ϕ. The electrostatic potential at a point, a, is defined as the work per unit charge required to bring a positive test charge reversibly from infinity to the point,

$$\phi(a) \equiv \frac{\delta w_{\infty \to a}}{dQ} \tag{31}$$

Note that potentials are defined with respect to a positive test charge, whereas electrochemical systems produce or consume negative charge (electrons). The test charge is of infinitesimal magnitude, so that it does not change the charge distribution in the system. The selection of infinity as the zero of electrostatic energy is arbitrary. Only differences in electrostatic potential between the various phases of the system are important.

The electrostatic potential at a distance r from a charge dQ is

$$\phi(r) = \frac{1}{4\pi\varepsilon_0} \frac{dQ}{r} \qquad (32)$$

where ε_0 is the permittivity of vacuum ($8.854 \times 10^{-12} \; C^2/N \, m$). Equation (32) has to be integrated over the existing charge distribution.

When a (real) charge is moved reversibly in an electrostatic potential, the work done increases U and G of the charged particle (with respect to its value at infinity) by $\phi(a) \, dQ$. The change in electrostatic energy for transferring dn moles of ions from infinity to a phase at a is $\phi(a)z\mathscr{F} \, dn$, where z is the charge on the ion (a small positive or negative integer) and $\mathscr{F}$ is the *Faraday constant*, the charge on 1 mol of hydrogen ions, $96,485 \; C/mol$. This change in the Gibbs free energy due to the position-dependent electrostatic potential must be added to the change in G due to the chemical potential of the ions:

$$dG_i = \mu_i \, dn_i + z_i \mathscr{F} \phi(a) \, dn_i \qquad (33)$$

The *electrochemical potential*, $\tilde{\mu}_i(a)$, of an ion at an electrostatic potential $\phi(a)$ is defined as

$$\tilde{\mu}_i(a) \equiv \left(\frac{\partial G}{\partial n_i} \right)_{T,P,n_j} = \mu_i + z_i \mathscr{F} \phi(a) \qquad (34)$$

and is the true measure of the escaping tendency of the ion from the phase a.

To find thermodynamic relationships for charged systems, we just replace chemical potentials in equations for uncharged systems by electrochemical potentials. For example, for equilibrium of a charged particle at a phase boundary, we have, in analogy to Eq. (24) of Chapter 6,

$$\tilde{\mu}_i(a) = \mu_i(a) + z\mathscr{F} \phi(a) = \tilde{\mu}_i(b) = \mu_i(b) + z\mathscr{F} \phi(b) \qquad (35)$$

Thus, when two phases are in contact, an ion that can exist in both phases will be transferred until

$$\phi(b) - \phi(a) = \frac{\mu_i(a) - \mu_i(b)}{z_i \mathscr{F}} \qquad (36)$$

For example, if a Pt rod containing some Zn is placed in contact with an aqueous $ZnCl_2$ solution, Zn^{+2} ions are transferred between the solution and the metal, until Eq. (36) holds. It takes only a minuscule transfer of Zn^{+2} to build up a very appreciable electrostatic potential, because electrons from the rod cannot transfer into the solution. Chemical potentials, which depend on concentrations, can therefore be calculated neglecting the ionic transfer. If the electrons could also transfer, much more of the rod would dissolve and equilibrium would have to be established by changing chemical potentials by changing concentrations.

For chemical reactions involving charged species in multiple (but contacting) phases, the condition for equilibrium is, in analogy to Eq. (30) of Chapter 7,

$$\sum_i \nu_i \tilde{\mu}_i = 0 \tag{37}$$

A *battery* (or *galvanic* or *voltaic cell*) is a device that uses oxidation and reduction reactions to produce an electric current. In an *electrolytic cell*, an external source of electric current is used to drive a chemical reaction. This process is called *electrolysis*. When the electric potential applied to an electrochemical cell is just sufficient to balance the potential produced by reactions in the cell, we have an electrochemical cell at equilibrium. This state also occurs if there is no connections between the terminals of the cell (open-circuit condition). Our discussion in this chapter will be limited to electrochemical cells at equilibrium.

The voltage produced by galvanic cells is due to the transfer of charged particles at a number of phase interfaces. In order for the battery to be able to produce an electric current, each phase must be able to conduct electricity, usually either by electron flow in solid or liquid metals or by ionic mobility in solutions. The voltage of the battery is measured at its *terminals*, which are metallic and permit electron flow. Each terminal is part of an *electrode*, at which an oxidation or reduction half-reaction occurs. The electrode at which oxidation occurs (in both galvanic and electrolytic cells) is called the *anode*, whereas that at which reduction occurs is called the *cathode*. The basic principle of the galvanic cell is the separation of these two reactions, so that electrons produced by oxidation must flow through an external circuit before causing reduction. For quantitative measurements, the terminals of the cell should be constructed from the same metallic material, so that the driving force for electrons at the terminals is due exclusively to an electrostatic potential difference within the cell, not to additional potential differences at the connecting wires.

10.8 Types of Electrodes

Most electrodes employed can be classified as being of the following types:

1. Gas electrode: Here, a gas is in equilibrium with its ions in solution. The most important example is the *hydrogen electrode* (or H^+/H_2 electrode), for which the electrode reaction is

$$2H^+_{(aq)} + 2e^- \leftrightarrow H_{2(g)} \tag{I}$$

Hydrogen gas is bubbled over a platinum surface that is coated with "platinum black," an electrolytically deposited coating of colloidal platinum, which is an excellent catalyst for the above equilibrium. The hydrogen electrode has been selected as the standard against which the potentials of other electrodes are measured. Equations of the type of reaction (I) are called half-cell reactions, because they include electrons. Reaction I is a reduction half-cell reaction.

2. Metal–metal ion electrode: Here, a metal is in equilibrium with its ion in solution. Metals such as Cu or Zn have an appreciable tendency to dissolve and form ions in water, but little tendency to react with water.

3. Amalgam electrode: Metals such as Ca or Na react with water and cannot be used directly as electrodes. They can be used, however, by dissolving them in liquid mercury, forming an *amalgam*. In the amalgam solution, the activity of the metal is reduced below what it is when pure; thus, its reaction with water is suppressed. Mercury does not dissolve in the aqueous solution.

4. Metal-insoluble salt electrode: Metal-insoluble salt electrodes are convenient for batteries. For example, the AgCl/Ag electrode consists of metallic silver, coated with the insoluble salt AgCl, and suspended in a solution or paste saturated with the salt. The electrode reaction is

$$AgCl_{(s)} + e^- \leftarrow Ag_{(s)} + Cl^-_{(aq)} \tag{II}$$

Note that the metal does not dissolve in the solution. An electrode of this type, which is often used because of its stability, is the *calomel electrode* shown in Fig. 4. This consists of a platinum wire inserted into liquid mercury, which is coated with mercurous chloride. The electrode reaction is

$$Hg_2Cl_2 + 2e^- \leftarrow 2Hg + 2Cl^- \tag{III}$$

5. Oxidation-reduction electrodes: These electrodes consist of an inert metal in a solution containing ions that can undergo oxidation-reduction reactions. An example is a platinum wire immersed in a solution containing Fe^{2+} and Fe^{3+} ions. Transfer of electrons into the solution is accompanied by reduction of some Fe^{3+} to Fe^{2+}.

6. Membrane electrode: The membrane electrode is a thin phase through which charge can be transported (by ion migration) so that electrochemical equilibrium can be maintained for at least one type of ion across the electrode. For example, pH is often measured in a solution (X) by means of a glass electrode between an HCl solution of known

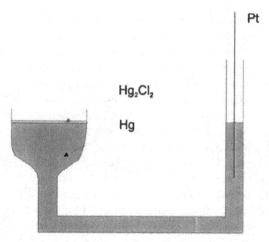

Figure 4   Calomel electrode.

$a(H^+)$ and the unknown solution. Assuming that the HCl solution is on the left-hand side of the glass membrane, the equilibrium condition is

$$\tilde{\mu}_{H^+}(HCl) = \tilde{\mu}_{H^+\,glass}(L)$$
$$\tilde{\mu}_{H^+}(X) = \tilde{\mu}_{H^+\,glass}(R) \tag{38}$$

where L and R are within the membrane on its left-hand and right-hand sides, respectively. Writing these expressions in terms of chemical and electrical potentials gives

$$\mu_{H^+}(HCl) + \mathscr{F}\phi(HCl) = \mu_{H^+}(glass, L) + \mathscr{F}\phi(glass, L)$$
$$\mu_{H^+}(X) + \mathscr{F}\phi(X) = \mu_{H^+}(glass, R) + \mathscr{F}\phi(glass, R) \tag{39}$$

We subtract the second equation from the first. In doing this, the terms on their right-hand sides cancel for the following reasons: The chemical environment on the two sides of the glass is similar enough that μ_{H^+} (glass, L) $= \mu_{H^+}$(glass, R), and, because the glass is weakly conducting, at equilibrium, ϕ(glass, L) $= \phi$(glass, R). It is not the hydrogen atoms that are transported through the glass to establish this equilibrium. The transport is by sodium and potassium ions that have some freedom to move through the pores of the silicon dioxide matrix of the glass. The potential difference introduced by the membrane is thus

$$\phi(HCl) - \phi(X) = \frac{\mu_{H^+}(X) - \mu_{H^+}(HCl)}{\mathscr{F}} \tag{40}$$

and is proportional to the difference in chemical potential of hydrogen ions in the two solutions. Because chemical potentials are proportional to logarithms of hydrogen ion activity, the potential difference across the electrode is proportional to the difference of pH's of the two solutions. In Sec. 10.12, we show how this type of membrane is incorporated into a pH-measuring electrode.

10.9 Electrochemical Cells

When drawing a galvanic cell, by convention, the anode is drawn on the left and the cathode on the right. In Fig. 5, an electrochemical cell using a hydrogen electrode as the anode and an AgCl/Ag electrode as the cathode is shown. The electrolyte is aqueous HCl and both terminals are Pt.

The balanced *electrochemical reaction* (so termed because it includes the electrons) occurring in the cell is obtained by subtracting reaction (I) from twice reaction (II) (to balance the number of electrons):

$$H_{2(g)} + 2AgCl_{(s)} + 2e^-(Pt_R) \rightarrow 2H^+_{(aq)} + 2Ag_{(s)} + 2Cl^-_{(aq)} + 2e^-(Pt_L)$$

$$(IV)$$

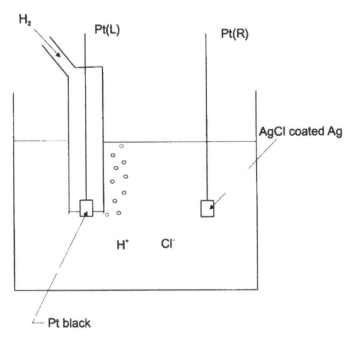

Figure 5 $H_2/H^+-Ag/AgCl$ electrochemical cell.

Here, it is explicitly indicated that electrons are supplied to AgCl/Ag at the right electrode and removed from H^+/H_2 at the left electrode. In the external circuit, electrons move from the anode at the left to the cathode at the right.

Rather than using a diagram such as that in Fig. 5, to describe an electrochemical cell, a *standard simplified diagram* is used. Vertical lines separate the various phases in the cell. For the separation between two liquid phases (by a porous barrier), a dotted or dashed vertical line is used. The terminals of the cells are placed on the ends of the diagram, with the anode on the left. Any metals attached to the terminals are written next to them. Gas or insoluble materials in contact with the metals are written next, and the electrolytic solution of the cell is described in the center of the diagram. To completely define the cell, the concentrations or activities of solutions and the pressures of gases are included. The simplified diagram for the cell illustrated in Fig. 5 is therefore

$$Pt \mid H_2 \text{ (gas, 1.0 atm)} \mid HCl \text{ (aqueous, } a) \mid AgCl \mid Ag \mid Pt \qquad (V)$$

where a is the activity of the HCl solution.

These diagrams may be a little misleading, because not every vertical line corresponds to a unique interface. In the above example, Pt, H_2 gas, and the HCl solution are all present at the same interface, as are Ag, AgCl, and the HCl solution. Thus, there are only two phase boundaries in the cell. By convention, we define the voltage of a galvanic cell as $\phi_R - \phi_L$. In the open-circuited condition, no current flows through the cell and $\phi_R - \phi_L$ is called the *potential*, $\mathscr{E}$, of the cell. With $a = 1.0$ at 298K, the cell above has an open-circuit voltage of 0.22 V. When the terminals are connected through a resistance, electrons flow from the anode to the cathode in the external circuit. In the solution, H^+ ions migrate from the anode to the cathode and Cl^- ions migrate in the opposite direction. The concentration of Ag^+ ions is negligible in the solution (because it reacts with Cl^- ions to form AgCl), so that migration of Ag^+ ions to the hydrogen electrode, followed by their direct chemical reduction can be ignored. The charge transferred by the cell for an extent of reaction ξ is 2ξ, by Eq. (5) of Chapter 7, because 2 mol of electrons are produced by the reaction as written in reaction (IV).

Another simple cell is the *Daniell cell*, diagramed as

$$Cu \mid CuSO_4 \text{ (aqueous, } m) \mid ZnSO_4 \text{ (aqueous, } m) \mid Zn \qquad (VI)$$

This cell has two electrolytic solutions, aqueous $CuSO_4$ and aqueous $ZnSO_4$, which are separated by a porous barrier. The barrier stops convective mixing of the two solutions. With the circuit open, there can be no net ionic flow through this barrier. However, ions of different size (e.g., Cu^{2+} and Zn^{2+}) have different rates of diffusion through the pores of the barrier. A *junction potential* must therefore develop at the barrier to counter the tendency for diffusion to produce a net current. Liquid junctions are indicated by broken vertical lines in the cell diagram. Junction potentials are of magnitude 5–20 mV. They can be reduced to just a few millivolts, by using a *salt bridge*, a tube filled with a gel containing a concentrated solution of KCl, between the two electrolyte solutions. A salt bridge is represented by a double broken line between the two electrolyte solutions, because there are two liquid junction interfaces between the solution in the salt bridge and the electrolytes of the cell. At reasonable concentrations of electrolyte solutions, the junction potential at each interface is mainly determined by the high concentration of K^+ and Cl^- ions. (Since these are the dominant ions diffusing at the interfaces.) K+ and Cl^- ions have approximately equal diffusion rates, making the junction potentials very small. In addition, the potentials at the two interfaces are in the opposite direction, so that they almost cancel.

The design criteria for each electrochemical cell depend on the use that will be made of the cell. For example, for an inexpensive disposable dry-cell flashlight battery, a carbon electrode surrounded by a moist paste of MnO_2 and graphite in $ZnCl_2$–NH_4Cl is used. The outer electrode is Zn foil. This cell is diagramed as

$$Zn \mid Zn(NH_3)_2{}^{2+} \mid Mn_2O_3 \mid MnO_2 \mid C \qquad\qquad (VII)$$

During operation of this cell, Zn and MnO_2 are consumed. The cell cannot be recharged.

For a rechargeable automobile battery, the lead–sulfuric acid cell is employed:

$$Pb \mid PbSO_4 \text{ (s)} \mid H_2SO_4 \text{ (aq)} \mid PbSO_4 \text{ (s)} \mid PbO_2 \text{ (s)} \mid Pb \qquad (VIII)$$

The cell is reversible because PbO_2 and $PbSO_4$ remain attached to the electrodes. Because this cell produces a voltage of ~2 V, six cells are connected in series for the standard 12-V automobile battery.

A fuel cell is an electrochemical cell in which the reagents of the cell reaction are continuously supplied and the products are continuously removed. The most commonly used fuel cell is the hydrogen–oxygen cell with a NaOH electrolyte and graphite electrodes:

$$C \mid H_2 \text{ (g)} \mid NaOH \text{ (aq)} \mid O_2 \text{ (g)} \mid C \qquad\qquad (IX)$$

The electrodes are often impregnated with a Pt catalyst to accelerate the electrode reactions and make more current available from the cell.

10.10 Thermodynamics of Electrochemical Cells

For comparison with theoretical analysis, the voltage of the cell is measured with an instrument that draws negligible current from the cell, by opposing the voltage generated by the cell with an external voltage. The voltage under these conditions is called $\mathscr{E}$, the potential of the cell. Under these conditions, the ionic and electronic transfers at each electrode reach equilibrium and the cell is reversible.

By raising or lowering the externally applied voltage an infinitesmal amount, the cell reactions can be made to go in either direction. An additional requirement for reversibility is that the cell has no liquid junctions. Diffusion at such junctions will proceed in the open-circuit condition and cannot be reversed by changing the externally applied voltage. Thus, the $H^+/H_2-AgCl/Ag$ cell in reaction (V) is reversible, whereas the Daniell cell in reaction (VI) is not reversible. This will generally limit our considerations to cells with a single liquid phase. Cells consisting of nonadjacent liquid phases can be treated as the sum of cells with single liquid phases.

The cell voltage is, by convention, written as a reduction potential, and because oxidation occurs at the left-hand electrode,

$$\mathscr{E} \overset{rev}{\equiv} \phi_R - \phi_L \tag{41}$$

The condition for equilibrium of the cell is by Eq. (37),

$$\sum_j v_j \tilde{\mu}_j = 0 \tag{42}$$

This sum is over all species in the electrochemical reaction [e.g., reaction (IV)], including the electrons, which exist at different potentials at the two electrodes. Writing the reaction in terms of chemical and electrical potentials,

$$\sum_i v_i(\mu_i + z_i \mathscr{F} \phi_i) + \sum_{elec} n(\mu_{elec} + \mathscr{F} \phi_{elec}) = 0 \tag{43}$$

In the first summation, only the ions in the reaction are charged, and all of these exist in a single liquid phase, with a single electrical potential, in a reversible system. In the second summation, assuming that the same metal is used for the two terminals, the chemical potentials of the electrons are the same at these terminals. Equation (43) then becomes

$$\sum_i v_i \mu_i + \mathscr{F} \phi \sum_{ions} v_i z_i = n\mathscr{F}(\phi_L - \phi_R) \tag{44}$$

The first summation on the left-hand side of this equation is over all the species in the chemical reaction of the cell. The second term is zero, due to electroneutrality of chemical reactions. Using Eq. (41), Eq. (44) becomes

$$\mathscr{E} = -\frac{\sum_i v_i \mu_i}{n\mathscr{F}} = -\frac{\Delta_{rxn} G}{n\mathscr{F}} = \frac{A}{n\mathscr{F}} \tag{45}$$

where A is the affinity of the chemical reaction of the cell.

From Eq. (45), we see that the cell potential is the negative of the Gibbs free-energy change per mole of electrons transferred in the cell reaction. A spontaneous cell reaction with a large negative free-energy change (a large affinity) gives a large positive cell potential. Defining the *standard potential of the cell*, $\mathscr{E}^0$, as the open-cell voltage with all reactants in their standard states,

$$\mathscr{E}^0 = -\frac{\Delta_{rxn}G^\circ}{n\mathscr{F}} \tag{46}$$

we have

$$\mathscr{E} = \mathscr{E}^0 - \frac{RT}{n\mathscr{F}} \ln \prod_i a_i^{\nu_i} = \mathscr{E}^0 - \frac{RT}{n\mathscr{F}} \ln Q_a \tag{47}$$

where Q_a is the proper reaction quotient, introduced in Chapter 7. Equation (47) is the *Nernst equation*, which gives the activity dependence (and thus the concentration dependence) of the voltage of an electrochemical cell. Although, strictly speaking, the above development does not hold for cells with liquid junctions, if appropriate liquid junction potentials are included, the Nernst equation holds to a high degree of accuracy for such cells in the open-circuit condition. It is inappropriate, however, to use the equation for cells through which current is passing and in which many nonequilibrium factors contribute to the potential.

Equations (45)–(47) show how the Gibbs free-energy change of many reactions involving ionic species can be measured by studying them in electro-chemcal cells. The enthalpy change of these reactions can be obtained from the temperature dependence of the potential. Combining the Gibbs–Helmholtz equation with Eq. (46),

$$\Delta_{rxn}H^\circ = \left(\frac{\partial(T^{-1}\Delta_{rxn}G^\circ)}{\partial(T^{-1})}\right)_P = -n\mathscr{F}\left(\frac{\partial(T^{-1}\mathscr{E}^0)}{\partial(T^{-1})}\right)_P \tag{48}$$

Knowing $\Delta_{rxn}G^\circ$ and $\Delta_{rxn}H^\circ$, the entropy of the cell reaction can be calculated from

$$\Delta_{rxn}S^\circ = \frac{\Delta_{rxn}H^\circ - \Delta_{rxn}G^\circ}{T} \tag{49}$$

Alternatively, $\Delta_{rxn}S^\circ$ can be obtained from

$$\Delta_{rxn}S^\circ = -\left(\frac{\partial\Delta_{rxn}G}{\partial T}\right)_P = n\mathscr{F}\left(\frac{\partial\mathscr{E}^0}{\partial T}\right)_P \tag{50}$$

10.11   Standard Electrode Potentials

Although potentials of half-cell electrode reactions cannot be measured, their consideration is extremely useful. For example, consider the two cells in series, as indicated in reaction (X),

$$Cu \mid CuSO_4 \text{ (aqueous, } m) \mid HCl \text{ (aqueous, } 1.0 \ m) \mid H_2 \text{ (gas, } 1.0 \text{ bar)} \mid Pt$$

$$(X)$$

connected to

$$Pt \mid H_2 \text{ (gas, } 1.0 \text{ bar)} \mid HCl \text{ (aqueous, } 1.0 \ m) \mid ZnSO_4 \text{ (aqueous, } m) \mid Zn$$

The first cell consists of a Cu/Cu^{2+} electrode combined with a H^+/H_2 electrode. It is connected in series with a second cell consisting of a H_2/H^+ electrode combined with a Zn^{2+}/Zn electrode. The combination

$$HCl \text{ (aqueous, } 1.0 \ m) \mid H_2 \text{ (gas, } 1.0 \text{ bar)} \mid Pt - Pt \mid H_2 \text{ (gas, } 1.0 \text{ bar)}$$
$$\mid HCl \text{ (aqueous, } 1.0 \text{ m)} \qquad (XI)$$

consists of hydrogen electrodes connected together with opposite polarities. The voltages of these electrodes cancel, and this part of the cell diagram in reaction (X) can be omitted. The voltage of reaction (X) must, therefore, just be that of the Daniell cell given in reaction (VI). Generalizing this reasoning leads to a particularly concise way of tabulating electrochemical data. For example, calling $\mathscr{E}_R$ the potential of the cell XII

$$Pt \mid H_2 \text{ (gas, } 1.0 \text{ bar)} \mid HCl \text{ (aqueous, } 1.0 \ m) \mid ZnSO_4 \text{ (aqueous, } a) \mid Zn$$

$$(XII)$$

and $\mathscr{E}_L$ the potential of cell XIII

$$Pt \mid H_2 \text{ (gas, } 1.0 \text{ bar)} \mid HCl \text{ (aqueous, } 1.0 \ m) \mid CuSO_4 \text{ (aqueous, } a) \mid Cu$$

$$(XIII)$$

we have for the potential of cell X (or of cell VI),

$$\mathscr{E} = \mathscr{E}_R - \mathscr{E}_L \tag{51}$$

In particular, at unit activity of the $CuSO_4$ and $ZnSO_4$ solutions,

$$\mathscr{E}^0 = \mathscr{E}_R^0 - \mathscr{E}_L^0 \tag{52}$$

We call $\mathscr{E}_R$ and $\mathscr{E}_L$ *electrode potentials* and $\mathscr{E}_R^0$ and $\mathscr{E}_L^0$ *standard electrode potentials* or *standard reduction potentials*. Notwithstanding these names, these quantities really are not the potentials of single electrodes, but rather the measured potentials of cells XII and XIII. These measured potentials contain three contributions: the potential of the electrode reaction of interest, written as a reduction; the potential of the standard hydrogen electrode; and a metal–metal potential resulting from dissimilar metals at the two electrodes. By arbitrarily setting the potential of the H^+/H_2 electrode to zero,[11] the measured potentials of cells XI and XII become just those of the reduction reactions, $Zn^{2+} + 2e \rightarrow Zn$ and $Cu^{2+} + 2e \rightarrow Cu$, each combined with the metal–metal potential of its terminals, respectively. For example, the potential of cell XII becomes

$$\mathscr{E}^0 = \mathscr{E}^0_{Zn^{2+}/Zn} + \mathscr{E}^0_{Pt/Zn} \tag{53}$$

For cell X, we have

$$
\begin{aligned}
\mathscr{E}^0 &= \mathscr{E}^0_R - \mathscr{E}^0_L \\
&= \mathscr{E}^0_{Zn^{2+}/Zn} + \mathscr{E}^0_{Pt/Zn} - \left(\mathscr{E}^0_{Cu^{2+}/Cu} + \mathscr{E}^0_{Pt/Cu} \right) \\
&= \mathscr{E}^0_{Cu/Cu^{2+}} + \mathscr{E}^0_{Zn^{2+}/Zn} + \left(\mathscr{E}^0_{Cu/Pt} + \mathscr{E}^0_{Pt/Zn} \right) \\
&= \mathscr{E}^0_{Cu/Cu^{2+}} + \mathscr{E}^0_{Zn^{2+}/Zn} + \mathscr{E}^0_{Cu/Zn} \tag{54}
\end{aligned}
$$

which, except for the junction potential, is that of the Daniell cell (VI).

Because any two oxidation–reduction reactions can be combined to make a cell, the tabulation of standard electrode potentials becomes a very efficient way of calculating cell potentials under standard conditions. As indicated by Eq. (54), if the electrode reactions involve the metals of the cell terminals, the metal–metal potential due to the cell terminals is automatically included in the result. A short table of standard electrode potentials is given in Table 2.

After calculating $\mathscr{E}^0$, the Nernst equation can be used to adjust the potential to nonstandard conditions.

Example 7. Calculate the open-circuit voltage of the Daniell cell (VI) at 298 K for the following: (Neglect the liquid-junction potential.)

(a) Under standard conditions
(b) With $a(ZnSO_4) = 1.0$ and $a(CuSO_4) = 0.05$

Solution:

(a) The electrode reactions and their potentials are as follows

$$
\begin{aligned}
Zn^{2+} + 2e &\rightarrow Zn \quad -0.7618 \text{ V} \\
Cu &\rightarrow Cu^{2+} + 2e \quad -(0.3419 \text{ V}) \\
&\qquad\qquad\quad \mathscr{E}^0 = -1.1037V
\end{aligned}
$$

Equation (46) shows that under standard conditions, this cell, with a positive $\Delta_{rxn}G$, is not spontaneous. Copper does not spontaneously replace zinc ions in solution. With the zinc electrode as anode (i.e., with zinc replacing copper ions in solution), the cell

TABLE 2   Electrode (Reduction) Potentials at 298 K

Half-Reaction	Volts
$K^+ + e \rightarrow K$	−2.931
$Na^+ + e \rightarrow Na$	−2.71
$Mg^{2+} + 2e \rightarrow Mg$	−2.372
$2H_2O + 2e \rightarrow H_{2(g)} + 2OH^-$	−0.828
$Zn^{2+} + 2e \rightarrow Zn$	−0.7618
$Fe^{2+} + 2e \rightarrow Fe$	−0.447
$PbSO_4 + 2e \rightarrow Pb + SO_4{}^{2-}$	−0.359
$2H^+ + 2e \rightarrow H_{2(g)}$	0
$AgCl + e \rightarrow Ag + Cl^-$	+0.2223
$Hg_2Cl_2 + 2e \rightarrow 2Hg + 2Cl^-$	+0.2691
$Cu^{2+} + 2e \rightarrow Cu$	+0.3419
$O_2 + 2H_2O + 4e \rightarrow 4OH^-$	+0.401
$Cu^+ + e \rightarrow Cu$	+0.518
$Fe^{3+} + e \rightarrow Fe^{2+}$	+0.771
$Ag^+ + e \rightarrow Ag$	+0.7996
$Cl_{2(g)} + 2e \rightarrow 2Cl^-$	+1.3583
$PbO_2 + SO_4{}^{2-} + 4H^+ + 2e \rightarrow PbSO_4 + 2H_2O$	+1.6913

has $\mathscr{E}^0 = +1.1037$ V, and is spontaneous. If a load is placed across the terminals of the cell, electrons will flow from the Zn electrode to the Cu electrode in the external circuit.

(b) The overall cell chemical reaction is $Cu + Zn^{2+} \rightarrow Cu^{2+} + Zn$. Because the activities of the metals are unity,

$$\mathscr{E} = \mathscr{E}^0 - \frac{RT}{2\mathscr{F}} \ln\left(\frac{a_{Cu^{2+}}}{a_{Zn^2}}\right)$$

$$= -1.1037 \text{ V} - \frac{8.314 \text{ J} \mid 298 \text{ K}\mid \text{mol}}{\text{mol K} \mid 2 \mid 96,485 \text{ C}} \ln(0.05) = -1.0652 \text{ V}$$

Example 7a is the idea behind the *electrochemical series*, a list of elements whose ions will displace the ions of another element from solution, as shown in Table 3. At comparable concentrations, ions of elements with larger negative reduction potentials (more easily oxidized) will displace those of elements with smaller negative reduction potentials from solution. In the short electrochemical series shown in Table 3, an element will displace the ions of any element lower in the list from aqueous solution. In some cases, however, the displacement reactions may be slow due to kinetic factors.

The addition of voltages in part (a) of Example 7 is really a form of addition of Gibbs free energy (Hess' law). Because in the overall cell chemical reaction, the electrons cancel, both electrode reactions must involve the same

TABLE 3 Electrochemical Series

Potassium	Lead
Sodium	Hydrogen
Magnesium	Copper
Zinc	Mercury
Iron	Silver

number of electrons, and electrode potentials can be added directly. This is shown in Example 8, where the Gibbs free energies of the electrode reactions are calculated.

Example 8. Calculate $\mathscr{E}^0$ and ΔG° of the cell $Cu \mid CuCl_2$ (aq) $\mid AgCl$ (s) $\mid Ag$.

Solution: The electrode reactions, standard electrode potentials, and Gibbs free energies are as follows:

Reaction	$\mathscr{E}^0(V)$	$\Delta G^\circ = -n\mathscr{F}\mathscr{E}^0$
$Cu \rightarrow Cu^{2+} + 2e$	-0.3419	$2(0.3419)\mathscr{F}$
$2(AgCl + e \rightarrow Ag + Cl^-)$	0.2223	$-2(0.2223)\mathscr{F}$
$2AgCl + Cu \rightarrow Cu^{2+} + 2Ag + 2Cl^-$		$2(0.1196)\mathscr{F}$

Because two electrons are transferred in the reaction, $\mathscr{E}^0 = -\Delta G^0/n\mathscr{F} = -0.1196$ V, as could be calculated directly, without calculating free energies.

In calculating the potential of a half-cell from those of other half-cells, electrons do not necessarily cancel and the calculation must be performed using free energies, as illustrated in Example 9.

Example 9. Calculate $\mathscr{E}^0$ of the Fe^{3+}/Fe electrode.

Solution:

Reaction	$\mathscr{E}^0$ (V)	$\Delta G^\circ = -n\mathscr{F}\mathscr{E}^0$
$Fe^{3+} + e \rightarrow Fe^{2+}$	0.771	$-(0.771)\mathscr{F}$
$Fe^{2+} + 2e \rightarrow Fe$	-0.447	$2(0.447)\mathscr{F}$
$Fe^{3+} + 3e \rightarrow Fe$		$0.123\mathscr{F}$

Therefore, $\mathscr{E}^0 = -\Delta G^\circ/3\mathscr{F} = -0.041$V.

Example 10. Calculate $\mathscr{E}^0$ of the cell Pt | H$_2$ (gas) | HCl (aqueous) |Cl$_2$ (gas) | Pt. Use your calculated value to obtain ΔG_f° of aqueous Cl$^-$ ions.

Solution:

Reaction	$\mathscr{E}^0$ (V)	$\Delta G^\circ = -n\mathscr{F}\mathscr{E}^0$ (kJ/mol)
H$_2$ (gas) $\rightarrow$ 2H$^+$ + 2e	0	0
Cl$_2$ (gas) + 2e $\rightarrow$ 2Cl$^-$	1.3583	-262.11
H$_2$ (gas) + Cl$_2$ (gas) $\rightarrow$ 2H$^+$ + 2Cl$^-$		-262.11

Because ΔG_f° of H$_2$ (gas), Cl$_2$ (gas), and H$^+$ are all zero,[11] $\Delta G^\circ = 2\Delta G_f^\circ(\text{Cl}^-)$, $\Delta G_f^\circ(\text{Cl}^-) = -131$ kJ/mol.

Example 10 illustrates how thermochemical data for aqueous ions may be obtained from measurements in electrochemical cells. The problem of measuring cell potentials in the standard state, which is a hypothetical state, will be discussed in section 10.12. The temperature variation of the voltage of such cells would provide ΔH_f° of aqueous ions, through the use of Eq. (48).

10.12 Applications of Electrochemical Cells

10.12.1 Measurements of Thermodynamic Quantities

Equations (47)–(50) indicate how thermodynamic quantities can be obtained from cell potentials measured under standard conditions. However, standard states are *hypothetical* states (e.g., infinitely dilute behavior at 1.0 m concentration), which cannot be prepared in the cell. As a result, an extrapolation procedure is used to find $\mathscr{E}^0$ from measured cell voltages as a function of concentration. From Eq. (47), we write the dependence of $\mathscr{E}$ on the concentration of the electrolyte in the form

$$\mathscr{E}(m_i) + \frac{RT}{n\mathscr{F}} \ln(\textstyle\prod_i m_i^{v_i}) = \mathscr{E}^0 - \frac{RT}{n\mathscr{F}} \ln(\textstyle\prod_i \gamma_i^{v_i}) \tag{55}$$

The left-hand side of this equation can be calculated from measurements of cell voltage as a function of concentration. The second term on the right-hand side becomes zero at infinite dilution. However, because no meaningful measurements can be made at zero concentration of reactants, we must extrapolate the equation to infinite dilution using the known concentration behavior of activity coefficients. In approaching infinite dilution, it is sufficient to use the Debye–Huckel limiting law. Substituting the infinitely dilute limit of Eq. (19) into the last term of Eq. (55), we find the concentration dependence of the right side of Eq. (55) to be

$$\mathscr{E}^0 - \frac{RT}{n\mathscr{F}} \sum_i \nu_i \ln \gamma_i = \mathscr{E}^0 + \frac{RT}{n\mathscr{F}} \sum_i \nu_i z^2 A I^{1/2}$$

$$= \mathscr{E}^0 + I^{1/2}\left(\frac{RT}{n\mathscr{F}} \sum_i \nu_i z^2 A\right) \qquad (56)$$

The only concentration dependent term is $I^{1/2}$. Thus, plotting the left-hand side of Eq. (56) versus $I^{1/2}$ (versus $m_i^{1/2}$, if the electrolyte i is the only contributor to the ionic strength) gives, in the limit of low concentrations, a straight line with intercept $\mathscr{E}^0$. From Eq. (46), ΔG° of the cell reaction is obtained. By repeating the procedure at different temperatures, ΔH° and ΔS° can be determined from Eqs. (48)–(50).

10.12.2 Measurements of Activity Coefficients

Once $\mathscr{E}^0$ is determined by the extrapolation procedure of Sec. 10.11 or from a table of electrode potentials (which implies that someone else has done the extrapolation), measurement of $\mathscr{E}$ as a function of concentration allows determination of Q_a as a function of concentration of the cell electrolyte from Eq. (47). Often, the electrolyte consists of a single solute, whose activity can then be determined from Q_a. For example, for the cell in Fig. 5, assuming the activities of the H_2 gas and condensed phases are unity,

$$Q_a = a_{H^+}^2 a_{Cl^-}^2 = a_{HCl}^2 = a_\pm^4$$

10.12.3 Determination of pH

The pH may be measured by means of a hydrogen gas electrode in the solution of interest (X), connected by a salt bridge to a reversible electrode, such as the AgCl/Ag electrode or the calomel electrode:

$$\text{Pt} \mid H_2\, a_{H_2} \mid X\, a_{H^+} \mid \text{aqueous HCl}\, a_{Cl^-} \mid \text{AgCl} \mid \text{Ag} \mid \text{Pt} \qquad \text{(XIV)}$$

The potential of this cell is

$$\mathscr{E}_X = \mathscr{E}^0 + \mathscr{E}_{JX} - \frac{RT}{\mathscr{F}}(\ln a_{H^+}(X) + \ln a_{Cl^-} - \tfrac{1}{2}\ln a_{H_2}) \qquad (57)$$

where $\mathscr{E}^0$ is the standard potential of the AgCl/Ag electrode and $\mathscr{E}_{JX}$ is the junction potentials in this cell. In order to account for the junction potential and variations in the activity of H_2 and Cl^-, measurements are made by comparison with the potential generated by the same electrodes and salt bridge using a standard solution, for which the measured potential is

$$\mathscr{E}_S = \mathscr{E}^0 + \mathscr{E}_{JS} - \frac{RT}{\mathscr{F}}(\ln a_{H^+}(S) + \ln a_{Cl^-} - \tfrac{1}{2}\ln a_{H_2}) \tag{58}$$

The junction potential, being dominated by the concentrated solution in the salt bridge, can be taken as the same in these two measurements, giving

$$\mathscr{E}_X - \mathscr{E}_S = -\frac{RT}{\mathscr{F}}(\ln a_{H^+}(X) - \ln a_{H^+}(S)) = \frac{2.303RT}{\mathscr{F}}(pH_X - pH_S) \tag{59}$$

or

$$pH_X = pH_S + \frac{\mathscr{F}}{2.303RT}(\mathscr{E}_X - \mathscr{E}_S) \tag{60}$$

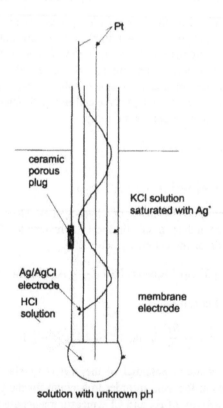

Figure 6 Membrane-type pH electrode.

The standard solution (which can be purchased from a chemical supply company) is chosen to have a pH close to that of the unknown solution. pH_S is calculated (by the company), using extended Debye–Huckel theory, to good accuracy.

Usually, rather than using a hydrogen gas electrode, a glass membrane electrode is used for the measurement. As discussed in Sec. 8, the potential across such a membrane can be proportional to the difference in pH's of the solutions on each side of the membrane. One design for a membrane-type pH electrode, which incorporates a Ag/AgCl reference electrode in a tube concentric to the membrane electrode, is shown in Fig. 6. The electrode is immersed in the solution whose pH is to be measured, with the solution level above the porous plug.

Questions

1. Show in two dimensions a typical arrangement of four positive ions and four negative ions in solution. How does your arrangement differ from that in a crystal?

2. Give a physical reason for the Debye length:

   (a) increasing with the relative dielectric constant of the solvent
   (b) increasing with the temperature of the solution
   (c) decreasing with the ionic strength of the solution

3. Show that for the Debye–Huckel theory [Eq. (21)] $\gamma_\pm$ is always less than 1.0 and that in concentrated solutions, $\gamma_\pm$ is independent of ionic strength.

4. Which of the following properties of $Ca(NO_3)_2$ do not change when a 0.03 m solution of $Ca(NO_3)_2$ is made 1.0 m in NaCl: m_i, $m_\pm$, and $\gamma_\pm$.

5. Show that for the Davies equation [Equation (25)], $\gamma_\pm$ may be greater than 1. Sketch the dependence of $\gamma_\pm$ on ionic strength for the Davies equation.

6. Show that when $v_+ = v_- = 1$, $m_\pm = m_i$, and when $v_+ = v_- = p$, $m_\pm = pm_i$.

7. In terms of the molality of the solution, what is the activity of a NaCl solution for the following conditions:

   (a) In the limit of infinite dilution
   (b) At finite concentration

8. Which of the following increase $\gamma_\pm$ above the DHLL value:

   (a) Inclusion of the $\sqrt{I_c}$ term in the denominator [Eq. (24)]
   (b) Inclusion of a linear term [Eq. (25)]
   (c) Ion pairing
   (d) Hydration of ions

9. Which of the following represents an ion pair in aqueous solution?

   (a) HF
   (b) $[Fe(CN)]^{2+}$
   (c) $[CaCl]^+$

10. Do you expect the formation of ion pairs to increase or decrease γ_i? Why?

11. For which of the following systems do you expect ion pairing to be the most important? Why?

(a) 0.1 m $(CH_3)_3NHCl$ in benzene
(b) 0.1 m $CaCl_2$ in water
(c) 0.3 m NaCl in water

12. What will happen if a piece of Zn metal is dropped into an essentially infinite volume of distilled water? Will all of the Zn dissolve? If not, why? Explain from a macro and a micro perspective.

13. For the Daniell cell, given by reaction (VI), what are the electrode reactions? Sketch the cell. How is current transported within the cell? Assuming that the smaller Zn^{2+} ions tend to diffuse faster, what is the direction of the junction potential?

14. What are the chemical reactions occurring in a lead–sulfuric acid car battery?

(a) During starting the car
(b) During charging the battery, while the car is driving

Which is the positive terminal of the battery?

15. The cells of automobile batteries used to include a capped opening for adding water that evaporated. The amount of charge stored on such cells could be determined with a *hydrometer*, a simple device that measured the density of the electrolyte in the cell. Explain why the density of the electrolyte is related to the charge state of the cell.

16. What is wrong with the following argument? If the terminals of an electrochemical cell are constructed from the same metal, the chemical potential of electrons [species i in Eq. (36)] at the terminals, which depends only on T, P and concentrations, are the same. From Eq. (36), the electromotive force of the cell is therefore zero!

17. Write down the standard simplified diagram for the electrochemical cell obtained when the pH electrode shown in Fig. 6 is placed in a solution with hydrogen ion activity, $a_{H^+}(X)$.

Problems

1. What is $m_\pm$ for each of the following solutions:

(a) 0.03 m NaCl
(b) 0.02 m $CaCl_2$
(c) 0.01 m $MgSO_4$
(d) 0.03 m $Ca_3(PO_4)_2$

2. Write $\gamma_\pm$ in terms of γ_+ and γ_- for each of the solutions in Problem 1.

3. Show that the definition in Eq. (13) gives $\phi \rightarrow 1$ in the limit of the infinitely dilute solution of a strong electrolyte. See Problem 2 of Chapter 9.

4. Derive Eq. (21) from Eq. (19) for an electrolyte that produces v_+ positive ions of charge z_+ and v_- negative ions of charge z_-.

5. Show that for a volatile electrolyte, such as HCl, the partial pressure above an aqueous solution, in the limit of infinite dilution, is proportional to m^2, and not to m, as given by Henry's law.

6. Use both the Debye–Huckel limiting law [Eq. (23)] and Eq. (24) to calculate the mean ionic activity coefficient of $CaCl_2$ in a solution prepared by adding 0.01 mol of $CaCl_2$ to the following:

    (a)   1.0 L of water
    (b)   1.0 L of 0.5 m $NaNO_3$ solution
    (c)   Repeat part (b) using the Davies formula [Eq. (25)].

7. Calculate the activity of a 0.03 m $ZnCl_2$ solution using the following:

    (a)   The Debye–Huckel limiting law [Eq. (23)]
    (b)   The Debye–Huckel theory [Eq. (24)]
    (c)   The Davies equation [Eq. (25)]

8. Using the result of Example 10, calculate the partial pressure of HCl in equilibrium with a 1.0 m aqueous HCl solution at 25°C.

9. Use the DHLL to calculate m_{H^+} and the pH for an aqueous solution at 25°C that is

    (a)   0.001 m in acetic acid
    (b)   0.001 m in acetic acid and 0.3 m in NaCl

Why is there more of a difference in the pH for these two cases than was found for Examples 4 and 5?

10. Use the DHLL to calculate m_{H^+} and the pH for a 0.1 m aqueous solution of chloroacetic acid at 25°C. $K_a = 1.40 \times 10^{-3}$ for chloroacetic acid. Compare your results with those calculated assuming an ideally dilute solution.

11. K_a for the reaction CO_2 (gas) $+ H_2O \leftrightarrow H^+ + HCO_3^-$ is 4.45×10^{-7}. What is m_{H^+} and the pH of water that is in equilibrium with 320 ppm of CO_2 in the atmosphere at 1.0 bar total pressure? You can neglect the dissociation of HCO_3^- into CO_3^{2-}.

12. Find the fraction of ion pairs in 0.15 m aqueous $CaSO_4$ at 25°C.

13. Show that for the case that all reactants and products are in the same phase, Eq. (37), for the equilibrium condition for a reaction involving charged species, reduces to Eq. (30) of Chapter 7.

14. For calculating the electrostatic potential of a charged metallic sphere, the charge on the sphere can be considered to reside at its center. What is the charge (in both Coulombs and moles of electrons) on a 1.0-cm-radius sphere so that the electrostatic potential at the surface of the sphere is 1.0 V?

15. Show that the potential difference across a membrane [Eq. (40)] between two solutions of different hydrogen ion activities is a linear function of the pH difference between the two solutions.

16. Write down the half-reactions for cell VII.

17. What is the voltage of the cell in reaction (V) when $P_{H_2} = 680$ torr and the molality of HCl is 0.05? See Example 2c for $\gamma_\pm$ of HCl.

18. Use Table 2 to calculate ΔG° for the reaction $2Fe^{3+} + 2Cl^- \rightarrow 2Fe^{2+} + Cl_2$.

19. The voltage for the cell in reaction (VIII) at $1.0\,m$ H_2SO_4 has been measured to be $\mathscr{E}$ (volts) $= 1.91737 + 5.61 \times 10^{-5}T + 1.08 \times 10^{-6}T^2$, with T (in °C) from 0–60°C. (HS Harned, WJ Hammer. J Am Chem Soc 57:33, 1935.

    (a) Write down the cell reaction.
    (b) Calculate ΔG, ΔH, and ΔS for the cell reaction at 25°C.
    (c) Calculate $\gamma_\pm$ for $1.0\,m$ H_2SO_4 at 25°C.

20. In a *hydrogen economy*, hydrogen for transportation fuels is produced by the electrolysis of water.

    (a) Write down the electrode reactions for producing H_2 and O_2 in an alkaline water hydrolysis cell.
    (b) Show the polarity of the external current source and indicate which electrode is the cathode.
    (c) What is the minimum voltage required for the external current source in order to electrolyze water?

21. For the cell Pt | H_2 (gas, 1.0 atm) | HCl (aqueous, m) | Hg_2Cl_2 (solid) | Hg (1) | Pt:

    (a) Write the half-cell reactions and the overall chemical reaction.
    (b) Express the open-circuit potential in terms of $\mathscr{E}^0$ and the activities of the components of the cell.

22. From the data in Table 2, find $\mathscr{E}^0$ for a Cu^{2+}/Cu^+ electrode.

Notes

1. From CP Jargodski, F Potter. Mad about Physics. New York: Wiley, 2001.
2. Because $v_\pm$ is independent of concentration, $dm_\pm = d(v_\pm m_i) = v_\pm\,dm_i$.
3. Nobel laureate in 1936.
4. Conceptually, a is the sum of the radii of the positive and negative ions (because such ions are more likely to be at closest approach); practically, a is usually taken as a parameter, which is optimized by fitting to experimental data.
5. The dielectric constant of water, 78.4, is very large. Deviations from ideally dilute behavior are much greater in less polar solvents than they are in water.
6. The symbol p is used as a shorthand notation for $-\log_{10}$.
7. This is not the official definition, which is a practical definition based on the measured potentials of some exactly defined electrochemical cells in which the hydrogen ion activity is calculated by an extended Debye–Huckel theory.
8. The results of the theory are not overly sensitive to small changes in this criterion.

9. In addition to the charge on a phase, polarization, which is the orientation and distribution of molecules and ions in and adjacent to the phase, can influence the escaping tendency of charged particles from the phase.
10. Under some conditions, low concentrations of solvated electrons can exist in solution.
11. This is equivalent to setting the Gibbs free energy of formation of $1.0\,m$ ideally dilute aqueous H^+ equal to zero and the chemical potential of Pt metal equal to zero.

11

Surfaces

Beauty is only skin deep.

Proverb

The surface of a phase has a different environment than its bulk. Surface tension is a measure of the work that must be done to increase the surface area of a phase. For curved surfaces, pressure is higher on the concave side of the surface than on the convex side. As a result, liquids rise in capillaries that they wet and the vapor pressure of droplets is greater than that of flat surfaces. The contact angle between condensed phases depends on the attractive interactions at their interface. When these interactions are large, there will be spreading (of a liquid over a solid) or mixing (of two liquids). Surface excess concentration measures how a solute concentrates at the surface of a solution. Positive excess concentration lowers the surface tension of the solvent. Substances that strongly concentrate at an interface are called surfactants. Gases can chemisorb or physisorb on solids. The Langmuir isotherm describes adsorption of a monolayer on a surface, whereas the Brunauer, Emmet, and Tetter (BET) isotherm describes multi-layer adsorption. The latter can be used to estimate the surface area of a solid. Statistical mechanics can be used to calculate thermodynamic properties of adsorption and treat gas–surface

308

equilibria. Colloids are comprised of particles containing 10^5–10^9 atoms. Their properties are dominated by surface effects.

11.1 The Surface Region

So far, we have considered only homogeneous phases of matter, where concentrations and other properties are uniform throughout the phase. The surface of a phase or, more exactly, the *interfacial region*, extending a few molecular diameters from the surface, is clearly different from the bulk of the phase. Energies and concentrations (for multicomponent phases) may vary in this region. Molecules in the interfacial region experience reduced interactions with other molecules in the phase and may begin to interact with molecules in the adjacent phase.

Because systems usually interact with their surroundings only at their surface, the properties of the surface assume an importance disproportionate to the fraction of molecules in this region. The interfacial region is so small that it makes a negligible contribution to the bulk properties of the phase, unless the phase is composed of very small particles. Conversely, for those properties which depend on the surface (e.g., adhesion, wetting, foam formation, etc.), very small additions of materials can result in very large effects, if the added substance concentrates at the surface. Changing surface concentrations can, therefore, often be a very cost-effective way of modifying the properties of materials.

11.2 The Surface of the Single-Component Condensed Phase

In thermodynamics, we focus on the most important variables needed to describe a system. Although we are interested in the size of a system (or of a phase), we usually do not concern ourselves with the shape of the system. One way in which the shape of a system does influence its thermodynamic properties is through its surface area. The surface of a phase is a different environment than its bulk region. Molecules on the surface of a material do not experience attractive interactions to other molecules in all directions and, therefore, have higher energy than molecules in the bulk of the material. Energy is increased when the surface area of a condensed system (usually a liquid) is increased at constant volume and temperature. Because the Helmholtz free energy has T and V as its natural variables, we can immediately write

$$dA = \left(\frac{\partial A}{\partial T}\right)_{V,\sigma} dT + \left(\frac{\partial A}{\partial V}\right)_{T,\sigma} dV + \left(\frac{\partial A}{\partial \sigma}\right)_{T,V} d\sigma =$$
$$- S\, dT - P\, dV + \gamma\, d\sigma \tag{1}$$

where the intensive variable γ is called the *surface tension* of the condensed phase. We are more likely to deal with condensed phases in air or in the presence of their vapors, rather than in vacuum. In such cases, to be exact, we should employ properties such as γ_{LG} or γ_{LV}, the liquid–air or liquid–vapor interfacial tensions in Eq. (1). However, at usual laboratory pressures, gas density is so low that these quantities are independent of the nature of the gas and the use of just γ, the surface tension of the condensed phase, is appropriate. We define partial derivatives of thermodynamic properties with respect to surface area, with volume and temperature constant, as the surface thermodynamic property (written lowercase, with superscript σ), so that

$$\gamma \equiv \left(\frac{\partial A}{\partial \sigma}\right)_{T,V} \equiv a^{\sigma} \tag{2}$$

that is, the surface tension is the surface Helmholtz free energy. We can also write

$$dU = d(A + TS) = T\,dS - P\,dV + \gamma\,d\sigma \tag{3}$$

showing that $\gamma = (\partial U/\partial \sigma)_{S,V}$.

Because $\gamma\,d\sigma$ is an additional work term, using Eqs. (39) and (40) of Chapter 4, we define the enthalpy and Gibbs free energy as

$$H = U + PV - \gamma\sigma \tag{4}$$

and

$$G = U - TS + PV - \gamma\sigma. \tag{5}$$

The differential forms of these equations are

$$dH = T\,dS + V\,dP - \sigma\,d\gamma \tag{6}$$

and

$$dG = V\,dP - S\,dT - \sigma\,d\gamma \tag{7}$$

With these definitions, the natural variables for the Gibbs free energy (P, T, and γ) are all intensive functions.

A useful Maxwell relation can be derived from Eq. (1)[1]

$$\left(\frac{\partial S}{\partial \sigma}\right)_{V,T} \equiv s^{\sigma} = -\left(\frac{\partial \gamma}{\partial T}\right)_{V,\sigma} \tag{8}$$

where s^{σ} is the surface entropy. Because $U = A + TS$,

$$\left(\frac{\partial U}{\partial \sigma}\right)_{V,T} \equiv u^{\sigma} = \left(\frac{\partial A}{\partial \sigma}\right)_{V,T} + T\left(\frac{\partial S}{\partial \sigma}\right)_{V,T} = \gamma - T\left(\frac{\partial \gamma}{\partial T}\right)_{V,\sigma} \tag{9}$$

where u^{σ} is the surface energy.

Even though it is not the Gibbs or Helmholtz free energy, it is useful to consider the F function, $F \equiv U + PV - TS$, with differential

$$dF = V\,dP - S\,dT + \gamma\,d\sigma \qquad (10)$$

This gives Maxwell relations

$$\left(\frac{\partial S}{\partial \sigma}\right)_{P,T} = -\left(\frac{\partial \gamma}{\partial T}\right)_{P,\sigma} \quad \text{and} \quad \left(\frac{\partial V}{\partial \sigma}\right)_{P,T} = \left(\frac{\partial \gamma}{\partial P}\right)_{\sigma,T} \qquad (11)$$

To the degree that a single phase of one component has only two degrees of freedom, $(\partial V / \partial \sigma)_{P,T}$ is zero. Even with highly accurate measurements, we expect this quantity and, from Eq. (11), the pressure dependence of the surface tension to be very small.

Note also that

$$\gamma = \left(\frac{\partial F}{\partial \sigma}\right)_{P,T} \qquad (12)$$

We can also write

$$\left(\frac{\partial U}{\partial \sigma}\right)_{P,T} = \left(\frac{\partial F}{\partial \sigma}\right)_{P,T} + T\left(\frac{\partial S}{\partial \sigma}\right)_{P,T} - P\left(\frac{\partial V}{\partial \sigma}\right)_{P,T} \qquad (13)$$

Dropping the very small last term and using Eq. (11), we get

$$\left(\frac{\partial U}{\partial \sigma}\right)_{P,T} = \gamma - T\left(\frac{\partial \gamma}{\partial T}\right)_{P,T} \qquad (14)$$

which is analogous to Eq. (9).

11.3 Surface Tension of Single Components

Because intermolecular interactions of surface molecules are intermediate between those in the vapor and those in the bulk, we can expect surface tensions to be roughly proportional to energies of vaporization at the same temperature. Water is one of the exceptions to this rule. It has a particularly large value of γ, due to surface molecules being much less hydrogen-bonded than molecules in the bulk. The surface tensions of some common liquids is given in Table 1. These values are measured in air, saturated with vapor at 1.0 atm total pressure.

One of the greatest difficulties in making surface tension measurements is ensuring the purity of the substance studied. Minute amounts of impurity, by concentrating at the surface, can make an appreciable difference in measured surface properties. Small amounts of reactive or condensable impurities in the gas phase can also greatly influence measured surface tensions.

TABLE 1   Surface Tension of Liquids at 298 K

Liquid	γ (10^{-3} J/m^2)	$d\gamma/dT$ $(10^{-3} \text{ J/m}^2 \text{ K})$
Ethyl alcohol	22.7	
Water	72.7	−0.15
n-Hexane	18.4	−0.10

Surface tensions of liquids decrease with increasing temperature and must equal zero at the critical temperature, where there is no longer a difference between the liquid and gaseous phase.[2] As a result, the temperature dependence of the surface tension is often expressed by the following empirical equation:

$$\gamma = \gamma_0 \left(1 - \frac{T}{T_c}\right)^n \tag{15}$$

A value of $n = 1$ is often employed, although Guggenheim has found $n = 11/9$ gives a better fit for organic liquids.

Equations (8) and (15) indicate that the surface entropy of liquids is positive. This is because extending the surface creates an additional environment into which molecules can partition. When, in Eq. (15), $n = 1$ is employed, the surface energy is independent of temperature (problem 3). In practice, this is found not to hold when approaching the critical temperature, where the surface energy is also found to approach zero.

There is no reason to expect that as the dimensions of a phase are reduced to the point where it contains a small number of molecules, its surface tension remains constant. This question has obvious relevance to important problems, such as condensation of droplets from supersaturated vapors. Although a number of authors have considered this problem, in these days of supercomputers, it is probably best handled by considering individual intermolecular interactions rather than a bulk property, such as surface tension.

Solids also have "surface tension" because molecules on the surface of a solid particle are subject to fewer attractive forces than molecules in the bulk of the solid. Measurements of the surface tension of solids (usually called the surface energy) are difficult because solids are rarely pure and smooth on the molecular scale.

As shown in Example 1, surface effects only contribute appreciably to thermodynamic quantities for very small particles. This is because relatively few of the molecules of a macroscopic sample lie close enough to the surface to experience interactions noticeably different from those in the interior.

Example 1. For what diameter droplet size does the extra energy of surface molecules become 1% of the bulk cohesive energy of *n*-hexane at 298 K? For hexane, $\rho = 0.66$ g/mL and $\Delta_{vap}H = 31.9$ kJ/mol.

Solution: The extra energy of surface molecules per mole equals $u^\sigma \sigma_m$, where σ_m is the surface area per mole:

$$\mu^\sigma = \gamma - T \frac{d\gamma}{dT} = 18.4 \times 10^{-3} + 298(0.1 \times 10^{-3})$$
$$= 48.2 \times 10^{-3} \text{ J/m}^2$$

Taking the energy of vaporization of *n*-hexane,

$$\Delta_{vap}U = \Delta_{vap}H - RT = 31{,}900 - 8.314(298) = 29{,}400 \text{ J/mol},$$

as a measure of its cohesive energy, the molar area of σ_m required to give this energy is determined by

$$48.2 \times 10^{-3} \frac{\text{J}}{\text{m}^2} \sigma_m = 0.01 \left(\frac{29{,}400 \text{ J}}{\text{mol}} \right) \quad \text{or} \quad \sigma_m = 7000 \frac{\text{m}^2}{\text{mol}}$$

The surface-to-volume ratio is

$$4\pi r^2 / 4\pi r^3 / 3 = 3/r$$
$$= \frac{7000 \text{ m}^2}{\text{mol}} \frac{\text{mol}}{86 \text{ g}} \frac{0.66 \text{ g}}{\text{mL}} \frac{10^6 \text{ mL}}{\text{m}^3}$$
$$= 5.4 \times 10^7 \text{m}^{-1}$$
$$d = 2r = 1.1 \times 10^{-7} \text{ m}$$

11.4 Processes Involving One Interface

Even though surface effects may not be an appreciable contribution to total energy, there are phenomena for which they are dominant and which permit the surface tension to be measured.

11.4.1 Stretching of a Film

Consider a film of liquid on a wire balance as shown in Fig. 1. A force is applied to a frictionless, movable wire, stretching the film, which adheres to the wire. The process occurs at constant temperature and volume of the liquid. The reversible work done as the movable wire moves a distance dx (at constant T and V) is equal to the increase in the Helmholtz free energy of the system:

$$\delta w_{rev} = f \, dx = dA = \gamma \, d\sigma = 2L\gamma \, dx \tag{16}$$

where the factor of 2 comes from the film having two sides. The surface tension can therefore be measured with such a balance as $\gamma = f/2L$. f is the force

Figure 1   Wire balance for stretching a liquid film.

required to keep the movable wire at equilibrium. It is useful to also think of the surface tension as the force per unit length, directed parallel to the surface, which "pulls" to reduce the surface area (hence, "tension"). An oppositely directed force must be applied to keep the surface from contracting. In the absence of surface tension, zero force would be required to move the frictionless wire.

11.4.2 Falling Droplet

Water dripping from a tube will form a droplet supported by surface tension at the tube–water boundary, until the surface tension force can no longer support the droplet. Just before the droplet falls, as shown in Fig. 2, the gravitational force on the droplet is balanced by the surface tension force at the edge of the tube:

$$\gamma \pi d = mg \tag{17}$$

With suitable calibration, this phenomenon can be used to measure surface tension.

Figure 2   Falling droplet.

11.4.3 Cavities in Liquids

A cavity in a liquid filled with an insoluble gas will attain an equilibrium radius, where the pressure in the cavity is larger than that in the liquid. To see this, consider Fig. 3.

We imagine a slight increase, dr, in the radius of a cavity in a liquid. The system is the constant volume of a liquid between the cavity and an outer boundary whose radius increases as the cavity volume increases. The system is held at constant temperature, and because it has constant volume, Eq. (1) becomes

$$dA = \gamma \, d\sigma \tag{18}$$

If the system is at equilibrium, dA is the net work that must be done on the system for the change in r. This is (because $dV_l = 0$)

$$dA = \gamma \, d\sigma = -P_l \, dV_g + P_g \, dV_g \tag{19}$$

or

$$P_g - P_l = \gamma \left(\frac{d\sigma}{dV_g} \right) \tag{20}$$

For a spherical cavity, $\sigma = 4\pi r^2$, $d\sigma = 8\pi r \, dr$, $V = \frac{4}{3}\pi r^3$, and $dV = 4\pi r^2 \, dr$. Therefore, $d\sigma/dV_g = 2/r$, and

$$P_g - P_l = \frac{2\gamma}{r} \tag{21}$$

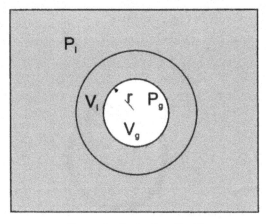

Figure 3 Cavity in a liquid.

This is for a spherical cavity. For a cavity with two principle radii of curvature, r_1 and r_2, Eq. (21) becomes[3]

$$P_g - P_l = \gamma\left(\frac{1}{r_1} + \frac{1}{r_2}\right) \tag{22}$$

This is known as the equation of Young and Laplace.

For water, with a surface tension of $73\,mJ/m^2$, Eq. (21) gives $P_g - P_l = 0.0015$ bar for a 1-mm-radius cavity and 1.5 bar for a 1-μm cavity (a very large pressure difference indeed!). Note that the surface acts like a membrane under tension, increasing the pressure on the concave side of the membrane. For a bubble, as shown in Fig. 4, there are two surfaces acting as membranes to increase the pressure of the gas inside the bubble over ambient pressure. The inside to outside pressure difference is thus twice that of Eq. (21):

$$P_{in} - P_{out} = \frac{4\gamma}{r} \tag{23}$$

11.5 Processes Involving More Than One Interface

11.5.1 Contact Angle and Spreading

When three phases are coexisting along an edge, we have three interfacial tensions. Assuming that one of the three phases is a gas (or a vapor), we will write these as γ_α, γ_β, and $\gamma_{\alpha\beta}$. The edge of the mobile α phase is at equilibrium under the tensions directed along the three interfaces. Equating the horizontal components of the forces in Fig. 5, we have

$$\gamma_\beta = \gamma_{\alpha\beta} + \gamma_\alpha \cos\theta \tag{24}$$

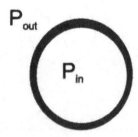

Figure 4   Pressure difference in a bubble.

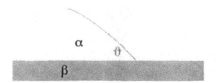

Figure 5 Interface among three phases.

or

$$\cos\theta = \frac{\gamma_\beta - \gamma_{\alpha\beta}}{\gamma_\alpha} \tag{25}$$

θ is called the (equilibrium) *contact angle*; $\cos\theta > 1$ represents a non-equilibrium situation and occurs if

$$\gamma_\beta - \gamma_{\alpha\beta} > \gamma_\alpha \tag{26}$$

or

$$\gamma_\alpha + \gamma_\beta - \gamma_{\alpha\beta} > 2\gamma_\alpha \tag{27}$$

$2\gamma_\alpha$ is called the *cohesive energy*, w_{coh}^α, of the α phase. It is the energy required to separate the α phase and produce two α-vapor interfaces. $\gamma_\alpha + \gamma_\beta - \gamma_{\alpha\beta}$ is called the *adhesive energy*, $w_{adh}^{\alpha\beta}$, of the α–β interface. It is the energy required to separate the α–β interface into separate α and β interfaces. We define the *spreading coefficient*, $S^{\alpha\beta}$, as

$$S^{\alpha\beta} \equiv \gamma_\beta - \gamma_\alpha - \gamma_{\alpha\beta} = w_{adh}^{\alpha\beta} - w_{coh}^\alpha \tag{28}$$

If $S^{\alpha\beta} > 0$, in the horizontal case, the liquid α will spread over the surface β. If β is a solid, α will completely wet it and enter its pores. This is called *complete wetting*, and is very useful (e.g., if we are using α to lubricate β). If the surface is not horizontal, the liquid will rise until the additional force of gravity balances the forces on the liquid edge. If β is also a liquid, α and β will mix and become a solution.

11.5.2 Wilhelmy Plate

When a plate is partially immersed in a liquid, the liquid adheres to the plate with contact angle θ, as shown in Fig. 6.

In order to suspend the plate in the liquid, the external force must balance the surface tension forces in addition to the gravitational and buoyancy forces on

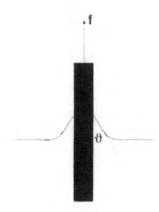

Figure 6 Plate suspended in a liquid.

the plate. Surface tension forces are exerted on the plate at the solid–liquid and solid–gas interfaces. The net downward surface tension force is

$$f_{st} = C(\gamma_{SG} - \gamma_{SL}) \tag{29}$$

where C is the circumference of the plate. Note that the direction of these forces on the plate are opposite to their directions on the mobile liquid interface. From Eq. (25),

$$f_{st} = C\gamma_{LG}\cos\theta \tag{30}$$

When the surface of the liquid is drawn up on the solid, we say that the liquid "wets" the solid. This results from a substantial attraction between the liquid and the surface of the solid, which is indicated by a small value γ_{SL}, the solid–liquid surface tension, and a large downward net surface tension force. A device that measures the force on a plate, such as shown in Fig. 6, is called a Wilhelmy balance.

11.5.3 Capillary Rise

Because the pressure is lower on the convex side of an interface, liquids will rise up in small tubes (capillaries), whose surface they "wet", with *contact angle* θ less than 90°, as shown in Fig. 7.

In the open capillary tube shown, the pressure in the gas phase at a and b are the same. Just below a in the liquid, we are on the convex side of the interface, so the pressure is lower by Eq. (21). At the bottom of the tube, the pressure must

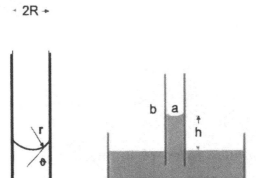

Figure 7 Capillary rise.

be the same inside and outside the tube, and so the liquid in the tube must rise a distance h, with

$$\frac{2\gamma}{r} = \rho g h \tag{31}$$

where ρ is the density of the liquid and r is the radius of curvature of the liquid interface. In the case of *perfect wetting*, when the adhesive energy between the liquid and the tube is greater than the cohesive energy of the liquid, the contact angle $\theta = 0°$ and the radius of curvature of the liquid surface, r, becomes equal to the radius of the tube, R. In this case, the capillary rise is

$$h = \frac{2\gamma}{\rho g R} \tag{32}$$

Capillary rise is responsible for water being drawn into a sponge or a cloth. When the cloth is coated with a waterproofing material, such as silicone, the adhesive forces are reduced, the contact angle is greater than 90°, and there is no longer a tendency for water to be drawn into the cloth.

Example 2. What pressure above atmospheric must be applied in a 1.0-mm-diameter tube placed just below the surface of water at 25°C, in order to blow a bubble. (Assume perfect wetting of the tube.)

Solution: The excess pressure applied must be sufficient to counteract the tendency of water to rise in the tube plus that needed to form a cavity in the water.

$$\Delta P = \rho g h + \frac{2\gamma}{R} = \frac{4\gamma}{R} \quad \text{[using Eq. (31) for perfect wetting]}$$

$$\Delta P = \frac{4 \mid 72.7 \times 10^{-3} \text{ J/m}^2}{5 \times 10^{-4} \text{ m}} = 5.8 \times 10^2 \text{ Pa} = 4.3 \text{ torr}$$

11.6 Thermodynamics of Immersion

Heat is evolved when a finely divided solid is immersed in a liquid. At constant temperature, this heat can be taken as $\Delta_{im} U$ for the process. Writing $\Delta_{im} U$ for the formation of unit area of surface, we have

$$\Delta_{im} U = u^\sigma_{SL} - u^\sigma_S \tag{33}$$

Note that there are no internal surfaces in the liquid before immersion. Substituting Eq. (9) gives

$$\Delta_{im} U = \left[\gamma_{SL} - T\left(\frac{\partial \gamma_{SL}}{\partial T}\right)_{V,\sigma} \right] - \left[\gamma_S - T\left(\frac{\partial \gamma_s}{\partial T}\right)_{V,\sigma} \right] \tag{34}$$

Applying Eq. (25),

$$\Delta_{im} U = T\left(\frac{\partial(\gamma_L \cos\theta)}{\partial T}\right)_{V,\sigma} - \gamma_L \cos\theta \tag{35}$$

In order to obtain measurable heats of immersions, it is necessary to work with powdered solids with very small particle size.

11.7 Effect of Surface Curvature on Vapor Pressure

In this section, we will derive the vapor pressure of a spherical droplet of radius r. This vapor pressure is greater than that of a flat surface because the liquid is on the concave side of the surface and is thus at a higher pressure than the surroundings. As we saw in Eq. (61) of Chapter 6, a higher pressure increases the escaping tendency of a condensed phase. The effect of surface curvature is only appreciable for very small droplets.

From Eq. (21), we have

$$dP_{liq} = -\frac{2\gamma}{r^2} dr \tag{36}$$

which gives a change of the chemical potential of the liquid of

$$d\mu_{liq} = V_m \, dP_{liq} = -\frac{2\gamma V_m}{r^2} \, dr \tag{37}$$

Equating this to the chemical potential of the vapor,

$$d\mu_g = RT \frac{dP}{P} = -2\gamma V_m \frac{dr}{r^2} \tag{38}$$

Integrating from an infinite radius, where the vapor pressure is P_∞,

$$\ln\left(\frac{P}{P_\infty}\right) = \frac{2\gamma V_m}{RTr} = \frac{2\gamma M}{RT\rho r} \tag{39}$$

In the last step, we have written the molar volume in terms of the molecular weight and liquid density.

Example 3. At 25°C the surface tension of water is 73×10^{-3} J/m^2 and its vapor pressure is 23.76 torr. What is the vapor pressure at 25°C of water droplets of radius 1.0 μm? of radius 10 nm?

Solution:
For 1.0 μm = 10^{-6} m:

$$\ln\left(\frac{P(10^{-6} \text{ m})}{P(r = \infty)}\right) = \frac{2(72.0 \times 10^{-3} \text{ J/m}^2)(0.018 \text{ kg/mol})}{(10^3 \text{ kg/m}^3)(10^{-6} \text{ m})(8.314 \text{ J/mol K})(298 \text{ K})}$$

$$= 0.00104$$

$$P(10^{-6} \text{ m}) = (23.76 \text{ torr}) \times \exp(0.00104)]$$

$$= (23.76 \text{ torr})(1.00104) = 23.78 \text{ torr}$$

For 10 nm = 10^{-8} m:

$$\ln\left(\frac{P(10^{-8} \text{ m})}{P(r = \infty)}\right) = 0.104(26.36 \text{ torr})$$

$$P(10^{-8} \text{ m}) = (23.76 \text{ torr})[\exp(0.104)] = 26.36 \text{ torr}$$

The higher vapor pressure of very small droplets allows air to achieve a considerable *supersaturation* (partial pressure of water in air greater than the vapor pressure of water) before liquid droplets begin to form. In fact, most cloud droplets form on *nucleation centers*, which may be dust particles or minute droplets of sulfuric or nitric acid. Cloud seeding involves adding nucleation centers (usually iodide salts), in an attempt to encourage precipitation from supersaturated air.

Another consequence of the effect of surface area on vapor pressure is that large regions of liquids will grow at the expense of small droplets. Also, in establishing the liquid–vapor equilibrium, the vapor will condense on flat surfaces or on large droplets of liquid, rather than form new microdroplets.

11.8 Thermodynamics of Solution Surfaces

We have defined solutions as homogeneous phases, with uniform concentrations throughout. Clearly, the surface of a solution provides a different environment than its bulk, and we should expect intensive properties (concentrations as well as intensive thermodynamic properties) to vary in this region. The mechanical and thermal variables, P and T, however, can be taken as uniform throughout the solution. It should be emphasized that the surface region of the solution is very thin, just a few molecular diameters thick. Bulk properties of the solution will, thus, only be affected by the surface if the solution is composed of very small droplets.

In treating interfacial (if) regions, we will follow the method of Gibbs and replace the nonuniform interfacial region by a two-dimensional *Gibbs surface phase* with uniform properties. Properties of this phase are called *surface excess properties* and their calculation is illustrated for the surface excess concentration of component i in Fig. 8. Here, the actual interfacial region, the region where properties vary, extends from z_1 to z_2 and is replaced by the surface phase located at position z_0, with the uniform bulk α and β phases extended up to this position.

The number of moles of i must be the same in the actual and the Gibbs model system:

$$\sigma \int_{z_1}^{z_2} c_i^{if} \, dz = \sigma c_i^{\alpha}(z_0 - z_1) + \sigma c_i^{\beta}(z_2 - z_0) + n_i^{\sigma} \tag{40}$$

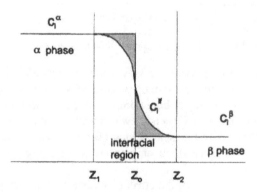

Figure 8 The Gibbs surface.

where σ is the area of the surface and n_i^σ is the *surface excess amount* of i. Rewriting gives

$$
\frac{n_i^\sigma}{\sigma} \equiv \Gamma_i^\sigma
$$

$$
= \left(\int_{z_0}^{z_2} c_i^{if} \, dz - c_i^\beta ((z_2 - z_0)) \right) - \left(c_i^\alpha (z_0 - z_1) - \int_{z_1}^{z_0} c_i^{if} \, dz \right) \tag{41}
$$

where Γ_i^σ is the *surface excess concentration* of i. Note that Γ_i^σ is the difference between the shaded area to the right of z_0 and the shaded area to the left of z_0. Surface excess concentrations have units of moles per unit area. z_0, the position of the Gibbs interface, can be arbitrarily located anywhere in the interfacial region. Therefore, surface excess properties are arbitrary. It is customary, however, to select z_0 so that one surface excess concentration, usually that of the solvent, is zero (i.e., for the solvent, the shaded areas above and below the concentration curve just cancel). We will assume that this convention has been adopted in positioning the interface. In Fig. 8, the surface excess concentration is slightly positive, representing an accumulation of component i in the interfacial region. A negative value for Γ_i^σ denotes a deficiency of component i in the interfacial region.

We consider the change in the Gibbs free energy for arbitrary additions of the various components to the Gibbs surface phase. To do this, we generalize Eq. (7) to an open system:

$$
dG = V \, dP - S \, dT - \sigma \, d\gamma + \sum_i \mu_i \, dn_i \tag{42}
$$

Because the surface tension, γ, is an intensive property, it can depend on P, T and concentrations, but not on the area of the surface phase.

We now apply Eq. (42) to a process in which we start with a surface phase of very small area and add components to the phase in the proper ratio to keep its surface excess concentrations constant as we increase its area. Pressure and temperature are also held constant in this process. Because $\mu_i = \mu_i(P, T, c_i)$ and $\gamma = \gamma(P, T, c_i)$, μ_i and γ are constant in this process and Eq. (42) becomes

$$
dG = \sum_i \mu_i \, dn_i \tag{43}
$$

Integrating this equation from a very small area to the final area of the interface gives

$$
G - G_0 = \sum_i \mu_i (n_i - n_{i,0}) \tag{44}
$$

In the limit of the initial area of the surface approaching zero, this becomes

$$
G = \sum_i \mu_i \, n_i \tag{45}
$$

Following our development of the Gibbs–Duhem equation in Chapter 8, we now apply Eq. (45) to a process at constant T and P, where components are added to the surface in an arbitrary ratio. In this process, the chemical potentials will vary:

$$dG = \sum_i \mu_i \, dn_i + \sum_i n_i \, d\mu_i \tag{46}$$

Equation (42), for the same process, becomes,

$$dG = -\sigma \, d\gamma + \sum_i \mu_i \, dn_i \tag{47}$$

Subtracting Eq. (47) from Eq. (46) and rearranging gives

$$d\gamma = -\sum_i \frac{n_i}{\sigma} \, d\mu_i = -\sum_i \Gamma_i d\mu_i \tag{48}$$

For a two-component solution, because z_0 has been positioned to make the surface excess concentration of solvent zero, Eq. (48) becomes

$$d\gamma = -\Gamma \, d\mu \tag{49}$$

where we have dropped the subscript for the single solute.

Because from Eq. (1) of Chapter 9, $\mu = \mu^0 + RT \ln a$, Eq. (49) becomes

$$\frac{d\gamma}{d \ln a} = -RT\Gamma \tag{50}$$

In practice, the isothermal dependence of surface tension on concentration of solute in an aqueous solution is of three different types, shown in Fig. 9.

With type 1 solutes, surface tension in aqueous solution mildly increases with concentration. Because activities generally increase with concentration, from Eq. (50), these solutes have a negative surface excess concentration (i.e., they are depleted in the surface layer). Inorganic electrolytes show this behavior. In the bulk solution, these ions are stabilized by interacting with the extended ionic environment of the solution. In the surface layer, this environment is limited in extent in one direction.

Type 2 solutes moderately decrease surface tension in aqueous solution and, thus, have positive surface excess concentrations. This class of solutes includes organic molecules with polar groups that give them some water solubility. Short-chain organic acids, amines, and alcohols are of this type.

Very small amounts of type 3 solutes produce a dramatic lowering of the surface tension of aqueous solutions. A substance with this property is called a *surfactant*. Usually, they consist of hydrocarbon chains, with $n = 10$–20 connected to polar groups (such as –OH, –CN, –COOH, –COOR, or –CONH$_2$) or ionic groups (such as –SO$_3^-$, –OSO$_3^-$, or –NR$_3^+$). They exclusively concentrate at the water–air interface. A monolayer of this type of

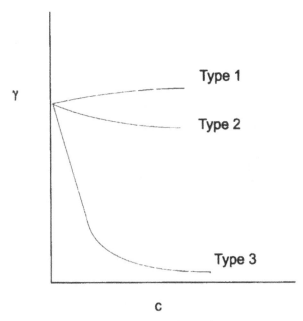

Figure 9   Concentration dependence of surface tension.

solute is called a *Langmuir layer.*[4] After a complete monolayer of a surfactant is added, further addition produces a new phase and no further decrease in surface tension.

Soaps and detergents are examples of a surfactant. At the interface between water and a greasy phase, the hydrocarbon is attracted to the grease and the polar group to the water. The detergent molecules thus accumulate at the interface between these two phases and, by reducing surface tension at the interface, promote mixing (and the removal of grease from clothing).

11.9   Properties of Surface Films

The dependence of the surface tension of a Langmuir layer containing a given amount of surfactant, as a function of its area, reveals many of the properties of the surfactant. (At constant temperature, such measurements are analogous to determining the P–V isotherm of a gas.) One device for making such measurements is the Langmuir balance, the essential principles of which are shown in Fig. 10.

Here, a monolayer film is prepared by dissolving a surfactant molecule in a volatile organic solvent. (The surfactant is assumed nonvolatile.) The solution is

fixed barrier

moveable barrier

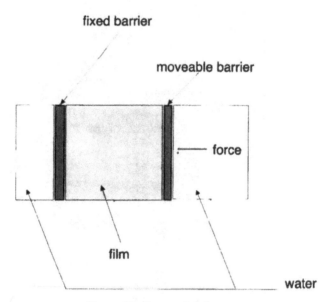

force

film

water

Figure 10   Langmuir balance.

added to the surface of clean water in a trough, in the region between a fixed and movable barricade. The surfaces are coated with a material, such as Teflon, that has negligible interaction with the surfactant. The force needed to hold the movable barricade at different positions, corresponding to different film areas, is recorded. As shown, the force per unit length of the barricade, the *surface pressure*, π, is given by

$$\pi = \gamma_0 - \gamma \tag{51}$$

where γ_0 is the surface tension of pure water and γ is that of the surfactant layer.

The shape of the typical π versus σ isotherm observed for a surfactant is shown in Fig. 11.

At large areas, as the area is decreased the concentration of surfactant is increased and the surface tension is decreased. From Eq. (51), the surface pressure increases in this region (but only slowly). At a critical area, σ_0, surface pressure rises quite abruptly if additional compression is attempted. This area, σ_0, called the *Pockels point*,[5] is considered to be the area at which the surfactant molecules are in contact in a monolayer. Any further compression requires a much larger force to "buckle" the monolayer. The surface tension is negative in this region. For a group of acids, $CH_3(CH_2)_nCOOH$, with $n = 14-24$, Langmuir found no change in the Pockels point area, indicating that these molecules take up very similar space in the monolayer. This suggest that at the Pockels point, the

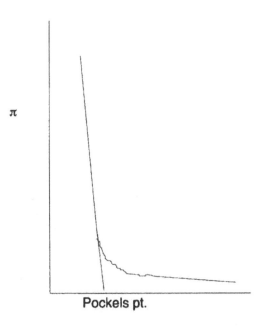

Figure 11 Isotherm for a surfactant.

organic chains of the different acids are sticking up out of the solution, so that it is the cross-sectional area of the chain that is important.

We can also analyze the part of π versus σ isotherm at large σ. When the film's area is very large, the surfactant molecules are sufficiently far apart that we can assume that their behavior is ideally dilute. In this case, the activity in Eq. (50) can be written in the usual ideally dilute solution approximation, $a = c = n/\sigma, d \ln c = dn/n$. We use moles/area for the interfacial concentration. Equation (50) then becomes

$$d\gamma = -RT\Gamma \, \frac{dn}{n} \tag{52}$$

For a surfactant, the concentration of solute in the aqueous layer is essentially zero. All of the surfactant is in the interfacial region and $\Gamma = n/\sigma$. Equation (52) then becomes

$$d\gamma = -\frac{RT}{\sigma} \, dn \tag{53}$$

Integrating from $n = 0$, $\gamma = \gamma_0$,

$$\gamma_0 - \gamma = \frac{nRT}{\sigma} = \pi \qquad (54)$$

(i.e., the two-dimensional analog of the ideal gas law).

11.10 Adsorption on Solids

Adsorption on solids is an important step in the industrially important process of heterogeneous catalysis. Adsorption, which takes place on the surface (including that of the pores) of the solid, should be distinguished from absorption, which occurs throughout its bulk. The latter is illustrated by the taking up of water by anhydrous calcium chloride.

In adsorption, we call the gas or solution solute the *adsorbate* and the solid the *adsorbent*. *Monolayer adsorption* involves up to one layer of adsorbate on the adsorbent, whereas *multilayer adsorption* involves more than one layer of adsorbate. Adsorption may be either *physical adsorption* (*physisorption*), where the adsorbate is bound to the surface by relatively weak physical forces ($\Delta H_{desorp} < 40$ kJ/mol) or *chemical adsorption* (*chemisorption*), where the binding forces are stronger ($\Delta H_{desorb} > 40$ kJ/mol).[6] Because chemical adsorption involves chemical-type bonds between adsorbate and adsorbent, it is limited to the first monolayer on the surface. Physical adsorption can involve multiple layers and physical adsorption can occur on top of chemisorbed layer.

In adsorption, the amount adsorbed on a surface is most fundamentally characterized by the fraction of surface sights, ϑ_i, that are occupied by i adsorbate molecules. Because the surface area is often not well known, adsorption isotherms are often reported as the amount adsorbed on the surface per gram of adsorbent. In gas adsorption, this amount is traditionally given as the volume of adsorbate at standard temperature and pressure (STP), v. This volume depends on temperature and on the pressure of the adsorbate in the gas phase. $v(T, P)$ is an equation of state for the surface and, when reported at a constant T, is known as an adsorption isotherm.

In order for the adsorbent surface to be well characterized, it must be extremely clean. Surface scientists have developed a number of techniques for producing clean surfaces. These include heating, cleaving crystals, and bombarding with high-energy ions. All of these processes must be carried out under ultrahigh-vacuum conditions, to avoid immediate contamination of the surface.

The simplest adsorption isotherm is that of Langmuir, which applies to surface coverage less than a monolayer. The isotherm assumes equilibrium between gas-phase adsorbate molecules, **A**, and molecules adsorbed on a surface site:

$$\mathbf{A + S \leftrightarrow AS}; \qquad K_1 = \frac{\vartheta_1}{\vartheta_V P} \tag{I}$$

S is a vacant surface site because the adsorbate cannot be adsorbed on a site **AS** that is already occupied. The activities of vacant and occupied surface sites are proportional to the fraction of the sites that are vacant, ϑ_V, and occupied by one adsorbate molecule, ϑ_1. Because in the Langmuir isotherm, only monolayer adsorption is allowed $\vartheta_V + \vartheta_1 = 1$ and

$$K_1 = \frac{\vartheta_1}{(1 - \vartheta_1)P} \tag{55}$$

where P is the pressure of the gas-phase adsorbate. Equation (55) can be written as

$$\vartheta_1 = \frac{K_1 P}{1 + K_1 P} \tag{56}$$

For adsorption from solution, a solution concentration variable replaces the gas-phase pressure. The amount adsorbed on the surface is proportional to the fraction of the surface sites occupied, giving the following for gas adsorption:

$$v = \frac{bK_1 P}{1 + K_1 P} \tag{57}$$

where b is a proportionality constant. The Langmuir isotherm is shown in Fig. 12.

At low pressure, surface coverage is proportional to pressure, whereas at high pressure, the surface coverage approaches unity, regardless of the temperature. Data are usually analyzed by plotting (or regression analysis) of v^{-1} versus P^{-1} because

$$\frac{1}{v} = \frac{1}{b} + \frac{1}{bK_1}\frac{1}{P} \tag{58}$$

An isotherm that considers multilayer adsorption was proposed by Brunauer, Emmet, and Teller in 1938 (BET isotherm). The importance of this isotherm is that, when its use is appropriate, it provides an estimate of the surface

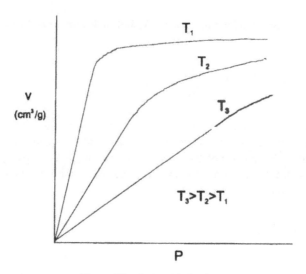

Figure 12   Langmuir isotherm.

area of an absorbent. For multilayer adsorption, we have to include the following, in addition to reaction (I):

$$A + AS \leftrightarrow A_2S; \qquad K = \frac{\vartheta_2}{\vartheta_1 P} \qquad \qquad (II)$$

$$A + A_2S \leftrightarrow A_3S; \qquad K = \frac{\vartheta_3}{\vartheta_2 P} \qquad \qquad (III)$$

$$A + A_{n-1}S \leftrightarrow A_nS; \qquad K = \frac{\vartheta}{\vartheta_{n-1} P} \qquad \qquad (IV)$$

where ϑ_n is the fraction of surface site on which there is a layer of an adsorbate n molecules thick. The justification for using the same equilibrium constant for all of these process is that, in each of them, the newly adsorbed adsorbate molecule is interacting directly with other adsorbate molecules. This interaction is taken as that for liquefaction of **A**.

$$A(gas) \leftrightarrow A(liq); \qquad K = \frac{1}{P_{vap}} \qquad \qquad (V)$$

where P_{vap} is the vapor pressure of **A** at the temperature of the isotherm.
  We can express all of the fractions in terms of ϑ_1:

$$\vartheta_n = \vartheta_{n-1}(KP) = \vartheta_1(KP)^{n-1} \qquad \qquad (59)$$

The sum of the fractions of surface sites with all numbers of adsorbate molecules must sum to unity:

$$1 = \vartheta_v + \sum_i \vartheta_i = \vartheta_v + \sum_i \vartheta_1(KP)^{i-1} = \vartheta_v + \frac{\vartheta_1}{1 - KP} \tag{60}$$

Since

$$1 + KP + (KP)^2 + \cdots + (KP)^n + \cdots = \sum_i (KP)^{i-1}$$
$$= \frac{1}{1 - KP} \tag{61}$$

The total number of molecules adsorbed on the surface per unit mass of adsorbent is

$$N = c_s \sum_i i\vartheta_i = c_s\vartheta_1 \sum_i i(KP)^{i-1} \tag{62}$$

where c_s is the number of surface sites per unit mass of adsorbent. This can be written as

$$\frac{N}{c_s} = \frac{\vartheta_1}{(1 - KP)^2} \tag{63}$$

since

$$\sum_i i\,(KP)^{i-1} = \frac{d(\sum_i (KP)^i)}{d(KP)}$$
$$= \frac{d[-1 + 1/(1 - KP)]}{d(KP)} \tag{64}$$
$$= \frac{1}{(1 - KP)^2}$$

From Eq. (60), using $\vartheta_v = \vartheta_1/K_1P$, from reaction (I),

$$1 = \frac{\vartheta_1}{K_1P} + \frac{\vartheta_1}{1 - KP}$$
$$= \frac{\vartheta_1[1 + (K_1 - K)P]}{K_1P(1 - KP)} \tag{65}$$

which when substituted into Eq. (63) gives

$$\frac{N}{c_s} = \frac{K_1P}{(1 - KP)[1 + (K_1 - K)P]} \tag{66}$$

Substituting $K = 1/P_{vap}$ from reaction (V) and rearranging gives

$$\frac{N}{c_s} = \frac{K_1 P_{vap} P}{(P_{vap} - P)[1 + (K_1 P_{vap} - 1)P/P_{vap}]} = \frac{v}{v_m} \tag{67}$$

where we have substituted the ratio of the amount of material adsorbed to the amount that would be adsorbed for a monolayer (both measured as volumes at STP) for the number of molecules adsorbed per surface site. Equation (67) is rearranged to

$$\frac{P}{v(P_{vap} - P)} = \frac{1}{K_1 P_{vap} v_m} + \frac{(K_1 P_{vap} - 1)}{K_1 P_{vap} v_m} \frac{P}{P_{vap}} \tag{68}$$

A linear plot (or regression) of $P/v(P_{vap} - P)$ versus P/P_{vap} therefore allows the determination of K_1 and v_m from the slope and intercept.

Although a number of assumptions of the derivation of the BET isotherm have been disputed, the isotherm does usually give a good fit to experimental data up to $P/P_{vap} \approx 0.35$. Obtaining the surface area of adsorbent from v_m requires a model for the size and shape of the adsorbate molecules and their arrangement on the monolayer surface. For small adsorbate molecules, such as Ar and N_2, a spherical shape and a close-packed arrangement is assumed. A surface area for a N_2 molecule of $16.2 \, Å^2$ has been generally agreed upon.

Data for the adsorption of N_2 on 45.6 g of copper at 90 K are shown in Fig. 13a. In Fig. 13b, these data are plotted in the form of Eq. (68). Linear regression gives a slope of 0.227 and an intercept of 2.338×10^{-3}, both in units of $(cm^3$ at $STP)^{-1}$.

Example 4. Use the plot of Fig. 13b to estimate the surface area per gram of the copper sample used in the experiment. The vapor pressure of N_2 at 90 K is 2710 torr.

Solution: From the slope and intercept of Eq. (68),

$$K_1 P_{vap} = 1 + \frac{\text{Slope}}{\text{Intercept}} = 1 + \frac{0.227}{2.338 \times 10^{-3}} = 98$$

$$v_m = \frac{1}{K_1 P_{vap} \text{ intercept}} = \frac{1}{98 \, 2.338 \times 10^{-3}} \, cm^3 \text{ at STP} = 4.36 \, cm^3 \text{ at STP}$$

$$\frac{4.36 \, cm^3 \text{ at STP}}{45.6 \text{ g Cu}} \frac{6.02 \times 10^{23} \text{ molec}}{22{,}400 \, cm^3 \text{ at STP}} \frac{16.2 \times 10^{-20} \, m^2}{\text{molec}}$$

$$= 0.416 \, \frac{m^2}{\text{g Cu}}$$

It should be noted that the "surface area of a solid" is not a very well-defined quantity. Much like the "length of the coastline of Maine," it depends on

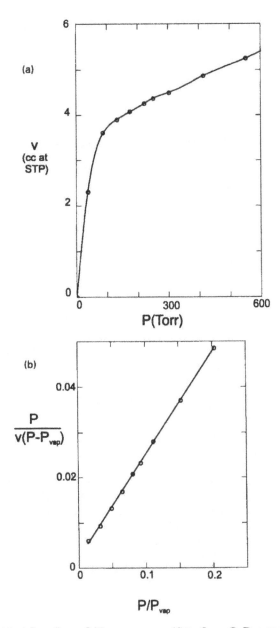

Figure 13 (a) Adsorption of N_2 on copper (data from S Brunauer, PV Emmett. Determination of the surface area of adsorbents. J Am Chem Soc 59:2683, 1972); (b) Adsorption of N_2 on copper; BET plot.

the scale of the measuring device used for the measurement. In Example 4, that scale was that of the N_2 molecule (i.e., approximately 4 Å). If the area of the same solid had been measured using a much larger molecule, whose size precluded it coating the finer details of the surface, a somewhat smaller surface area would have been found.

11.11   Statistical Mechanics of Adsorption

11.11.1   Statistical Approach for the Langmuir Isotherm

Physical understanding of the assumptions underlying the various isotherms can be increased by deriving them by statistical mechanics. This will be done only for the Langmuir isotherm. At equilibrium, the chemical potential of the adsorbate in the gas phase and on the surface must be equal:

$$\mu_g = \mu_{surf} \tag{69}$$

For μ_g, we use Eq. (38) of Chapter 5 with U_0 as the desorption energy and the molecular partition function written as a product of a translational partition function [Eq. (72) of Chapter 5] and q_{int}, which accounts for the internal degrees of freedom of the molecule:

$$
\begin{aligned}
\mu_g &= U_0 - RT \ \ln \ q_{int} \frac{V_m}{N_A h^3} (2\pi m k T)^{3/2} \\
&= RT \ \ln\left[q_{int} \frac{V_m}{N_A h^3} (2\pi m k T)^{3/2} \ \exp\left(-\frac{U_0}{RT}\right)\right]
\end{aligned}
\tag{70}
$$

For the surface, we calculate the Helmholtz free energy from Eq. (45) of Chapter 5: $A = -RT \ \ln Q$. We assume that surface molecules are distinguishable (by their position) and noninteracting, so that the system partition function is a product of N molecular partition functions. However, because we are not interested in which of the N out of a total of M surface sites are occupied, we must include a degeneracy factor of $M!/N!(M - N)!$. The energy of a molecule on the surface is taken as zero.

$$Q_{surf} = \frac{M!}{N!(M - N)!} q_{surf}^N \tag{71}$$

Using Stirling's approximation and making the assumption that

$$q_{surf} = q_{int} \tag{72}$$

(i.e. that the molecules on the surface have all of their internal degrees of freedom identical to gas-phase molecules, but have lost their translational degrees of freedom), for the adsorbed molecules we get

$$A = -kT[M \ln M - M - N \ln N + N - (M - N) \ln (M - N)$$
$$+ (M - N) + N \ln q_{int}] \tag{73}$$

The chemical potential on the surface is obtained from

$$\mu_{surf} = \left(\frac{\partial A}{\partial n}\right)_{M,T} = N_A \left(\frac{\partial A}{\partial N}\right)_{M,T} \tag{74}$$

Note that here M is the number of surface sites, which takes the place of volume, the usual natural variable for A.

Inserting Eq. (73) into Eq. (74) gives

$$\mu_{surf} = RT[\ln N - \ln(M - N) + \ln q_{int}]$$
$$= RT \ln\left(\frac{Nq_{int}}{(M - N)}\right) \tag{75}$$

Equating Eqs. (70) and (75) and cancelling q_{int} gives

$$\frac{N}{M - N} = \frac{N_A h^3}{V_m (2\pi mkT)^{3/2}} \exp\left(\frac{U_0}{RT}\right) \tag{76}$$

Because $\vartheta = N/M$, Eq. (76) becomes

$$\frac{\vartheta}{1 - \vartheta} = \frac{h^3 P}{kT(2\pi mkT)^{3/2}} \exp\left(\frac{U_0}{RT}\right) = b(T)P \tag{77}$$

Equation (77) may be written as

$$\vartheta = \frac{bP}{1 + bP} \tag{78}$$

which is identical in form to Eq. (56).

11.11.2 Statistical Approach to the Thermodynamics of Adsorption

In analogy to what we did for solutions, we define the change in a thermodynamic function upon adsorption as

$$\Delta_{ads} X = X - \sum_i n_i X^*_{m,i} = \sum_i n_i (\bar{X}_i - X^*_{m,i}) \tag{79}$$

$X^*_{m,i}$ is the molar thermodynamic property of a pure component (adsorbate or adsorbent) and $\bar{X}_i$ is the partial molar property of the component, defined as

$$\bar{X}_i \equiv \left(\frac{\partial X}{\partial n_i}\right)_{T,P,n_j} = \left(\frac{\partial X}{\partial n_i}\right)_{T,P,\vartheta} \tag{80}$$

$\Delta_{ads}X$ is an integral change and involves contributions as the fractional coverage varies from $0 \rightarrow \vartheta$. Usually, we are interested in how thermodynamic properties vary with coverage and want the differential change

$$\Delta_{ads,dif}X_i \equiv \left(\frac{\partial \Delta_{ads}X}{\partial n_i}\right)_{T,P,\vartheta} = \bar{X}_i - X^*_{m,i} \tag{81}$$

We will illustrate the method by outlining the calculation of the (differential) entropy of adsorption. For S^*_m, we use the Sackur–Tetrode expression, Eq. (76) of Chapter 5, with a term added for the internal degrees of freedom:

$$S^*_m = R \ln q_{int} + \frac{5}{2}R + R \ln\left(\frac{(2\pi M)^{3/2} R^{5/2} T^{5/2}}{N_A^4 h^3 P}\right) \tag{82}$$

Assuming that the adsorbed molecules have lost their three translational degrees of freedom, we calculate S from Eq. (32) of Chapter 5, taking the energy term as zero. The partition function for the molecules on the surface is given by Eqs. (71) and (72):

$$S = k \ln\left(\frac{M!}{(M-N)!N!}q_{int}^N\right) \tag{83}$$

Using Sterling's approximation for the factorials,

$$S = k[M \ln M - (M-N) \ln(M-N) - N \ln N + N \ln q_{int}] \tag{84}$$

From Eq. (80), the partial molar entropy of the adsorbate is

$$\bar{S} = \left(\frac{\partial S}{\partial n}\right)_{T,P} = N_A\left(\frac{\partial S}{\partial N}\right)_{T,P} = R[\ln(M-N) - \ln N + \ln q_{int}] \tag{85}$$

Combining the first two terms and writing them in terms of $\vartheta = N/M$, gives

$$\bar{S} = R \ln\left(\frac{1-\vartheta}{\vartheta}\right) + R \ln q_{int} \tag{86}$$

The entropy of adsorption then becomes

$$\Delta_{ads}S = R \ln\left(\frac{1-\vartheta}{\vartheta}\right) - \frac{5}{2}R - R \ln\left(\frac{(2\pi M)^{3/2} R^{5/2} T^{5/2}}{N_A^4 h^3 P}\right) \tag{87}$$

The first term, called the (differential) *configurational entropy*, results from the different possible ways the adsorbed molecules can occupy the surface sites. The last two terms, which are negative, are the Sackur–Tetrode entropy due to the lost translational degrees of freedom of the gaseous molecules.

11.12 Colloids

Colloids are systems with size intermediate between the microscopic and macroscopic realms. They have been defined as particles with a characteristic dimension between a micron and a few nanometers or, alternatively, as entities containing between 10^3 and 10^9 atoms. With their small size, colloids have a very large surface-to-volume ratio and surface interactions are dominant in determining their stability. Some properties of systems in the colloidal size range are as follows:

1. They cannot be removed from solution by filters, but they can by semipermeable membranes.
2. They do not rapidly sediment (settle out) from solution under the influence of gravity,[8] but they often do in an ultracentrifuge.[9]
3. While individual dispersed particles are not visible, the suspensions can scatter light (the Tyndall effect).

Colloidal suspensions are everywhere around us. We encounter them in everything from washing our clothes to painting our homes to eating our morning cereal (Question 12). A large fraction of our bodies are colloidal materials. Colloids are used extensively in a variety of industrial processes. Their importance is illustrated by the original name of the *Journal of Physical Chemistry* being the *Journal of Physical and Colloid Chemistry*.

Colloidal suspensions can be classified in a number of ways. Most obvious is by the phase of the homogeneous *dispersing medium* and that of the *dispersed particles*, as shown in Table 2.

Colloids can also be classified by the nature of the interaction between the dispersed particles and the homogeneous dispersing medium. If this interaction is attractive, the system is known as a *lyophilic colloid*. If the interaction is repulsive, it is called a *lyophobic colloid*. Because water is the most common dispersing medium, these terms usually become *hydrophilic* and *hydrophobic*. If all the dispersed particles are roughly the same size, the colloid is called *monodispersed*; if they cover a range of sizes, it is called *polydispersed*. The dispersed particles may be *macromolecules* or *aggregates* and their shape may be *globular* (roughly spherical) or *fibrous*. The function of biological macromolecules is intimately related to their dispersed shape. Determining the relationship

TABLE 2   Classification of Colloidal Systems

Medium	Particle	Name	Example
Gas	Liquid	Fog	Fog
Gas	Solid	Smoke	Smoke
Liquid	Gas	Foam	Beer foam
Liquid	Liquid	Emulsion	Milk
Liquid	Solid	Sol	White paint
Solid	Liquid	Gel	Skin
Solid	Solid	Suspension	Metallic alloy

between macromolecular structure, shape and function is a very active field of biochemical research.

Because in lyophilic colloids there is an attractive interaction between the particles and the medium, the interface between these two phases has a low surface energy and the particles are solvated. There is little reason for particles to clump together under these circumstances; thus, the dispersed particles are usually individual macromolecules. The shape of the macromolecule depends on the relative magnitude of the forces within the molecule to those between the molecule and the medium. If the former is larger, the particle will be globular, if the latter is larger, the particles will be extended. In the latter case, there may be some interactions between different dispersed molecules, producing semisolid properties (a gel).

Lyophobic colloids (sols) may be prepared by grinding crystalline materials or running an electric arc between metallic electrodes, both in the dispersing medium. More commonly, they are prepared by precipitating the solid from a strongly supersaturated solution, which produces a large number of precipitation nuclei. Because there is little attractive interaction between the particles and the medium, attractive forces between the particles would soon lead to their aggregation (*flocculation*). This tendency, however, is counterbalanced by repulsive electrical forces between the particles.

The origin of the electrical forces between colloidal particles is the difference in chemical potentials for some charged species at the particle–medium interface. As shown in Eq. (36) of Chapter 10, this results in charge transfer, which produces an electric potential difference between the particle and the medium. The net charge on the colloidal particle influences the distribution of ions in the surrounding medium, much as we have discussed for the Debye–Huckel theory of ionic activities. Thus, if the particle has a net positive surface charge, there will be a predominance of negative charge in its vicinity in the medium. This produces what is commonly called an *electrical double layer*.

Because the charge distributions around similar particles have the same sign, they will repel each other, counteracting the influence of the attractive interactions. The Debye length, a measure of the distance from the charged particle over which electrical imbalances persist, is, by Eq. (18) of Chapter 10, inversely proportional to the square root of the ionic strength of the medium. Adding electrolytes to the medium thus reduces the influence of the repulsive electrical forces and often flocculates the colloid. To increase the stability of colloids, they can be dialyzed after their formation to remove excess electrolyte. Some observations require for their explanation a tightly held layer of adsorbed ions on the particle, called the *Stern layer*, as well as the diffuse cloud of ions in the medium.

Surfactant molecules in solution can form *association colloids* (called *micelles*) when the concentration of the surfactant is above a *critical micelle concentration*. This behavior only occurs above a given temperature, called the *Kraft temperature*. Below this temperature, the surfactant shows normal solubility behavior. In Fig. 14, a two-dimensional cut through a micelle, according to the most popular model, is shown.

This micelle is comprised of surfactant molecules consisting of long hydrocarbon tails attached to an anionic lyophilic group. Typically, there are 50–100 molecules in the micelle. Some counterions in the medium are adsorbed on the aggregate, whereas others form the diffuse ionic environment. Some workers believe that there is considerable penetration of the medium into the micelle. Micelles are important for detergent action, with oily dirt particles dissolved in the hydrocarbon interior of the micelle.

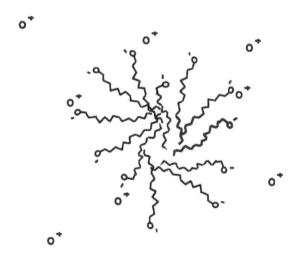

Figure 14 A micelle.

Questions

1. Explain how certain insects are able to walk on water, when their density is greater than that of water.

2. Arrange the following liquids in order of increasing surface tension: ether, water, and mercury.

3. Tables of surface tension often list values in dynes per centimeter, millinewtons per meter, or grams per second squared. How is each of these related to the units used in Table 1?

4. Equation (32) predicts that the larger the γ, the higher up into a capillary a liquid will be drawn. However, it is found that liquids with very high surface tension will not be drawn into capillaries at all. Rationalize this in light of the derived equation. (Hint: The surface tension is a measure of the cohesive energy of a liquid.)

5. Show that Young's equation [Eq. (24)] can be derived by requiring that the net horizontal component of force on the interface is zero.

6. Two bubbles, one of radius 1 cm and the other of radius 2 cm, stick together. In which direction does the interface between the two bubbles curve?

7. Devise a method to accurately use two Wilhelmy plates (see Fig. 6) to measure the surface tension of a Langmuir film. Do not forget about the gravitational and buoyancy forces. Such a device is called a Wilhelmy balance.

8. How would you expect the solubility of a solid to depend on crystal size? What happens as a newly formed precipitate ages in contact with its supernatant liquid?

9. Consider the physical assumptions used to derive the BET isotherm. List three ways in which these assumptions might be in error.

10. The Langmuir isotherm describes chemisorption, whereas the BET isotherm describes physisorption. Why is more appropriate to estimate surface areas using the latter isotherm?

11. Sketch the dependence of the differential and integral configurational entropies on the fraction of a monolayer covered.

12. How do colloids play a role in washing our clothes, painting our homes, and eating our morning cereal?

13. To make Jello, powder containing gelatin (a protein), sugar and flavoring is dissolved in hot water and then placed in the refrigerator to cool. Describe, from a microscopic perspective, what happens in this process.

14. If a $0.1\,M$ mixture of AgI is prepared in colloidal form, no measurable osmotic pressure is observed even though the colloidal particles do not pass through a semipermeable membrane. Explain.

15. Explain why detergent does not clean very well below their Kraft temperatures.

16. For 1% by mass of each of the following in water: sugar, starch, and colloidal gold, which has the largest osmotic pressure?

Problems

1. Calculate the diameter of circular plates that a 150-lb. man would have to wear on his feet in order to walk on water. You can assume that the plates are perfectly nonwettable.

2. We add 1.0 nmol of an organic substance to 1 L of water in a 15-cm-diameter container. The substance concentrates in the top 2.0-nm layer of the solution. What is its concentration in this layer?

3. Show that if $n = 1$ in Eq. (15), the surface energy is temperature independent.

4. Show that if $(\partial \gamma / \partial P)_T = 0$, then $(\partial \gamma / \partial T)_V = (\partial \gamma / \partial T)_P$.

5. For a metal plate, 1 cm $\times$ 3 cm $\times$ 10 cm with density 7.0 g/cm^3, suspended 5 cm. deep in water and making a contact angle of 30° with the water surface, compare the gravitational force, buoyancy force, and the vertical component of the surface tension force on the plate.

6. Show that in the case that a liquid makes a contact angle ϑ with the wall of a capillary tube of radius R, the capillary rise in the tube is $h = (2\gamma \cos \vartheta)/\rho g R$.

7. Calculate the height to which water is raised in capillary tubes of diameters of 1.0 mm and 0.1 mm, assuming that the water completely wets the surface of the tube.

8. What pressure above atmospheric must be applied in a tube of diameter 1.0 mm, placed 3.0 cm below the surface of water at 25°C in order to blow a bubble? (Assume perfect wetting of the tube.)

9. What would the radius of a water droplet have to be at 25°C in order for it to have a vapor pressure of 25 torr?

10. Find an expression for the surface entropy, $s_\sigma \equiv (\partial S / \partial \sigma)_T$, for a substance whose surface tension follows the temperature dependence of Eq. (15). You can assume $(\partial S / \partial \sigma)_{T,V} \approx (\partial S / \partial \sigma)_{T,P}$. Comment on the sign of s_σ.

11.* For the case in which, in the absence of gravity, a liquid droplet adsorbs onto a surface as a hemispherical droplet, what is the ratio of the adhesive energy to the cohesive energy?

12. The surfactant molecule hexadecanol, $C_{16}H_{33}OH$, has been used to retard evaporation from water reservoirs. If the cross-sectional area of this molecule is 0.25 nm^2, what mass of hexadecanol would have to be added to a reservoir of area 10^4 m^2 in order to completely cover the surface and retard the evaporation of water?

13. Derive the Langmuir isotherm by equating the rate of desorption of adsorbate from the surface (proportional to the fraction of surface sites occupied) to the rate of adsorption (proportion to the pressure of adsorbate in the gas phase).

14.   For a gas–solid system following the Langmuir isotherm, how would you plot $\vartheta(P)$ so that a fit to a straight line would provide $b(T)$? From Eq. (77), how would you plot $b(T)$ so that a fit to a straight line would provide U_0?

15.   From Eq. (75), sketch the chemical potential of adsorbed molecules as a function of ϑ, the fraction of surface sites occupied. What values does μ_{surf} approach as $\vartheta \to 0$ and $\vartheta \to 1$? Is μ_{surf} an "escaping tendency" from the surface phase in the sense that it is just related to the rate that molecules escape from the surface?

16.   Show that, following the derivation of the Langmuir isotherm, for the case of two different species competing for monolayer adsorption sites,

$$\vartheta_1 = \frac{b_1 P_1}{1 + b_1 P_1 + b_2 P_2}, \qquad \vartheta_2 = \frac{b_2 P_2}{1 + b_1 P_1 + b_2 P_2}$$

17.   Show that Eq. (55) can be written in the form of Eq. (56).

18.*   Monodispersed colloidal gold particles allowed to sit in perfectly quiescent aqueous suspension at 25°C form a sedimentation equilibrium in which their number density decreases by a factor of 2 in 10 cm of elevation. What is the average number of gold atoms in each colloidal particle?

19.[M]   On the same surface as used in Example 4, Brunauer and Emmett also obtained the following data for adsorption of CO_2 at −78°C:

P (torr)	65	122	175	212	260	305	420	522	600	710
v (cm^3)	1.70	2.55	3.15	3.50	3.78	4.05	4.60	5.22	5.78	6.65

The sublimation pressure of CO_2 at −78°C is 780 torr. What area should be assigned to solid CO_2 on the Cu surface in order to obtain the same surface area as calculated in Example 4.

Notes

1. The derivative at constant V and σ means at constant surface-to-volume ratio.
2. Note that the energy of vaporization also goes to zero at the critical point.
3. AW Adamson. Physical Chemistry of Surfaces. New York, Interscience, 1960, pp 4–6.
4. After Irving Langmuir (1881–1957), Nobel laureate in chemistry, 1932.
5. After Agnes Pockels, an amateur scientist, who did many of the early experiments on surface tension in her kitchen.
6. Of course, 40 kJ/mol is an arbitrary cutoff. For particular purposes, we might decide to modify this cutoff between physisorption and chemisorption.
7. See Example 4 of Chapter 1.
8. Suspensions of colloidal gold prepared by Michael Faraday (1791–1867) are still on exhibit at the British Museum.
9. An ultracentrifuge typically provides an acceleration 10^5 times that of gravity.

12

Steady-State Systems

What an organism feeds upon is negative entropy.

Erwin Schrodinger[1]

The time invariance of steady-state systems depends on flows of energy, matter, and charge between the system and the surroundings. Intensive properties of such systems are not uniform but can be represented as functions of position. The net transfer of a conservative quantity, such as energy, through the boundaries of the system must be zero. Entropy is a nonconservative property and is always generated in systems undergoing flow (irreversible) processes. In processes involving steady-state systems, entropy generation is most conveniently measured as the entropy increase of the surroundings. Formulas are developed for the rate of entropy generation per unit volume in heat transfer, mechanical work, electric current, diffusion, chemical reaction, and phase transfer, in terms of flows and their conjugate forces. The flows are expressed as linear functions of the forces, including terms involving nonconjugate forces. The symmetries of the coefficients relating flows to forces, including Onsager's relation, are discussed. Forces with no conjugate flows adjust so that entropy production is a minimum in the steady state. A number of irreversible processes are considered.

12.1 Steady-State Systems

Up to this point, our thermodynamic analyses have dealt with systems at equilibrium or in transition between equilibrium states. Systems in equilibrium states have no tendency to change with time and retain this characteristic when the influence of their surroundings is removed. Steady-state systems also have no tendency to change with time. However, in the steady state, the static condition depends on the surroundings. If the interaction with the surroundings is removed, the system will change with time as it progresses toward a different static state, namely that of equilibrium. Steady-state systems may be open or closed, but not isolated. They are of great practical importance; to a good approximation, they encompass almost the entire biological world and many industrial processes.[2]

A steady-state system is not in equilibrium with the surroundings and may not be in internal mechanical, thermal or material equilibrium. Differences in pressure, temperature, and chemical and electrical potential between the system and surroundings are the driving forces for the bulk flow of material[3] and the flow of energy, species, and charge, respectively, between the system and the surroundings. These driving forces create nonuniform thermodynamic properties within steady-state systems. Thus, although steady-state systems are unchanging in time, they are not uniform in space.

Because we have adopted the philosophy that intensive systems properties, such as temperature and pressure, are measured in the surroundings, we should consider under what conditions it is appropriate to assign such thermodynamic properties to steady-state systems. Usually, we will be able to regard a steady-state system as composed of numerous smaller systems, over each of which intensive properties can be treated as uniform, with negligible error. If these smaller systems are large enough for these properties to be statistically determined with negligible fluctuations, we can treat intensive properties as functions of positions in a steady-state system. This approach will fail if systems are far from equilibrium and properties vary abruptly with position, such as occur in some regions of flames. Although properties vary abruptly at phase boundaries, these can be handled by considering the system as composed of separate subsystems for different phases. Extensive thermodynamic properties, such as U, can be handled by treating related intensive properties, such as the molar energy, U_m or the *energy per unit volume*, u. The intensive variables are functions of position, and the energy of the entire system can be obtained by a volume integral of u over the system.

12.2 Conservative and Nonconservative Properties

Extensive quantities, such as energy, entropy, and charge, can be transported through the boundaries of both closed and open systems; chemical species can be transported only through the boundaries of open systems. In analogy to what was done for heat and work, we will use the symbol δ to indicate a small amount of these quantities transported, with the convention that positive transport indicates transport into the system. Some quantities, such as chemical species and entropy, can also be generated within a system. We will use the symbol d_{int} to indicate a small amount of a quantity generated within a system. As usual d without a subscript indicates the change of quantity of a variable in the system. For any extensive property, X, we can write

$$dX = \sum_i \delta X_i + d_{int}X \overset{ss}{=} 0 \tag{1}$$

where a summation is included because we may want to treat transport that occurs at different boundaries separately.[4] We have equated dX to zero because we are treating only steady-state systems, which have time-invariant properties. Rearranging gives

$$d_{int}X \overset{ss}{=} -\sum_i \delta X_i \tag{2}$$

the net X that is produced in a steady-state system must be transported out of the system.

Some extensive quantities can be neither created nor destroyed. They are called *conservative quantities*, and the d_{int} of these are zero. Among these are total mass, charge, and energy, and the mass of each element.[5] For these quantities,

$$\sum_i \delta X_i \overset{ss}{\underset{cons}{=}} 0 \tag{3}$$

Each δX_i of a conservative quantity is matched by a corresponding decrease in X in the surroundings; thus, for conservative quantities,

$$\sum_i dX_{sur,i} \overset{ss}{\underset{cons}{=}} 0 \tag{4}$$

(Because we deal exclusively with steady-state systems in this chapter, the "ss" designation will henceforth be omitted in equations.)

When chemical reactions occur in a system, the number of moles of the various compounds (but not the elements) may be *nonconservative*. Another very important nonconservative quantity is entropy, the production of which is the hallmark of real processes. In a steady-state system,

$$dS = d_{int}S + \sum_i \delta S_i = 0 \tag{5}$$

$$dS_{univ} = dS + \sum_i dS_{sur,i} = d_{int}S + \sum_i \delta S_i + \sum_i dS_{sur,i} > 0 \tag{6}$$

Although, in general, transports are not reversible, we can idealize them as being reversible from the point of view of the system. This is because, by definition, we are not interested in the details of the processes that occur in the surroundings. Thus, we can imagine heat transfer to occur from a heat reservoir, which is constructed of a material with infinite heat conductivity, so that it maintains a uniform temperature as heat is withdrawn from it. Moreover, the reservoir is in contact with the boundary of the system sufficiently long so that the boundary is at the temperature of the reservoir. In this case, the decrease of the entropy of the reservoir is exactly equal to the entropy transported to the system:

$$\delta S_i = -dS_{sur,i,rev} \tag{7}$$

Because this can be idealized to occur for each transport to the system, we can write

$$d_{int}S + \sum_i (\delta S_i + dS_{sur,i}) = d_{int}S > 0 \tag{8}$$

because each term in the summation is zero. Entropy is always produced in systems undergoing flow processes.

Calculation of the entropy produced in systems undergoing different flow processes (called *irreversible processes*) is key for considering steady-state systems. In order to measure the entropy produced in the system, we think of it as transported to the surroundings in a reversible manner and measure the entropy changes in the surroundings. From Eqs. (5) and (7),

$$dS_{int} = \sum_i dS_{sur,i,rev} \tag{9}$$

The entropy generated in a steady-state system is the entropy added to the surroundings, if this entropy is transported in a reversible manner. This is a convenient operational principle, because the surroundings can be idealized as composed of reservoirs, each with uniform intensive properties.

12.3 Entropy Generation in Some Simple Processes in Steady-State Systems

12.3.1 Heat Flow

We consider a system of uniform area and thickness dx, between and in contact with heat reservoirs at temperatures T and $T + dT$, as shown in Fig. 1.

Because energy is conserved, the heat flows into and out of the system are equal. Using Eq. (9), we have, for the entropy generated in the system,

$$d_{int}S = \sum_i dS_{sur,i,rev} = \frac{\delta q}{T + dT} - \frac{\delta q}{T} = \delta q \left(\frac{1}{T + dT} - \frac{1}{T} \right) \tag{10}$$

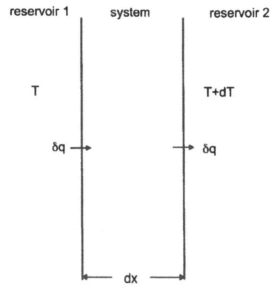

Figure 1 Heat transfer.

which can be written as

$$d_{int}S = \delta q \, d\left(\frac{1}{T}\right) = \delta q \frac{d(T^{-1})}{dx} \, dx \tag{11}$$

Of particular interest is Θ, the *rate of entropy generation per unit volume of the system*, which must always be positive:

$$\Theta \equiv \frac{1}{V} \frac{d_{int}S}{dt} > 0 \tag{12}$$

For heat transfer, this becomes

$$\Theta_q = \frac{1}{A \, dx} \frac{\delta q}{dt} \frac{d(T^{-1})}{dx} \, dx = J_q \frac{d(T^{-1})}{dx} \tag{13}$$

where J_q is the *heat flux* (heat flow per unit area per unit time). Readers with some exposure to vector calculus can generalize Eq. (13) to three dimensions as

$$\Theta_q = \mathbf{J}_q \cdot \nabla(T^{-1}) \tag{14}$$

where ∇ refers to the gradient of a scalar function.[6] In this derivation, we have considered heat reservoirs, so that temperatures upstream and downstream of the system are constant during the heat flow. However, in a steady-state system, the temperatures upstream and downstream of *any small part* of the system are constant. The results of Eqs. (13) and (14) thus hold for all parts of such systems.

The forms of Eqs. (13) and (14) are called *bilinear*, giving the entropy generation as the product of a flux and the gradient of temperature. The flux must be related to the gradient, because Θ is always positive, requiring heat flux in the opposite direction of the temperature gradient $[\nabla(T^{-1}) = -(1/T^2)\nabla T]$. We define the quantity multiplying a flux in the bilinear expression for entropy as the *force conjugate to the flux*. $\nabla(T^{-1})$ is, therefore, the "force" driving heat flux.

The *dissipation*, Φ, is defined as

$$\Phi \equiv T\Theta \tag{15}$$

and is an alternative way of defining forces (the product of a force and flux being Φ). Forces defined in this manner will differ from those used here by a factor of T. The reader should be wary of such differences in pursuing further readings in the field. The dissipation function is the rate of destruction of free energy in the system by irreversible processes.

12.3.2 Mechanical or Electrical Energy

In many industrial processes, mechanical energy is continuously added to stir a system. Because we regard mechanical energy as completely nondegraded, its addition does not transport any entropy. At the steady state, however, the input of mechanical work δw must be balanced by an equal magnitude of heat flowing to the surroundings (assumed at constant temperature T). This gives

$$\Theta_{mech} \equiv \frac{1}{V}\frac{d_{int}S}{dt} = \frac{1}{V}\frac{dS_{sur,rev}}{dt} = \frac{1}{T}\frac{1}{V}\frac{\delta w_{mech}}{dt} \tag{16}$$

where $(1/V)(\delta w_{mech}/dt)$ is the rate per unit volume at which mechanical energy is added to the system. Expression of Θ_{mech} in terms of the product of a force and a flux requires specification of additional details of the mechanical system.

Exactly similar considerations hold for electrical energy, where the volumetric production of entropy in the system is

$$\Theta_{elec} = \frac{1}{T}\frac{1}{V}\frac{\delta w_{elec}}{dt} \tag{17}$$

$\delta w_{elec}/dt$ is the electrical power dissipated by a current flow through a drop in electrical potential $d\phi$ over the system of width dx:

$$\frac{\delta w_{elec}}{dt} = -Id\phi \tag{18}$$

Substituting into Eq. (17) gives

$$\Theta_{\text{elec}} = -\frac{1}{T}\frac{I}{A}\frac{d\phi}{dx} = -\frac{J_e}{T}\left(\frac{d\phi}{dx}\right) \tag{19}$$

where J_e is the flux of electric current. Θ_{elec} must be greater than zero, indicating that current flows in the direction of decreasing electric potential. In three dimensions, Eq. (19) becomes

$$\Theta_{\text{elec}} = -\frac{\mathbf{J_e}}{T}\cdot\nabla\phi, \tag{20}$$

where $\nabla\phi$ is the gradient of the electrical potential. The force driving the flow of electric charge is therefore $-(1/T)\nabla\phi$.

12.3.3 Diffusion

Figure 2 describes diffusion of a component through a system of width dx and constant area, A. The entropy generated in the system is by Eq. (9), using Eq. (19) of Chapter 6 to calculate the entropy change of the constant temperature and pressure reservoirs:

$$d_{\text{int}}S = \sum_i dS_{\text{sur},i,\text{rev}} = \frac{1}{T}(\delta q_{\text{tot},2} - \delta q_{\text{tot},1})_{\text{rev}} - \frac{1}{T}(\mu_i + d\mu_i - \mu_i)\delta n_i \tag{21}$$

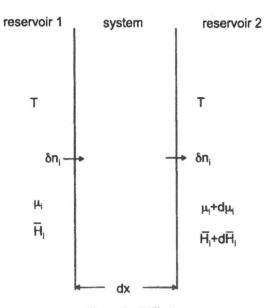

Figure 2 Diffusion.

Because P is constant, the first term is the enthalpy change of the reservoirs. However, because enthalpy is a conserved quantity in this case,[7] this term is zero by Eq. (4), and

$$d_{int}S = -\frac{1}{T}\left(\frac{\partial\mu_i}{\partial x}\right) dx\, \delta n_i \tag{22}$$

$$\Theta_i \equiv \frac{1}{V}\frac{d_{int}S}{dt} = -\frac{1}{T}\left(\frac{\partial\mu_i}{\partial x}\right)\frac{\delta n_i}{A\, dt} = -\frac{1}{T}\left(\frac{\partial\mu_i}{\partial x}\right)J_i \tag{23}$$

where J_1 is the flux of component i. In three dimensions, this becomes

$$\Theta_i = -\frac{J_i}{T}\cdot\nabla\mu_i \tag{24}$$

For charged particles (ions), there is an additional electrostatic energy $z\mathscr{F}\phi$ transported, which is incorporated by using the electrochemical potential, $\tilde{\mu}_i$, defined in Eq. (35) of Chapter 10. Equation (24) then becomes

$$\Theta_i = -\frac{J_i}{T}\cdot\nabla\tilde{\mu}_i = J_i\cdot\left(-\frac{1}{T}\nabla\tilde{\mu}_i\right) \tag{25}$$

The total entropy generation due to the diffusion of a number of species is

$$\Theta_{dif} = \sum_i J_i\left(-\frac{1}{T}\nabla\tilde{\mu}_i\right) \tag{26}$$

The force driving the diffusive flow of component i is therefore $-(1/T)(\nabla\tilde{\mu}_i)_T$, where we have indicated that this result is for a constant-temperature situation.

12.3.4 Chemical Reactions and Phase Transfer

When a chemical reaction occurs in a steady-state system, products must be transferred out of the system and reactants must be transferred into the system. To avoid additional entropy generation, we take these transfers as being transfers by diffusion to reservoirs at material equilibrium with the system. The entropy generated in the system by the chemical reaction is then, by Eq. (9), the entropy increase of the reservoirs:

$$dS_{int} = \sum_i dS_{sur,i} = \frac{1}{T}\sum_i \delta q_{tot,sur} - \sum_i \frac{1}{T}\mu_i\delta n_i = -\frac{1}{T}\sum_i \mu_i\, dn_i \tag{27}$$

As discussed in Section 12.2, we can drop the first summation in a process in which enthalpy is conserved. Following Eq. (5) of Chapter 7, we have $dn_i = \nu_i\, d\xi$, giving

$$dS_{int} = -\frac{1}{T}\sum_i \nu_i\mu_i\, d\xi = \frac{1}{T}A\, d\xi, \tag{28}$$

where $A \equiv -\sum_i \nu_i \mu_i$ is called the *affinity* of the chemical reaction. The volumetric rate of entropy production is then

$$\Theta_{react} = \frac{1}{T} A \frac{1}{V} \frac{d\xi}{dt} = \frac{A}{T} v_{react} \qquad (29)$$

where v_{react} is the *volumetric rate of the reaction* (defined as the rate of production of a product with unit stoichiometric coefficient). A/T is, therefore, the "force" driving the chemical reaction, and "motion" of the reaction is along the coordinate of ξ, the extent of the reaction. Because the entropy generated within a system must be positive, if a single reaction can occur in a system, v_{react} is positive [i.e., toward products, if the affinity of the reaction is positive ($\Delta_{react}G = \sum_i \nu_i \mu_i < 0$)]. This is a generalization of a principle derived in Chapter 7 for the special cases of reactions occurring at constant T and V or constant T and P.

With j chemical reactions occurring in a system, the total rate of entropy generation by chemical reactions is

$$\Theta_{react} = \sum_j (A_j/T)v_{react,j} \qquad (30)$$

If the reactions are independent, each term in the sum must be positive. However, if the reactions are "coupled," most simply by having reactants and products in common, only the total entropy generation must be positive. Thus, as is well known in biochemical systems, it is possible for reactions with negative affinity (positive free-energy change) to be "driven" to products by reactions with positive affinity.

Phase transfer ($\alpha \rightarrow \beta$) can be treated exactly as for chemical reactions, with the affinity of phase transfer equal to the difference in chemical potentials of the phases:

$$A_\phi = -(\mu_\beta - \mu_\alpha) = -\Delta_\phi G \qquad (31)$$

For charged particles (ions), the electrostatic energy of the transfer is incorporated by using the electrochemical potential defined in Eq. (35) of Chapter 10. The electrochemical affinity is

$$\bar{A}_\phi = -(\bar{\mu}_\beta - \bar{\mu}_\alpha) \qquad (32)$$

and is used in Eq. (30) or (31).

12.3.5 Simultaneous Flows

As an example of simultaneous flows, we will consider simultaneous diffusion and heat transfer. This is illustrated in Fig. 3.

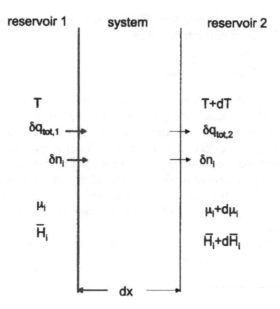

Figure 3 Simultaneous heat transfer and diffusion.

As in the previous examples, we calculate the entropy generated in the steady-state system by the entropy increase of the reservoirs:

$$
\begin{aligned}
d_{int}S &= dS_{res\,1,rev} + dS_{res\,2,rev} \\
&= -\frac{1}{T}\delta q_{tot\,1} - \frac{\bar{\mu}_i}{T}(-\delta n_i) + \left[\frac{1}{T} + d\!\left(\frac{1}{T}\right)\right]\delta q_{tot\,2} - \left[\frac{\bar{\mu}_i}{T} + d\!\left(\frac{\bar{\mu}_i}{T}\right)\right]\delta n_i
\end{aligned}
$$

$$(33)$$

Because enthalpy is conserved in heat transfer and diffusion, $\delta q_{tot\,1} = \delta q_{tot\,2} = \delta q_{tot}$, giving

$$
d_{int}S = \delta q_{tot}d(1/T) - \delta n_i d(\bar{\mu}_i/T) \tag{34}
$$

$$
\Theta \equiv \frac{1}{V}\frac{d_{int}S}{dt} = J_{q_{tot}}\frac{d(1/T)}{dx} - J_i\frac{d(\bar{\mu}_i/T)}{dx} \tag{35}
$$

or, in three dimensions,

$$
\Theta = \mathbf{J}_{q\,tot}\cdot\nabla\!\left(\frac{1}{T}\right) - \mathbf{J}_i\cdot\nabla\!\left(\frac{\bar{\mu}_i}{T}\right) \tag{36}
$$

We write $\mathbf{J}_{q\,\text{tot}} = \mathbf{J}_q + \bar{H}_i\mathbf{J}_i$ and use Eq. 12A in Appendix A to write a component of the gradient as

$$\frac{d}{dx}(T^{-1}\mu_i) = \left(\frac{\partial(T^{-1}\mu_i)}{\partial x}\right)_{T^{-1}} + \left(\frac{\partial(T^{-1}\mu_i)}{\partial T^{-1}}\right)_x \frac{dT^{-1}}{dx} = \frac{1}{T}\left(\frac{\partial\mu_i}{\partial x}\right)_{T^{-1}}$$
$$+ \bar{H}_i\frac{dT^{-1}}{dx} \tag{37}$$

where the Gibbs–Helmholtz equation, Eq. (29) of Chapter 4, has been used in this constant-pressure system. Substituting into Eq. (35) gives[8]

$$\Theta = \mathbf{J}_q \cdot \nabla\left(\frac{1}{T}\right) - \mathbf{J}_i \cdot \frac{1}{T}(\nabla\mu_i)_T \tag{38}$$

which is the superposition of entropy generations for heat flow and diffusion, Eqs. (14) and (24). We will accept as a general principle that when a number of simultaneous irreversible processes are occurring in a system, the entropy generation and dissipation can both be written as the sum of terms, one for each of the processes, and the forms of terms are analogous to that obtained when the processes are considered individually. Either Eq. (36) or Eq. (38) can be used to analyze systems undergoing heat transfer and diffusion simultaneously. The choice is usually based on which provides a simpler calculation.

12.4 The Phenomenological Equations Relating Flows and Forces

Although it may be interesting to compare the entropy generated by various irreversible processes in a given system, in order to make real progress with the theory, we must postulate relations between the flows and forces that we have discussed in the last section. Assuming that the flows depend on the forces, we can expand the functional relationship between a flow, J_k, and the forces, X_i, in a multivariable power series:

$$J_k = J_k(0) + \sum_i L_{ki}X_i + \text{h.o.} \tag{39}$$

Because there are no flows at equilibrium, where the forces are zero, $J_k(0) = 0$. We will explore the regime, not too far from equilibrium, where the higher-order (h.o.) terms can be ignored, giving

$$J_k = \sum_i L_{ki}X_i \tag{40}$$

Flow processes have been studied experimentally for quite some time and these phenomena have often been found to follow Eq. (40). For example, Ohm's law for the flow of charge is

$$\mathbf{J}_e = -k_e \nabla \phi \tag{41}$$

where k_e is the electrical conductivity. Because, from Eq. (20), $X_e = -(1/T)\nabla\phi$, we have

$$\mathbf{J}_e = -k_e \nabla \phi = L_{ee}\left(-\frac{1}{T}\nabla\phi\right) \tag{42}$$

or

$$L_{ee} = Tk_e \tag{43}$$

Similarly, for heat flow, we have Fourier's law,

$$\mathbf{J}_q = -k_q \nabla T \tag{44}$$

where k_q is the thermal conductivity. Because, from Eq. (14),

$$X_q = \nabla(1/T) = -(1/T^2)\nabla T,$$
$$\mathbf{J}_q = -k_q \nabla T = -L_{qq}\frac{1}{T^2}\nabla T \tag{45}$$

or

$$L_{qq} = T^2 k_q \tag{46}$$

For diffusion, we have Fick's law:

$$\mathbf{J}_i = -D_i \nabla c_i \tag{47}$$

where D_i is the diffusion coefficient. However, the force conjugate to the diffusion flow is, from Eq. (26), $X_i = -(1/T)(\nabla\mu_i)_T$. At uniform temperature, using a $1.0\,m$ standard state, this can be written as

$$X_i = -R\nabla \ln c_i = -\frac{R}{c_i}\nabla c_i \tag{48}$$

giving

$$\mathbf{J}_i = -D_i \nabla c_i = -L_{ii}\frac{R}{c_i}\nabla c_i \tag{49}$$

from which

$$L_{ii} = \frac{c_i D_i}{R} \tag{50}$$

For some irreversible processes, it is not obvious that fluxes are proportional to thermodynamic driving forces. For example, in chemical kinetics, the flux of a chemical reaction (its reaction rate) is expressed in terms of concentration, because it is proportional to the collision rate, whereas the thermodynamic driving force is expressed in terms of chemical potentials, which determine the affinity. However, it has been shown that near equilibrium, the kinetic and thermodynamic forms for the rates of chemical reactions become identical (i.e., both are proportional to the deviation of a concentration variable from its equilibrium value). For example, for the simple monatomic reversible reaction,

$$A \overset{k_f}{\underset{k_b}{\rightleftharpoons}} B \tag{51}$$

use of Eq. (40) gives

$$J_{react} = L_{react} X_{react} = L_{react} \frac{A}{T} = \frac{L_{react}}{T}(\mu_A - \mu_B) \tag{52}$$

Because at equilibrium the chemical potential of A and B are equal, this can be written as

$$\begin{aligned} J_{react} &= \frac{L_{react}}{T}[(\mu_A - \mu_{A,eq}) - (\mu_B - \mu_{B,eq})] \\ &= L_{react} R \left[\ln\left(\frac{c_A}{c_{A,eq}}\right) - \ln\left(\frac{c_B}{c_{B,eq}}\right) \right] \end{aligned} \tag{53}$$

Writing this in terms of the deviation of concentrations from their equilibrium values,

$$\Delta c_A = c_A - c_{A,eq} = -\Delta c_B = c_{B,eq} - c_B \tag{54}$$

gives

$$J_{react} = L_{react} R \left[\ln\left(1 + \frac{\Delta c_A}{c_{A,eq}}\right) - \ln\left(1 + \frac{\Delta c_B}{c_{B,eq}}\right) \right] \tag{55}$$

After expanding the logarithms in Taylor series,

$$J_{react} = L_{react} R \Delta c_A \left(\frac{1}{c_{A,eq}} + \frac{1}{c_{B,eq}} \right) \tag{56}$$

Problem 2 deals with showing that this is identical to the kinetic expression for the volumetric rate of the reaction if

$$L_{react} = \frac{k_f c_{A,eq}}{R} \tag{57}$$

This analysis suggests that, for a general reaction, the expression for L_{react} is $Rate_{eq}/R$.

12.5 Fluxes Produced by Nonconjugate Forces

Equation (40) is of the form of the multiplication of a second-order tensor by a vector. However, the equations of Ohm, Fourier, and Fick relate a flux only to its conjugate force (i.e., involve only the diagonal elements of the tensor L). The question immediately arises of whether the nondiagonal elements of this tensor can be nonzero (i.e., whether a flow can depend on nonconjugate forces). There is abundant evidence that, indeed, this does occur. In Chapter 5, thermal transpiration was presented as an example of a diffusive flow resulting from a temperature difference. Another example is the thermoelectric effect, on which the operation of thermocouples is based. Thus, when using Eq. (40), it is important to include the nonconjugate forces, which produce "coupling" between irreversible processes. In fact, it is in the consideration of such "nondiagonal terms" that steady-state thermodynamics provides its most interesting results.

Although nonconjugate forces must be considered, certain limitations can be placed on the coefficients in Eq. (40). This can be seen by considering the case of two forces and two flows[9]:

$$
\begin{aligned}
J_1 &= L_{11}X_1 + L_{12}X_2 \\
J_2 &= L_{21}X_1 + L_{22}X_2
\end{aligned}
\tag{58}
$$

Suppose process 1 is heat flow and process 2 is the rate of a chemical reaction: Eqs. (58) then become

$$
\begin{aligned}
J_q &= L_{11}\nabla\left(\frac{1}{T}\right) + L_{12}\frac{A}{T} \\
J_{react} &= L_{21}\nabla\left(\frac{1}{T}\right) + L_{22}\frac{A}{T}
\end{aligned}
$$

Now assume that we treat a case in which a chemical reaction is occurring in a system with uniform temperature, the first of Eqs. (59) becomes

$$
J_q = L_{12}\frac{A}{T}
\tag{60}
$$

However, A and T are scaler quantities, and because L_{12} is only a property of the medium, in an isotropic medium it must also be a scalar. Because there is no way to define a direction for the vector heat flow, $L_{12} = 0$ is required. *There can be no coupling between vector and scalar irreversible flow processes in isotropic media*.[10,11] This is known as the Curie–Prigogine principle.

A very important relationship between the coefficients in Eq. (40) was derived by Onsager.[12] Starting with the concept of *microscopic reversibility*,[13] Onsager treated the fluctuations around the equilibrium state. Then, reasoning that macroscopic irreversible flows should behave in the same way as these fluctuations, he showed that

$$L_{12} = L_{21} \tag{61}$$

(i.e., *the matrix of coupling coefficients is symmetric*).

Equation (58) indicates that in order for only a single flux, J_1, to result from a force X_1, either L_{21} must be zero or X_2 must be adjusted to keep J_2 zero. Because we are exploring the consequences of "cross coupling," we must assume the latter situation. An example of this was seen in Chapter 5, where we used statistical mechanics to investigate thermal diffusion. Here, we had a temperature gradient, which drives a heat flux through the system. This is the only flux occurring in the system at steady state, because a pressure gradient is established that balances the tendency of molecules to diffuse under the influence of the thermal gradient. This pressure gradient is an example of an *uncontrolled force*, as it is not exerted by action external to the system.

Prigogine[14] has proved an important theorem regarding uncontrolled forces in systems with single irreversible flows. To derive this theorem, we write the entropy generation in terms of forces as

$$\Theta = J_1 X_1 + J_2 X_2 = L_{11} X_1^2 + (L_{12} + L_{21}) X_1 X_2 + L_{22} X_2^2 \tag{62}$$

From this, we obtain

$$\left(\frac{\partial \Theta}{\partial X_2} \right)_{X_1} = (L_{12} + L_{21}) X_1 + 2 L_{22} X_2 = 2(L_{21} X_1 + L_{22} X_2) = 2 J_2 \tag{63}$$

because $L_{12} = L_{21}$. Thus, for a given value of the controlled force, X_1, the variation of entropy generation with respect to the uncontrolled force, X_2, is proportional to the flux conjugate to the uncontrolled force. When this flux is zero, $(\partial \Theta / \partial X_2)_{X_1} = 0$ (i.e., the entropy generation is an extremum). Because entropy generation is always positive, the single extremum is a minimum. *In steady-state systems, uncontrolled forces with zero conjugate fluxes adjust so that entropy production is a minimum.* This is the principle of minimum entropy production first developed by Prigogine.

12.6 Thermal Diffusion

In Chapter 5, we used an approach based on microscopic velocity distributions to investigate the ideal gas pressure difference established between two chambers held at different temperatures at steady state. The chambers were separated by a partition containing a hole small enough that effusive flow existed between the chambers. We will now apply steady-state thermodynamics to the same system. Because, in Eq. (62) of Chapter 5, we calculated the total energy flow between the chambers, we will analyze the problem using the approach of Eq. (36) (where the fluxes are taken as the mass flow and the *total* energy flow). We write Eq. (58) as

$$\mathbf{J}_{q \text{ tot}} = L_{qq} \nabla\left(\frac{1}{T}\right) - L_{qi} \nabla\left(\frac{\mu}{T}\right)$$
$$\mathbf{J}_i = L_{iq} \nabla\left(\frac{1}{T}\right) - L_{ii} \nabla\left(\frac{\mu}{T}\right) \tag{64}$$

In thermal transpiration, we have a flow of energy, but no flow of matter; therefore, $\mathbf{J}_i = 0$ and

$$\nabla\left(\frac{\mu}{T}\right) = \frac{L_{iq}}{L_{ii}} \nabla\left(\frac{1}{T}\right) \tag{65}$$

For a one-dimensional system, we can substitute the differential for the gradient:

$$d\left(\frac{\mu}{T}\right) = \mu \, d\left(\frac{1}{T}\right) + \frac{1}{T} \, d\mu = \frac{L_{iq}}{L_{ii}} \, d\left(\frac{1}{T}\right) = \frac{L_{qi}}{L_{ii}} \, d\left(\frac{1}{T}\right) \tag{66}$$

the last step resulting from the application of Onsager's relation [Equation (61)]. Substituting for the chemical potential,

$$(H_m - TS_m)\left(-\frac{1}{T^2}\right) dT + \left(\frac{1}{T}\right)(V_m \, dP - S_m \, dT) = -\frac{L_{qi}}{L_{ii} T^2} \, dT \tag{67}$$

$$\left(\frac{L_{qi}}{L_{ii}} - H_m\right) \frac{dT}{T^2} = -R \frac{dP}{P} \tag{68}$$

where use has been made of the ideal gas equation. The quantity $L_{qi}/L_{ii} \equiv Q^*$ is the *energy of transport*, the total energy transported per mole of gas that passes through the hole connecting the chambers. In Eq. (62) of Chapter 5, this was shown to be $2RT$. The molar enthalpy of these molecules is $(5/2)RT$, giving

$$\frac{1}{2} \frac{dT}{T} = \frac{dP}{P} \tag{69}$$

Upon integration, we get

$$\frac{P_1}{P_2} = \sqrt{\frac{T_1}{T_2}} \tag{70}$$

which is identical to the result obtained in Eq. (65) of Chapter 5.

A solution normally has uniform concentration. However, if a temperature gradient is established in a solution, a concentration gradient will result, a phenomenon that is called the *Soret effect*. For analysis, we employ Eq. (58), which in a binary solution leads to the coupled flux equations:

$$\mathbf{J}_q = L_{qq}\nabla\left(\frac{1}{T}\right) - L_{qi}\frac{1}{T}(\nabla\mu_i)_T - L_{qA}\frac{1}{T}(\nabla\mu_A)_T$$

$$\mathbf{J}_i = L_{iq}\nabla\left(\frac{1}{T}\right) - L_{ii}\frac{1}{T}(\nabla\mu_i)_T - L_{iA}\frac{1}{T}(\nabla\mu_A)_T \tag{71}$$

$$\mathbf{J}_A = L_{Aq}\nabla\left(\frac{1}{T}\right) - L_{Ai}\frac{1}{T}(\nabla\mu_i)_T - L_{AA}\frac{1}{T}(\nabla\mu_A)_T$$

where A is the solvent and i is the solute.

Using the Gibbs–Duhem equation [Chapter 8, Eq. (15)] in the form

$$(\nabla\mu_A)_T = -\frac{x_i}{x_A}(\nabla\mu_i)_T = -\frac{c_i}{c_A}(\nabla\mu_i)_T \tag{72}$$

and applying the second of Eqs. (71) to the case of no material flux through the system, gives, with the use of Onsager's relation,

$$L_{iq}\nabla\left(\frac{1}{T}\right) = \frac{1}{T}\left(L_{ii} - \frac{c_i}{c_A}L_{Ai}\right)(\nabla\mu_i)_T \tag{73}$$

We note that

$$\mathbf{J}_i - \frac{c_i}{c_A}\mathbf{J}_A = c_i\left(\frac{\mathbf{J}_i}{c_i} - \frac{\mathbf{J}_A}{c_A}\right) = c_i(\mathbf{v}_i - \mathbf{v}_A) \equiv c_i\mathbf{v}_i^{\text{rel}} \equiv \mathbf{J}_i^{\text{rel}} \tag{74}$$

where $\mathbf{v}_i^{\text{rel}}$ and $\mathbf{J}_i^{\text{rel}}$ are the velocity and flux of the solute relative to the solvent. Equation (73) can therefore be written as

$$L_{iq}\nabla\left(\frac{1}{T}\right) = \frac{1}{T}L_{ii}^{\text{rel}}(\nabla\mu_i)_T \tag{75}$$

where L_{ii}^{rel} is the relative coupling constant in the case of a pure heat flow.

Using a $1.0\,M$ standard state for chemical potentials, this becomes, for one-dimensional variation,

$$L_{iq}\frac{dT^{-1}}{dx} = L_{ii}^{\text{rel}}R\frac{d\ln c_i}{dx} \tag{76}$$

Making use of Onsager's relation and rearranging,

$$\frac{d\ln c_i}{dT^{-1}} = \frac{L_{qi}}{RL_{ii}^{\text{rel}}} = \frac{Q^*}{R} \tag{77}$$

which shows obvious similarity to the Clausius–Clapeyron equation [Eq. (44), Chapter 6].

12.7 Thermoelectric Effects

The most familiar thermoelectric effect is the generation of Ohmic heat due to current flow, discussed in Section 12.3.2. Less well known is the Thomson heat, produced or absorbed when a current flows in a temperature gradient. The *Thomson coefficient*, σ, can be defined as

$$\sigma \equiv -\frac{\delta q}{\delta Q \, dT} \tag{78}$$

It is the heat emitted per unit charge passing through unit temperature difference. This is a reversible effect; identical heat must be absorbed if the charge passes through the temperature gradient in the opposite direction.

There are also two well-known thermoelectric effects resulting from the joining of dissimilar materials (forming a junction): the *Seebech effect*, on which thermocouples are based, and the *Peltier effect*, used for thermopiles. The Seebech effect results when the two junctions of the dissimilar materials are held at different temperatures. The *Seebech coefficient*, ε, is defined as the open-circuit voltage generated per unit temperature differential of the two junctions:

$$\epsilon \equiv \left(\frac{\partial \phi}{\partial T}\right)_{J_{elec}=0} \tag{79}$$

The *Peltier effect*, on which thermopiles are based, is the heat released or taken up when a current flows through an isothermal junction. In a thermopile, the current flows in the opposite direction through the two junctions and, thus, transfers heat from one junction to the other. The *Peltier coefficient*, Π, is defined as the heat absorbed per unit charge flow:

$$\Pi \equiv \left.\frac{\delta q}{\delta Q}\right|_T = \left(\frac{\partial J_q}{\partial J_{elec}}\right)_T \tag{80}$$

Thomson applied the first law of thermodynamics to a virtual charge of magnitude δQ passing through the circuit at equilibrium shown in Fig. 4. Equating the reversible work needed to drive the charge to the heat emitted gives

$$\delta w_{rev} = \delta Q \, d\phi = \Pi(T + dT)\delta Q - \Pi(T)\delta Q + (\sigma_A - \sigma_B)\,\delta Q \, dT \tag{81}$$

or

$$\frac{d\phi}{dT} = \frac{d\Pi}{dT} + (\sigma_A - \sigma_B) \tag{82}$$

Note that when there is no temperature gradient, the Thomson effect vanishes and the Peltier effects cancel. By considering the test charge to flow in infinite time,

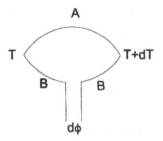

Figure 4 Thomson's circuit.

the current and the Ohmic heat go to zero and the potential becomes that for the open circuit, which is how the Seebech coefficient is defined. We get

$$\epsilon = \frac{d\Pi}{dT} + (\sigma_A - \sigma_B) \tag{83}$$

which is Thomson's equation. Additional results can be obtained from the equations for coupled heat and charge flows:

$$\mathbf{J}_q = L_{qq}\nabla\left(\frac{1}{T}\right) - \frac{L_{qe}}{T}\nabla\phi$$
$$\mathbf{J}_e = L_{eq}\nabla\left(\frac{1}{T}\right) - \frac{L_{ee}}{T}\nabla\phi \tag{84}$$

The Peltier coefficient is obtained by dropping the temperature gradient terms:

$$\Pi = \frac{\mathbf{J}_q}{\mathbf{J}_e}\bigg|_{\nabla T=0} = \frac{L_{qe}}{L_{ee}} \tag{85}$$

The Seebach coefficient results from setting the electric current equal to zero:

$$0 = -\frac{L_{eq}}{T^2}\left(\frac{\partial T}{\partial x}\right)_{J_e=0} - \frac{L_{ee}}{T}\left(\frac{\partial\phi}{\partial x}\right)_{J_e=0} \tag{86}$$

$$\epsilon \equiv \left(\frac{\partial\phi}{\partial T}\right)_{J_e=0} = -\frac{L_{eq}}{L_{ee}T} = -\frac{L_{qe}}{L_{ee}T} = -\frac{\Pi}{T} \tag{87}$$

where we have used Onsager's equation in the next to last step.

Questions

1. Can a steady-state system that is not in equilibrium with its surroundings be in internal mechanical, thermal, and material equilibrium?

2. Apply Eq. (1) to the total mass of carbon in the atmosphere

    (a) over a short time period (days)
    (b) over a long time period (decades).

3. How would you obtain the internal energy, U, of a steady-state system from u_m, the energy per unit mass as a function of position in the system?

4. Must each irreversible process occurring in a steady-state system produce entropy?

5. What are the forces conjugate to the flux of heat, electric current, diffusion and chemical reaction if the dissipation function is used to define the forces?

6. If mechanical energy enters a system at a shaft of radius R, which rotates at an angular velocity, ω, exerting a torque, τ, what is the mechanical energy flux added to the system? What is the force conjugate to this flux?

7. The second law of thermodynamics not only gives us a direction for time but also gives us a macroscopic explanation for the direction for irreversible processes in steady-state systems. For heat flow and diffusion, give a *microscopic* reason why the flows are in the direction opposite to the gradient of temperature and concentration, respectively.

8. If a temperature gradient is established across a single chamber, will there be a resulting pressure gradient, as we obtained for the case of thermal transpiration? Why are different results obtained in these two cases?

9. If a temperature gradient is established in a two-chamber system, with the chambers connected by an orifice through which Joule–Thomson type flow occurs, will there be a resulting pressure gradient? Explain your answer?

10. Discuss similarities and differences between Eq. (77) and the Clausius–Clapeyron equation and the phenomena underlying each of these.

11. A reservoir of saltwater and a reservoir of pure water are connected by a tube. The system is well insulated so that no heat transfer will occur between it and the surroundings or directly between the two reservoirs. Describe what will happen.

12. Which of the following must be zero: $L_{react,t}$, $L_{react,q}$, and $L_{q,t}$?

13. If a chemical reaction cannot couple with heat flow in an isotropic medium, how does the heat produced in an exothermic reaction get out of the system?

14. Give a *microscopic* explanation for the Seebech effect and for Thomson heat generation.

Problems

1. Show that enthalpy is not a conserved quantity in a steady-state system when an electric current is flowing. (Compare with Note 7.)

2. The chemical kinetics expression for the rate of the chemical reaction given in Eq. (51) is $J_B \equiv dc_B/dt = k_f c_A - k_b c_B$. Show that the requirement that the thermodynamic reaction

flux, Eq. (52), be the same as that given by chemical kinetics is that L_{react} be given by Eq. (57), for the case of reversible first-order reactions.

3.* Solve Eq. (58) for the forces in terms of J_1 and J_2 and express Θ in terms of the flows only. Show that, with Onsager's relation, $(\partial\Theta/\partial J_1)_{J_2} = 2X_1$ and $(\partial\Theta/\partial J_2)_{J_1} = 2X_2$.

4. A mixture of He and Ne is held in a one-dimensional temperature gradient. The mole fraction of He is 0.500, where the temperature is 373 K, and the mole fraction of He is 0.512, where the temperature is 288 K. Estimate the heat of transfer for diffusion of He through Ne.

Notes

1. E. Schrodinger. What Is Life? Cambridge: Cambridge University Press, 1974, p 76.
2. Assuming that the rate of change of the biological system or the industrial reactor can be neglected.
3. We will not consider bulk flow in this volume.
4. In cases in which there is continuous variation of the transport over the boundary of the system, integration, rather than summation, is required.
5. In the realm of chemistry, where nuclear transformations are not considered.
6. In Cartesian coordinates, with unit vectors, $\hat{i}$, $\hat{j}$, and $\hat{k}$,

$$\nabla f = \hat{i}\,\frac{df}{dx} + \hat{j}\,\frac{df}{dy} + \hat{k}\,\frac{df}{dz}$$

   Boldface type indicates vector quantities.
7. Because U is always conserved, $\Delta U_1 + \Delta U_2 = 0$ in a steady-state process. Because the boundaries of the system and pressure are unchanging $\Delta(PV)_1 + \Delta(PV)_2$ is also zero.
8. Constant T^{-1} implies constant T.
9. The forces and flows are written in boldface because they may be vector quantities, as is the case with heat flow and diffusion.
10. Such coupling is possible in anisotropic media, such as in active transport at membranes in biological systems.
11. In general, there can be no coupling between flow processes represented by tensors of different orders.
12. Lars Onsager, Nobel laureate.
13. Microscopic processes are reversible, because they depend on Newton's laws of motion, which contain only a *second* derivative with respect to time.
14. Ilya Prigogine, Nobel prize in chemistry, 1977.

Appendix A

Multivariable Calculus

A.1 Differentials

Consider an inclined plane in x-y-z coordinate space, as shown in Fig. 1. We can think of the plane as a meadow with the x axis pointed east and the y axis pointed north; the z coordinate is the elevation above sea level. Elevation is a state function of the east–north position in the field; we write $z(x, y)$. We seek the change of elevation (z) as we go from point i to point f in the meadow. This can be calculated by first traveling east, a distance Δx, ending up at point g. We indicate the easterly slope of the meadow as $(\Delta z/\Delta x)_y$. The subscript y indicates that the north–south position remains constant when measuring this slope. Our change in elevation upon arrival at point g is $\Delta z_1 = (\Delta z/\Delta x)_y \Delta x$. We then complete our journey to point f by traveling in the northerly (y) direction. Our change in elevation in the second part of the trip is $\Delta z_2 = (\Delta z/\Delta y)_x \Delta y$. The change in elevation for the total trip is the sum of that in the first and the second parts of the trip:

$$\Delta z = \left(\frac{\Delta z}{\Delta x}\right)_y \Delta x + \left(\frac{\Delta z}{\Delta y}\right)_x \Delta y \tag{1}$$

What happens if the meadow is not planar? At a given point, we can approximate a nonplanar meadow with the plane that is tangent to the meadow at

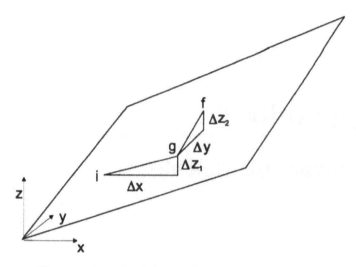

Figure 1 An inclined plane in the *x-y-z* coordinate space.

the point. If we stay very close to the original point, the tangent plane is a very good approximation to the meadow, and in the limit of infinitesimal displacement from the point, it is a perfect approximation and we can write

$$dz = \left(\frac{\partial z}{\partial x}\right)_y dx + \left(\frac{\partial z}{\partial y}\right)_x dy \tag{2}$$

where *dz* is called the *total differential* of the function $z(x, y)$. Quantities such as $(\partial z/\partial x)_y$, the *partial derivative of z with respect to x holding y constant*, are the slopes of the meadow in particular directions. Given a differential expression

$$dz = M(x, y) \, dx + N(x, y) \, dy \tag{3}$$

it must be identical to Eq. (2). Because *dx* and *dy* are arbitrary,

$$M(x, y) = \left(\frac{\partial z}{\partial x}\right)_y \quad \text{and} \quad N(x, y) = (\partial z/\partial y)_x \tag{4}$$

Example 1. If $z = x^3 \sin y$, what are $(\partial z/\partial x)_y$, $(\partial z/\partial y)_x$, and *dz*?

Solution: It is assumed that the reader can take ordinary derivatives of simple functions. In finding $(\partial z/\partial x)_y$, *y* is a constant, so *z* can be treated as Cx^3. Then,

$$\left(\frac{\partial z}{\partial x}\right)_y = 3x^2 C = 3x^2 \sin y$$

For $(\partial z/\partial y)_x$, we write $z = C' \sin y$ and $(\partial z/\partial y)_x = C' \cos y = x^3 \cos y$. Finally, $dz = 3x^2(\sin y)\, dx + x^3(\cos y)\, dy$.

If $z(x, y)$, then $x(y, z)$ and $y(z, x)$. Each of these functions permits the definition of two partial derivatives. What are the relationships between the six partial derivatives formed from three variables? If two partial derivatives hold the same variable constant, we only have two variables and can use the reciprocal rule from ordinary calculus:

$$\left(\frac{\partial z}{\partial x}\right)_y = \left(\frac{\partial x}{\partial z}\right)_y^{-1}, \qquad \left(\frac{\partial z}{\partial y}\right)_x = \left(\frac{\partial y}{\partial z}\right)_x^{-1}$$

and

$$\left(\frac{\partial y}{\partial x}\right)_z = \left(\frac{\partial x}{\partial y}\right)_z^{-1} \tag{5}$$

If z is held constant, we can set $dz = 0$ in Eq. (2):

$$0 = \left(\frac{\partial z}{\partial x}\right)_y dx_z + \left(\frac{\partial z}{\partial y}\right)_x dy_z \tag{6}$$

where we have placed a subscript on dx and dy to indicate that z is constant. Rearranging, we get

$$\frac{dy_z}{dx_z} \equiv \left(\frac{\partial y}{\partial x}\right)_z = -\left(\frac{\partial z}{\partial x}\right)_y \left(\frac{\partial z}{\partial y}\right)_x^{-1} = -\left(\frac{\partial y}{\partial z}\right)_x \left(\frac{\partial z}{\partial x}\right)_y \tag{7}$$

The last equation looks just like the chain rule for ordinary derivatives, except for the minus sign, which is often omitted by students. However, the need for the minus sign is fairly obvious. If z increases with x at constant y and with y at constant x, for z to remain constant, as we increase x, we must decrease y. We will call Eq. (7) the *chain rule for partial derivatives*.

If there is a relation among x, y, and z, any function of these variables can be written in terms of only two of the three variables, [e.g., $F(x, y)$ or $F(x, z)$]. If x, y, and z change, F will also change, and it will change the same amount whether we decide to consider it as a function of x and y or x and z.

$$dF(x, y) = \left(\frac{\partial F}{\partial x}\right)_y dx + \left(\frac{\partial F}{\partial y}\right)_x dy = dF(x, z) = \left(\frac{\partial F}{\partial x}\right)_z dx + \left(\frac{\partial F}{\partial z}\right)_x dz \tag{8}$$

We can apply Eq. (8) to a process at constant y $(dy = 0)$:

$$\left(\frac{\partial F}{\partial x}\right)_y dx_y = \left(\frac{\partial F}{\partial x}\right)_z dx_y + \left(\frac{\partial F}{\partial z}\right)_x dz_y \tag{9}$$

which gives

$$\left(\frac{\partial F}{\partial x}\right)_y = \left(\frac{\partial F}{\partial x}\right)_z + \left(\frac{\partial F}{\partial z}\right)_x \left(\frac{\partial z}{\partial x}\right)_y \qquad (10)$$

Equation (10) shows how to change partial derivatives when changing the variable held constant.

We will also be interested in $dF(x, y)$ when $y = y(x)$. We have

$$dF = \left(\frac{\partial F}{\partial x}\right)_y dx + \left(\frac{\partial F}{\partial y}\right)_x dy = \left(\frac{\partial F}{\partial x}\right)_y dx + \left(\frac{\partial F}{\partial y}\right)_x \frac{dy}{dx} dx \qquad (11)$$

This also gives

$$\frac{dF}{dx} = \left(\frac{\partial F}{\partial x}\right)_y + \left(\frac{\partial F}{\partial y}\right)_x \frac{dy}{dx} \qquad (12)$$

To extend Eq. (2) to functions of more than two variables, the equation for the total derivative must include a term for each variable, with the partial derivative for that variable holding all other variables constant. For $h(x, y, z)$,

$$dh = \left(\frac{\partial h}{\partial x}\right)_{y,z} dx + \left(\frac{\partial h}{\partial y}\right)_{x,z} dy + \left(\frac{\partial h}{\partial z}\right)_{x,y} dz \qquad (13)$$

which can be written as

$$dh = M(x, y, z)\, dx + N(x, y, z)\, dy + P(x, y, z)\, dz \qquad (14)$$

A.2 Integrals

Integrals of the form $\int_{x_1}^{x_2} F(x)\, dx$ are completely determined, although the integration may not be trivial for complicated functions $F(x)$. Integrals of the form $\int_{x_1,y_1}^{x_2,y_2} G(y)\, dx$ or $\int_{x_1,y_1}^{x_2,y_2} G(x,y)\, dx$ are not determined until the relation between x and y, $x(y)$, is specified. They are called *path integrals*.

Example 2.   Evaluate the integral $\int_{0,0}^{1,1} y\, dx$ for the following two paths: path A, $y = x$; path B, $y = 0 \rightarrow 1$, with $x = 0$, then $x = 0 \rightarrow 1$, with $y = 1$.

Solution:

Path A:

$$\int_0^1 x\, dx = \left.\frac{x^2}{2}\right|_0^1 = \frac{1}{2}$$

Path B:

$$0 + \int_0^1 1 \, dx = 1$$

Clearly, these are not the same.

The integral of a total differential, however, is the difference between two values of a state function and, therefore, cannot depend on the path of the integral. This holds even if this difference is evaluated by means of partial derivatives.

$$\int_{x_1,y_1}^{x_2,y_2} dz = z(x_2,y_2) - z(x_1,y_1) = \int_{x_1,y_1}^{x_2,y_2} M(x,y) \, dx + \int_{x_1,y_1}^{x_2,y_2} N(x,y) \, dy$$

$$(15)$$

Example 3. Evaluate $\int_{0,0}^{1,1} dz$, where $z = xy$, both directly and using partial derivatives, with the paths used in Example 2.

Solution: Directly, $\int_{0,0}^{1,1} dz = (1 \times 1) - (0 \times 0) = 1$.
Path A, $x = y$:

$$\int_{0,0}^{1,1} dz = \int_{0,0}^{1,1} y \, dx + \int_{0,0}^{1,1} x \, dy = \int_0^1 x \, dx + \int_0^1 y \, dy = \frac{1}{2} + \frac{1}{2} = 1$$

Path B, $y = 0 \rightarrow 1$, with $x = 0$, then $x = 0 \rightarrow 1$, with $y = 1$:

$$\int_{0,0}^{1,1} dz = \int_{\text{path}} y \, dx + \int_{\text{path}} x \, dy = [0 + 1] + [0 + 0] = 1$$

The direct method and the two paths give the same value.

An integral over one cycle of a cyclic process in indicated by $\oint$. If the integrand is a total differential,

$$\oint dz = z(x_1,y_1) - z(x_1,y_1) = 0 \tag{16}$$

A.3 Second Derivatives

There are four second partial derivatives of the function $z(x,y)$. These are

$$\left(\frac{\partial^2 z}{\partial x^2}\right)_y, \quad \left(\frac{\partial^2 z}{\partial y^2}\right)_x,$$

$$\left(\frac{\partial^2 z}{\partial x \, \partial y}\right) \equiv \left(\frac{\partial}{\partial x}\left(\frac{\partial z}{\partial y}\right)_x\right)_y$$

and

$$\left(\frac{\partial^2 z}{\partial y\,\partial x}\right) \equiv \left(\frac{\partial}{\partial y}\left(\frac{\partial z}{\partial x}\right)_y\right)_x$$

The latter two derivatives are known as *cross-derivatives*, and for any well-behaved function of x and y, they are equal[1]:

$$\left(\frac{\partial^2 z}{\partial x\,\partial y}\right) = \left(\frac{\partial^2 z}{\partial y\,\partial x}\right) \qquad (17)$$

Example 4. Find the four second derivatives of the function $z = x^3 \sin y$. Compare the two cross-second-derivatives.

Solution:

$$\left(\frac{\partial z}{\partial x}\right)_y = 3x^2 \sin y \quad \text{and} \quad \left(\frac{\partial z}{\partial y}\right)_x = x^3 \cos y$$

$$\left(\frac{\partial^2 z}{\partial x^2}\right)_y = 6x \sin y, \quad \left(\frac{\partial^2 z}{\partial y^2}\right)_x = -x^3 \sin y$$

$$\left(\frac{\partial^2 z}{\partial y\,\partial x}\right) = 3x^2 \cos y \quad \text{and} \quad \left(\frac{\partial^2 z}{\partial x\,\partial y}\right) = 3x^2 \cos y$$

The latter two derivatives are equal as stated.

For a total differential of the form of Eq. (3), Eq. (17) requires

$$\left(\frac{\partial M}{\partial y}\right)_x = \left(\frac{\partial N}{\partial x}\right)_y \qquad (18)$$

This useful equation is known as the cross-derivative rule. There are nine second partial derivatives of a function $h(x, y, z)$ of three variables. Calculating derivatives for a few of these functions can convince the reader that the cross-derivative rule also holds for such functions. Thus, using the notation of Eq. (14),

$$\left(\frac{\partial M}{\partial y}\right)_{x,z} = \left(\frac{\partial N}{\partial x}\right)_{y,z}, \qquad \left(\frac{\partial M}{\partial z}\right)_{x,y} = \left(\frac{\partial P}{\partial x}\right)_{x,y}$$

$$\left(\frac{\partial N}{\partial z}\right)_{y,x} = \left(\frac{\partial P}{\partial y}\right)_{z,x} \qquad (19)$$

Problems

1. For further justification of the chain rule for partial derivatives, consider the following example: Let z equal the money in your checking account, x be the number of times you go shopping for clothes in a month, and y the number of times you eat out in a month. What

are the signs of $(\partial z/\partial x)_y$ and $(\partial z/\partial y)_x$? What is the significance of $(\partial y/\partial x)_z$ and what is its sign (from your experience)? Are the signs of these three partial derivatives in agreement with the chain rule for partial derivatives?

2. Demonstrate Eq. (17) for the function

$$F = 3xy^2z + 2z \sin y + 4x \cos y.$$

Note

1. A proof of this can be found in most textbooks on multivariable calculus. Alternatively, the reader can just pick a few functions of two variables and compare the cross-second-derivatives, as is done in Example 4. The latter procedure has the advantage of providing some practice in taking partial derivatives.

Appendix B

Numerical Methods

This appendix will discuss a few numerical methods which might be useful to you in solving some of the problems at the end of the chapters.

B.1 Solving Equations

Some equations, such as that of van der Waals, are trivial to solve for some variables (e.g., P or T), but more difficult to solve others (V). As discussed in Chapter 1, the van der Waals equation may be rearranged into a cubic equation for V, and formulas are available to obtain the roots for such equations. Our objective here is to discuss more general methods for obtaining approximate solutions for such equations. One way to approach the solution of the general equation,

$$y = f(x, y) \tag{1}$$

is to rewrite it in the form

$$F(x, y) = y - f(x, y) = 0 \tag{2}$$

With a calculator, it is usually simple to try different values of y, for a given value of x, until the function $F(x, y)$ changes sign and then close in on the root of Eq. (2), until the required accuracy for y is obtained. Such a procedure is known as *trial and error*. Of course, one must be aware that sometimes Eq. (2) may have more than a single real root, as does the van der Waals equation in the two-phase region.

Sometimes, the function $f(x, y)$ can be written in the form

$$y = f(x, y) = g[h(x) + j(x, y)]$$ (3)

where h, g, and j are functions. If j is much smaller than h, it may be possible to neglect it completely and solve directly for y. If j is somewhat larger, this first value of y may be used in the right-hand side of Eq. (3) to calculate a corrected value for y. This procedure can be repeated until successive values of y are unchanged to the desired accuracy (convergence). This method is called *successive approximations* and often may be carried out faster than trial and error.

A successive approximation method that does not rely on the y-dependent part of F being small is the *Newton–Raphson* method. As seen in Fig. 1, if a value of y, y_1, gives $F(x, y_1)$, a closer solution to Eq. (2) can usually be obtained by setting

$$y_2 = y_1 - \frac{F(x, y_1)}{(\partial F(x, y)/\partial y)_{x, y=y_1}}$$ (4)

Computer programs (such as Mathcad) automatically implement the Newton–Raphson method of finding roots. Usually, such programs require the input of an

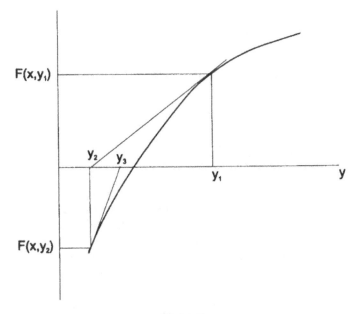

Figure 1

initial guess of the root. However, certain precautions are wise when using such programs. First, it is usually worthwhile to plot or tabulate the function, to see whether it has multiple roots, and to aid in choosing a guess that is sufficiently close to the root of interest, so that it will converge to that root. Second, in cases in which $(\partial F(x,y)/\partial y)_{x,y=y_1}$ is very small, the program may reach the convergence criterion on $F(x,y)$ without identifying the root with sufficient accuracy.

B.2 Fitting Data

We will only discuss the problem of fitting data points to a straight line. In earlier days, this was accomplished by placing a transparent piece of plastic with a straight edge over the data, so as to minimize the sum of the magnitudes of the deviations of the point from the edge.

For data influenced only by small random fluctuations, it can be shown that the proper criterion is a minimization of the sum of the squares of the deviations from the fitting line. Obtaining the best-fitting line by this criterion is known as *linear regression analysis*. Computer programs, such as Mathcad, as well as most scientific calculators can perform linear regression analysis. In case these are not available, the relevant formulas are as follows.

To fit a set of N points (x_i, y_i) to the straight line $y = mx + b$, first calculate D as

$$D \equiv N \sum_i x_i^2 - \left(\sum_i x_i \right)^2 \tag{5}$$

The slope and intercept of the best-fit straight line are then given by

$$m = \frac{N \sum_i (x_i y_i) - (\sum_i x_i)(\sum_i y_i)}{D} \tag{6}$$

$$b = \frac{(\sum_i y_i)(\sum_i x_i^2) - (\sum_i x_i y_i)(\sum_i x_i)}{D} \tag{7}$$

For best accuracy, it is necessary to keep twice as many significant figures in the calculations of the various sums as there are in the original dataset. Books on regression analysis will provide criteria for "goodness of fit" and for the accuracy with which the slope and intercept are known. It is very important to plot the original data points on the same graph as the straight-line fit, to see whether linear regression is appropriate or whether there is a *systematic* deviation of the data from the line. The latter is indicated by the points at the two extremes of the abscissa range tending to be above or below the line. If systematic deviation is observed, a fit to a polynomial function can be employed.

B.3 Numerical Integration

Numerical integration is needed when the functional form of the integrand is such as to preclude analytical integration or when the data to be integrated are obtained in tabular form as the result of measurements. In the latter case, the data have varying amounts of scatter, and linear or polynomial regression should often be use before (analytical) integration.

The simplest method of numerical integration is the trapezoidal method, where the abscissa is divided into intervals and each resulting area is estimated as the abscissa interval times the average of the initial and final ordinate in the interval:

$$\int_{x_0}^{x_{n+1}} f(x)\,dx = \sum_{i=0}^{n} (x_{i+1} - x_i)\frac{f(x_{i+1}) + f(x_i)}{2} \tag{8}$$

A numerical integration of a function using three trapezoids is shown in Fig. 2. The exact integral is the shaded area, and for the function chosen, each of the trapezoids overestimates the area under the curve. The accuracy of the method can, of course, be improved by dividing the range of integration into a larger number of intervals (i.e., using narrower trapezoids). In using the trapezoidal

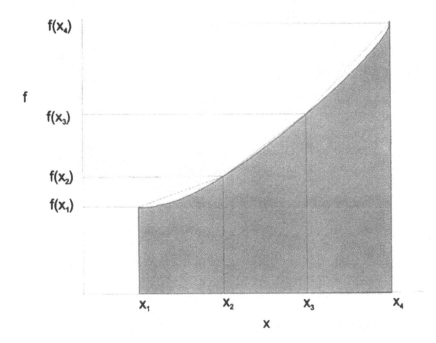

Figure 2 Numerical integration of a function using three variables.

method, the trapezoids do not all have to be of the same width. This is particularly convenient in integrating experimental data, which may not have been obtained at regular intervals of the independent variable.

A more accurate method of numerical integration is by the use of Simpson's rule. This method, however, requires that the integration range be divided into an even number of intervals of equal width h. This requires an odd number of points on the abscissa, which are numbered from 0 to n. Simpson's rule gives

$$\int_{x_0}^{x_n} f(x)\, dx = \frac{h}{3}(f_0 + 4f_1 + 2f_2 + 4f_3 + 2f_4 + \cdots + 4f_{n-1} + f_n) \tag{9}$$

Problems

1. Solve the equation $e^y = xy/(y + xe^{-xy})$ for $x = 5$, finding y with an error of less than 0.001, using both trial and error and successive approximation. Which takes less time?

2. Find the best least-squares straight-line fit to this data:

x	0	0.40	0.70	1.00	1.50	2.00
y	0.87	1.13	1.19	1.42	1.58	1.82

3. Integrate the function $f(x) = e^{-x^2}$, from $x = 1$ to $x = 2$, using the trapezoidal rule with x intervals of 0.25 and using Simpson's rule with the same interval.

Appendix C

Table of Thermodynamic Data

Selected Thermodynamic Properties at 298.15 K, $P^\circ = 1.0$ bar, $m^\circ = 1.0$ mol/kg

Compound	ΔH_f° (kJ/mol)	ΔG_f° (kJ/mol)	S° (J/K mol)	C_P° (J/K mol)
AgCl(s)	−127.07	−109.80	96.2	50.79
Ag$^+$(aq)	105.58	77.124	72.68	21.7
Ar(g)	0	0	154.734	20.786
Br(g)	111.88	82.429	174.912	20.786
Br$_2$(g)	30.907	3.14	245.35	36.0
Br$_2$(liq)	0	0	152.23	75.688
C(diamond)	1.897	2.900	2.38	6.1149
C(graphite)	0	0	5.740	8.527
CO(g)	−110.52	−137.15	197.56	29.12
CO$_2$(g)	−393.51	−394.36	213.6	37.1
CCl$_4$(g)	−103.0	−60.63	309.7	83.30
CCl$_4$(liq)	−135.4	−65.27	216.4	131.7
CF$_4$(g)	−925.0	−879.0	261.5	61.09
HCOOH(aq)	−425.43	−372.0	163.0	
HCOO$^-$(aq)	−425.55	−351.0	92.0	
CH$_2$O(g)	−117.0	−113.0	219.9	35.4
CH$_3$(g)	145.7	147.9	194.2	38.7

(*continued*)

Compound	ΔH_f° (kJ/mol)	ΔG_f° (kJ/mol)	S° (J/K mol)	C_p° (J/K mol)
$CH_3Cl(g)$	−80.83	−57.40	234.5	40.7
$CH_3OH(g)$	−201.0	−162.3	239.9	44.1
$CH_3OH(liq)$	−238.67	−166.4	127.0	81.6
$CH_4(g)$	−74.81	−50.75	186.15	35.31
$C_2H_2(g)$	227.4	209.9	200.9	44.0
$C_2H_4(g)$	52.26	68.12	219.5	43.56
$C_2H_5OH(g)$	−235.1	−168.6	282.6	65.44
$C_2H_6(g)$	−84.68	−32.9	229.5	52.63
$C_6H_6(liq)$	49.1	124.5	173.4	136.0
$C_6H_6(g)$	82.9	129.7	269.2	82.4
$CaO(s)$	−668.56	−604.04	39.7	42.80
$CaCl_2(s)$	−795.8	−748.1	104.0	72.59
$CaCO_3(s)$	−1206.9	−1128.8	92.9	81.88
$Cl(g)$	121.68	105.70	165.09	21.84
$Cl_2(g)$	0	0	222.96	33.91
$Cl^-(aq)$	−167.16	−131.26	56.5	
$HCl(aq)$	−167.16	−131.26	56.5	
$Cu(s)$	0	0	33.15	24.43
$Cu^{2+}(aq)$	64.77	65.52		
$H(g)$	217.96	203.26	114.60	20.786
$H_2(g)$	0	0	130.57	28.82
$H^+(aq)$	0	0		
$OH^-(aq)$	−229.99	−157.29		
$OH(g)$	39.0	34.2	183.6	29.89
$H_2O(liq)$	−285.83	−237.18	69.91	75.291
$H_2O(g)$	−241.82	−228.59	188.72	33.58
$He(g)$	0	0	126.040	20.786
$K^+(aq)$	−252.4	−283.2	102.0	22.0
$Fe(s)$	0	0	27.3	25.1
$Fe^{2+}(aq)$	−89.1	−78.87		
$Fe^{3+}(aq)$	−48.5	−4.6		
$Fe_2O_3(s)$	−824.2	−742.2	87.40	103.8
$Ne(g)$	0	0	146.219	20.786
$N_2(g)$	0	0	191.5	29.12
$NH_3(g)$	−46.11	−16.5	192.3	35.1
$NO(g)$	90.25	86.57	210.65	29.84
$NO_2(g)$	33.2	51.30	240.0	37.2
$NOBr(g)$	82.2	82.4	273.7	45.5
$N_2O_4(g)$	9.16	97.82	304.2	77.28
$NH_4Cl(s)$	−314.4	−203.0	94.6	84.1
$PCl_3(g)$	−287.0	−268.0	311.7	71.84
$PCl_5(g)$	−374.9	−305.0	364.6	112.8

Compound	ΔH_f° (kJ/mol)	ΔG_f° (kJ/mol)	S° (J/K mol)	C_P° (J/K mol)
$O_2(g)$	0	0	205.03	29.35
$O(g)$	249.2	231.7	161.1	21.9
$O_3(g)$	143.0	163.0	238.8	39.2
$NaCl(s)$	−411.15	−384.15	72.13	50.50
$Na^+(aq)$	−240.1	−261.9	59.0	46.4
$SO_2(g)$	−296.83	−300.19	248.4	39.9
$SO_3(g)$	−395.7	−371.1	256.6	50.67
$SO_4^{2-}(aq)$	−909.27	−744.63		
$Zn(s)$	0	0	41.6	25.4
$Zn^{2+}(aq)$	−153.9	−147.0		

Source: Data from: Handbook of Chemistry and Physics. 64th ed. and 77th ed. Boca Raton, FL: CRC Press, 1983, 1996.

Appendix D

Glossary of Symbols

If no units are given, the quantity is either without units or has varying units.

Symbol	Meaning	S.I. units
α	Thermal expansion coefficient, or elongation	K^{-1}
α_L	Linear thermal expansion coefficient	K^{-1}
γ	Surface tension,	$J\,m^{-2}$
	or heat capacity ratio,	
	or activity coefficient	
Γ_i	Surface excess concentration	$mol\,m^{-2}$
δ	Infinitesimal amount of	
Δ	Change in a quantity	
ϵ	Molecular energy, or	J
	strain, or	
	efficiency or coefficient of performance or	
	Seebach coefficient	$V\,K^{-1}$
ϵ_i	Energy of ith molecular state	J
ϵ_0	Permittivity of free space	$C^2\,s^2\,kg^{-1}\,m^{-3}$
$\epsilon_{r,A}$	Dielectric constant of solvent	
ξ	Extent of reaction	mol
θ	Contact angle	

(*continued*)

Symbol	Meaning	S.I. units
θ_f	Freezing-point depression	K
θ_b	Boiling-point elevation	K
ϑ	Fraction of surface sites occupied	
Θ	Volumetric rate of entropy generation	$J\,K^{-1}\,s^{-1}\,m^{-3}$
κ	Isothermal compressibility,	Pa^{-1}
	or number of polymer chains in sample	
π	Surface pressure	$J\,m^{-2}$
Π	Osmotic pressure, or	Pa
	Peltier coefficient	$J\,C^{-1}$
ρ	Density	$kg\,m^{-3}$
μ	Chemical potential	$J\,mol^{-1}$
$\tilde{\mu}$	Electrochemical potential	$J\,mol^{-1}$
μ_J	Joule coefficient	$K\,m^{-3}$
μ_{JT}	Joule–Thomson coefficient	$K\,Pa^{-1}$
ν	Number of chains per unit volume,	m^{-3}
	or total number of ions produced in ionization	
ν_i	Stoichiometric coefficient of component i in a chemical reaction	
ν_+	Number of positive ions produced in ionization	
ν_-	Number of negative ions produced in ionization	
σ	Diameter of hard-sphere molecule, or	m
	surface area, or	m^2
	Thomson coefficient	$J\,C^{-1}\,K^{-1}$
ϕ	Electrostatic potential, or	V
	fugacity coefficient, or	
	osmotic coefficient	
Φ	Dissipation	$J\,s^{-1}\,m^{-3}$
Ω	Number of configurations	
a	van der Waals gas constant, or	$N\,m^4\,mol^{-2}$
	linear term in stress–strain relation, or	Pa
	activity or	
	the number of stoichiometric restrictions in a system, or	
	the distance of closest approach of two ions	m
a'	Berthelot gas constant	$N\,m^4\,K\,mol^{-2}$
a''	Redlich–Kwong gas constant	$N\,m^4\,K^{1/2}\,mol^{-2}$
A	Helmholtz free energy, or	J
	constant in Debye–Huckel equation	
A	Affinity of a chemical reaction	$J\,mol^{-1}$
b	van der Waals constant, or	$m^3\,mol^{-1}$
	quadratic term in stress–strain relation	Pa
b'	Berthelot gas constant	$m^3\,mol^{-1}$
b''	Redlich–Kwong gas constant	$m^3\,mol^{-1}$

Symbol	Meaning	S.I. units
B	Second virial coefficient, or	$m^3\,mol^{-1}$
	constant in Debye–Huckel equation	
B'	Second virial coefficient	Pa^{-1}
c	Molecular speed, or	$m\,s^{-1}$
	number of component of a system	
c_i	Molarity of component i	$mol\,L^{-1}$
c_V	Specific heat (at constant volume)	$J\,g^{-1}\,K^{-1}$
c_P	Specific heat (at constant pressure)	$J\,g^{-1}\,K^{-1}$
C_V	Heat capacity at constant volume	$J\,K^{-1}$
C_P	Heat capacity at constant pressure	$J\,K^{-1}$
d	Infinitesimal change of	
D_i	Diffusion coefficient of component i	$m^2\,s^{-1}$
E	Young's modulus, or	Pa
	total energy	J
$\mathscr{E}$	Cell potential	V
f	Fugacity, or	Pa
	number of degrees of freedom of a system	
f_i	Fraction of molecules in ith state	
$f(v_i)\,dv_i$	Fraction of molecules with component of velocity	
	$v_i \leftrightarrow v_i + dv_i$	
$f(\epsilon)\,d\epsilon$	Fraction of molecules with energy $\epsilon \leftrightarrow \epsilon + d\epsilon$	
F	F Function	J
F_x	x Component of force	N
$F(c)\,dc$	Fraction of molecules with speeds $c \leftrightarrow c + dc$	
$\mathscr{F}$	Faraday's constant	$C\,mol^{-1}$
g_i	Degeneracy of states at ϵ_i	
G	Gibbs free energy	J
g	Acceleration due to gravity	$m\,s^{-2}$
H	Enthalpy	J
$\mathbf{H}$	Magnetic field	T
I_c	Ionic strength	$mol\,L^{-1}$
J_e	Flux of electric current	$C\,s^{-1}\,m^{-2}$
J_i	Flux of component i	$mol\,s^{-1}\,m^{-2}$
J_q	Heat flux	$J\,s^{-1}\,m^{-2}$
k	Boltzmann's constant	$J\,K^{-1}$
k_e	Electrical conductivity	$C\,s^{-1}\,m^{-1}\,V^{-1}$
k_q	Thermal conductivity	$J\,s^{-1}\,m^{-1}\,K^{-1}$
K	Kinetic energy, or	J
	equilibrium constant	
K_γ	Proper quotient of activity coefficients at equilibrium	
K_f	Freezing-point depression constant	$K\,kg\,mol^{-1}$
K_b	Boiling-point elevation constant	$K\,kg\,mol^{-1}$
K_{12}	Distribution coefficient between two solvents	

(*continued*)

Symbol	Meaning	S.I. units
$K_{m,i}$	Molality-based Henry's law constant	Pa kg mol^{-1}
l	Length (e.g., of a segment of a polymer chain), or generalized coordinate	m
i.s.	For an ideal solution	
id.s.	For an ideally dilute solution	
L	Length, or generalized force	m
L_D	Debye length	m
L_{ij}	Proportionality constant between flow i and force j	
m	Mass of a molecule	kg
m_i	Molality of component i	mol kg^{-1}
M	Molar mass, or mass of an object	kg mol^{-1} kg
n	Number of moles	mol
N_A	Avogadro's number	molecules mol^{-1}
N_i	Number of molecules in the ith state	
p	Number of coexisting phases in a system	
P	Pressure	Pa
P_c	Critical pressure	Pa
P_i	Partial pressure of component i	Pa
$P(r, N)\, dr$	Probability that a polymer chain of N segments has length $r \leftrightarrow r + dr$	
q	Heat, or molecular partition function	J
Q	Electric charge, or system partition function	C
Q_a	Proper coefficient of activity coefficients for a chemical reaction	
Q^*	Energy of transport	J mol^{-1}
r	Length of a polymer chain, or radius of curvature, or the number of independent reactions in a system	m m
R	Gas constant, or radius of a tube	J mol^{-1} K^{-1} m
S	Entropy	J K^{-1}
t	Temperature, or time	°C s
T	Temperature	K
T_B	Boyle temperature	K
T_c	Critical temperature	K
T_r	Reduced temperature	
U	Internal energy	J
v	Amount adsorbed on a adsorbate	cm^3

Symbol	Meaning	S.I. units
v_x	x Component of velocity	$m\ s^{-1}$
v_{react}	Volumetric rate of reaction	$mol\ s^{-1}\ m^{-3}$
V	Volume, or	m^3
	potential energy	J
V_c	Critical volume	m^3
$V_{m,r}$	Reduced molar volume	
x_i	Mole fraction of component i	
X	General thermodynamic variable	
$\bar{X}$	Partial molar thermodynamic variable	
X^E	Excess thermodynamic variable	
y_i	Mole fraction of component i in vapor above a solution	
z_i	Overall mole fraction of component i in system, or charge on an ion	
Z	Compression factor	
$\langle\ \rangle$	Average over molecules	
∇	Gradient of a scalar function	

Subscripts

ϕ	For a phase change
A	Of solvent
bd	For bond dissociation
c	Cold Reservoir
cycle	For a cyclic process
e	At equilibrium
elec	In an electrical process
exp	For expansion
ext	External
f	Of formation
fus	Fusion (melting)
gas	Of gas
h	Hot reservoir
im	For immersion
ind	Independent
int	Internal
liq	For liquid
m	Per mole
mech	In a mechanical process
melt	For melting
mix	For mixing
oth	Other (besides expansion)
rxn, react	For a chemical reaction

(*continued*)

Symbol	Meaning	S.I. units
rev	For a reversible process	
rot	Rotational	
solid	For solid	
sub	Sublimation	
sur	Of the surroundings	
surf	Of a surface	
tot	Total	
tr	Translational	
trans	For transfer between solvents	
univ	For the universe	
vap	Vaporization, or for vapor	
vib	Vibrational	
+	For positive ion	
−	For negative ion	
±	Mean ionic	

Superscripts

σ	Surface thermodynamic function	
°	In the standard state	
*	Of a pure substance	
rel	Solute relative to solvent	

Index

activity   164, 246
    coefficient   246–247, 258, 301
        determining from colligative
            properties   255
        determining from distribution
            coefficients   253
        determining from vapor
            pressure   249
        mean ionic   275–280
        of solute from that of solvent   256
adiabatic   38
    demagnetization   94
adsorption   228–237
    chemical (chemisorption)   328
    physical (physisorption)   328
    statistical thermodynamics of 335–337
affinity   192
allotrope   171
anode   288
azeotrope   259

battery   288

boiling-point
    diagram   235
    elevation   226–229
Bose-Einstein condensate   116n
bubble   316
bubble-point line   233

calorie   50
calorimetry
    adiabatic   185
    bomb   185
capillary rise   318–320
cathode   288
cavity in liquid   315
chain rule for partial derivatives   367
chemical potential   151–153
    from partition function   131
chemical reaction   179–207, 355
    direction   192
    entropy generation   351
    at equilibrium   199

[chemical reaction]
  equilibrium constant (*see*
    equilibrium, constant)
  extent of   180
  going to completion   190
  nomenclature   180
coefficient
  Joule   59
  Joule-Thomson   61
  Peltier   360
  second virial   17
  Seebach   360–361
  spreading   317
  stoichiometric   180
  thermal expansion   6, 26
    linear   27
  Thomson   360–361
cogeneration   92n
colligative properties   228
colloid   213, 337–339
  lyophilic   337
  lyophobic   337
component   197
  independent   197
concentration
  at equilibrium   199–204
  molality units   213
  molarity units   213
configurations, number of
  of an energy distribution   124
  of a polymer chain   141–144
  in space   121
constant
  critical   21
  gas   5
constraints   39
contact angle   316–318
cooling curve (*see* thermal analysis)
corresponding states   24
critical point   19
critical solution temperature   261
cross derivatives   370
Curie-Prigogine principle   356

Debye length   278
Debye-Huckel

[Debye-Huckel]
  limiting law   279
  theory   277–280
deformation, plastic   28
degrees of freedom   196–199
deviations
  from ideal solution
    behavior   245–266
    positive and negative   249
dew point line   233
dielectric constant   278, 284, 306n
differentials   365
diffusion   349–350, 354
  thermal   136, 357–359
dissipation   348
distillation   236
  fractional   236
distribution
  barometric   31
  binomial   143
  Boltzmann   127
  coefficient   232, 253
  over energy states   124
  Gaussian   143
  normal   143
  of solute between
    solvents   231
  in space   121
  velocity   132–134
droplet   314

effusion
  Graham's law of   11
electrochemical cell   291–303
  Daniell cell   292, 296
  standard diagram   292
  thermodynamics of   293–297
electrochemical potential   287
electrochemical series   298
electrode   288–291
  amalgam   289
  calomel   289
  hydrogen   288
  membrane   289, 302
  pH   303
  potential   296–300

electrolytic cell   288
energy
  adhesive   317
  bond (*see* enthalpy, bond dissociation)
  cohesive   317
  degradation   73
  electrical   45
  hierarchy   73
  internal   49
    from partition function   128, 131
  molecular interaction   105
  relativistic   181
  of transport   358
engine
  Carnot cycle   64–67
  heat   85–88
ensemble
  microcanonical   120
  canonical   130
enthalpy   53
  bond dissociation   188
  group   189
  from partition function   130
entropy
  absolute   97, 98t
  change
    of Carnot cycle engine   77
    for expansions   79–80
    for heating   80–81
    of mixing   109
    for open systems   152
    for phase change   81
    for stretching a rubber   83
    of the surroundings   76, 78
    of system   77–78
    of the universe   75
  from configurations   122
  generation   347
    in chemical reaction   351
    in diffusion   350
    in heat flow   347
    in mechanical or electrical
      processes   348–349
  meter   76–77
  from partition function   129, 131
  of pure, perfect crystals   97

equation
  Clapeyron   162
  Clausius-Clapeyron   162, 359
  Gibbs-Helmholtz   104, 353
  Nernst   295
  Sackur-Tetrode   138
  of state   6
    Berthelot   15, 23
    of condensed phases   26
    of ideal gas   5
    of real gases   12
    Redlich-Kwong   15, 23
    thermodynamic   104
    of two-phase region   166–168
    van der Waals   12–15, 22
  Tait   27
  of Young and Laplace   316
equilibrium   6
  chemical reaction   190–192
  condensed phases in   195
  constant   191, 193–195, 257
    for dissociation of ion pairs   284
    from partition function   206
    temperature dependence   204–206
  internal   38
  ionic   282–283
  material   102, 151, 155
  mechanical   6, 38, 43, 46
  phase   155–157, 161–163
    condensed   169
    three   170
  solid-liquid   236–240
  with surroundings   38
  ternary systems   266–269
  thermal   6, 38
equipartition theorem   139
eutectic   238–240
extraction, liquid-phase   231

field, gravitational   30, 43
film   313
  Langmuir   325
  surface   325–328
first law of thermodynamics   49–51
flocculation   338
fluctuation   124

force
  attractive 13, 17
  intermolecular 12
  nonconjugate 356
  repulsive 12, 17
  uncontrolled 357
freezing-point depression 226–229
fuel cell 293
fugacity 165–166
  coefficient 166
function (*see* thermodynamic
    function)

gas
  ideal 4
    model 8
  mole change in reaction 200
  nonideal 164–166
  van der Waals 12, 105
Gibbs free energy 100, 192–195, 301
  from partition function 130
Gibbs-Duhem relation 216
glass 174
  transition 174, 175

heat 40–43, 354
  diagram 84
  flow at steady state 346–348
  of formation 182
  of reaction 182
    estimation 187–190
  reservoir 41
  reversible 42
  of vaporization 163
heat capacity 57, 106
  at constant pressure 53
  at constant volume 52
  from equipartition theorem 140
  graph 54
  molar 52
  ratio 64
  table 56
Helmholtz free energy 99
  for adsorption 334–335
  from partition function 130, 131
  of rubber 145

immersion 320
integrals 368–369
interfacial (*see* surface)
ionic strength 278
ion pairs 284–286
ion solvation 286
isobar 6, 20
isochore 6
isoenthalp 61
isotherm 6, 20
  Brunauer, Emmet, and Teller 329–332
  Langmuir 329
    from statistical mechanics 334–335
isothermal compressibility 7, 26
isotope separation 34n

junction potential 292–293

kinetics 3

lapse rate 56
law
  Amagat's 26
  Dalton's 24
  Fick's 354
  Fourier's 354
  Graham's 11
  Henry's 224–225
    reference 247
  Hess's 182
  Ohm's 354
  Onsager's 356
  Raoult's 222, 224
    reference 246
layer
  electric double 338
  Stern 339
lever rule 234
limit, proportional 28
liquid crystal 173
  nematic 173

magic camera 119, 149n
Maxwell relations 103
micelle 339
  critical concentration 339

miscibility   260
mixtures, gas   24
model
    Bernoulli   8
    hard-sphere   14
molecular weight
    determination   229
    mass-average   231
    number-average   231
mole fraction   26, 213
multivariable calculus   365–371

nucleation center   321
numerical methods   372–376
    fitting data   374
    integration   375–376
    solving equations   372–374

osmotic coefficient   257
    for electrolytes   276
osmotic pressure   227–229

partial molar quantities (*see*
       thermodynamic functions,
       partial molar)
partition function
    molecular   126
    system   130
Peltier effect   360
peritectic point   265
perpetual motion machines   74
pH   282
phase   154, 197
    diagram   171
       acetone-chloroform   262
       benzene-ethyl alcohol   260, 261
       carbon dioxide   173
       ethyl ether-ethyl alcohol   259
       ideal binary system   234, 236
       triangular   267
       water   171, 172
       water-phenol   261, 263
    mesomorphic   154, 173–175
    metastable   172
    rule   196–199

[phase]
    stability regimes   158–159
    transformation   154
       first-order   159–160
       lambda   160–161
       second-order   160
plasma   154
Pockels point   326
polymer
    displacement length of chains
       141–144
    phase transitions in   175
    strain-induced crystallization   146
pressure
    critical   21
    internal   57, 104
    partial   24
    vapor   19, 159, 162–163
       effect of inert gas on   168
       diagram   222, 233
       droplet   320
       variation in gravitational field   30
process
    constant-pressure   46
    cyclic   40
    gedanken   46
    isothermal   48
    Joule   59, 119–121
    Joule-Thomson   59–62
    path of a   40
    reversible adiabatic   62–64
properties   5
    colligative   228, 255
    conservative vs.
       nonconservative   344–346
    extensive   5, 39
    intensive   5, 39
    molar   6
    of polyatomic ideal gas   138
    quality of   73
    uniform   40

quantum mechanics   3, 125, 140, 148–149
quotient, proper, of activity
       coefficients   193

rate of hitting wall   135
reaction (*see* chemical reaction)
rubber
  elongation   111
  equation of state   28
  ideal   111
    theory of elasticity   144–146
    thermodynamics of stretching   110

salt bridge   293
salting in   283
scientific creationism   78
second derivatives   369–370
second law of thermodynamics   72–78
Seebach effect   360
semipermeable membrane   227
solubility
  of gases   225t
solution   213
  ideal   220–222
  ideally-dilute   222–225
  interstitial solid   262
  ionic   274
  substitutional solid   262
Soret effect   359
specific heat   52, 53
speed
  average   134
  root-mean-square   11
  of sound   11
standard state
  condensed phase   163
  gas   107
  nonideal gas   165–166
statistics
  Boltzmann   124
  corrected Boltzmann   125
steady state   38, 135
  systems   344
Stirling's approximation   129
strain   28
strength
  ultimate   28
  yield   28
stress   28
supercritical fluid   21

supersaturation   321
surface
  area determination   332
  energy   310
  entropy   310
  excess concentration   323–324
  excess properties   322–324
  Gibbs phase   322
  pressure   326–328
  region   309
  tension   27, 310–312, 324
    measurement of   313–319
    of solids   312
  thermodynamics   309–311
surfactant   324–328
surroundings   38
system
  closed   38
  heterogeneous   40
  homogeneous   40
  isolated   38
  open   38
    thermodynamics of   151–154

Taylor series expansion   34n
temperature   4, 41–43
  absolute   4
  absolute zero   94–97
  Boyle   18
  critical   21
  ideal gas   4
  Joule-Thomson inversion   61, 62
  Kelvin   34n
  Kraft   339
  scales   42
theoretical plate   243
thermal analysis   239
thermochemistry   181–185
thermodynamic data   377–379
thermodynamic function
  excess   248
  F 107
  partial molar   214–217
    measurement of   217–219
    from partition function   128
    of monatomic ideal gas   136–138

[thermodynamic function]
  per unit area 310
  per unit volume 344
  state 39
third law of thermodynamics 94–97
tie line 234, 268
transpiration, thermal 136
triple point 170
two-phase region 20, 166–168, 233

vaporization 19
velocity distribution 132–134
virial expansion 17
volume
  critical 21

[volume]
  excluded 12
  reduced 23

wetting 317–319
Wilhelmy plate 317
work 43–48
  diagram 47
  function 99
  other 48
  reversible 45–47

Young's modulus 28

zeroth law of thermodynamics 41